Ecological Encyclopedia of The Marine Animals of The Indo-Pacific

Shin Nippon Kyoiku Tosho Co., Ltd.

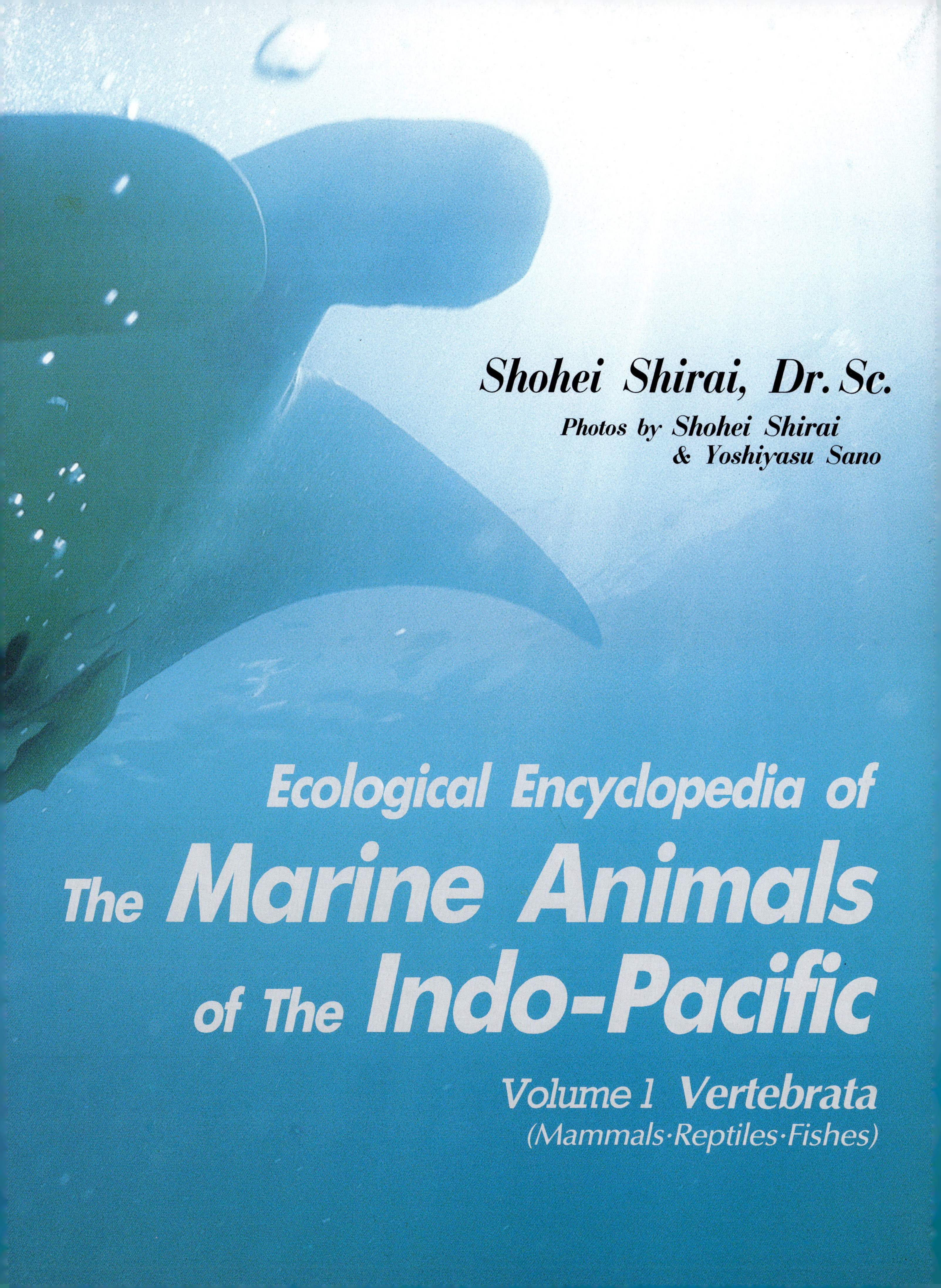

Shohei Shirai, Dr. Sc.

Photos by Shohei Shirai
& Yoshiyasu Sano

Ecological Encyclopedia of The Marine Animals of The Indo-Pacific

Volume 1 Vertebrata

(Mammals·Reptiles·Fishes)

Ecological Encyclopedia of the Marine Animals of the Indo-Pacific
Volume 1 (Vertebrata)

Published 1986 by Shin Nippon Kyoiku Tosho Co., Ltd.
203, Sanaizaka Heim, 11 Sanai-cho, Ichigaya,
Shinjuku-ku, Tokyo, Japan

First published 1986

Printed in Japan
ISBN4-88024-092-3 C1640

CONTENTS

Foreword

The coral reef enchanted many people. An acquaintance of mine long cherished a desire of painting the coral reef and sent a work after her heart to an exhibition for the first time when she was seventy.

Dr. Shohei Shirai began studying Scleractinian corals, the basics of the coral reef, at an early age and widened the sphere of his research to the whole marine life on and around the coral reef at the same time deepening the study on each subject.

This "Ecological Encyclopedia of the Marine Animals of the Indo-Pacific" embodies a part of Dr. Shirai's achievement.

Let me express my great respect for the author and heartfelt congratulations on this publication.

Tokiharu Abe, Dr. Sc.
ex-chairman of
the Ichthyological
Society of Japan,
Director of the Tsukiji
Fish Information Center

Foreword

The Indo-Pacific region—i.e. the tropical and subtropical Indian Ocean and Central and Western Pacific combined—is the largest biogeographical region of the world. It has by far the greatest number of marine animals and plants of any province.

Only one man has dared to bring a major part of this immense fauna and flora of the sea into one large work of three volumes. That man is Dr. Shohei Shirai. We have seen his skill in photographing marine life and in compilations that both the scientist and layman can read and enjoy. The recent work will have about 4,000 photographs taken by Shirai and Yoshiyasu Sano over a period of more than 20 years; the color reproduction is excellent. The species accounts are brief but informative, giving scientific and English common names, a diagnosis, distribution, and pertinent remarks. Volume 1 will cover marine mammals, reptiles, and fishes. As I am an ichthyologist, you can be sure that I will be among the first to order a copy.

John E Randall

John E. Randall, Ph. D.
Senior Ichthyologist
Division of Ichthyology
Bishop Museum, Honolulu

Introduction

It was in 1977 that the "Ecological Encyclopedia of the Marine Animals of the Ryukyu Islands" was first published.

It had been the purpose to make an ecological encyclopedia of Invertebrata, centering on the shells and the stony corals of Coelenterata which the writer majors in. However, the publisher wished to make one with all the marine animals including fishes. It was a hard task. The result was better than expected. Both in 1978 and 1980 it was reimpressed. The contents had been revised and augmented. Then in 1980 an revised edition was published. It was in 1983 that the fourth edition was produced. The encyclopedia is known to not only the Ryukyu Islands but also most universities, institutions, libraries, aquariums of Japan and divers, anglers, and the people who are interested in marine residents.

South-East Asian and Oceanic universities and institutions of fisheries show interest in it, too, and seem to be using it, though its Japanese explanation is difficult for them. A South-African university has asked for it.

South Africa is the homeground of its famous ichthyologist, Dr. J. L. B. Smith known for his study on Coelacanth. The people who are interested in fishes must know that many tropical fishes live along the coast of South Africa through Dr. Smith' study, but the writer had never expected that the encyclopedia of the marine animals of the Ryukyu Islands would be of any use there. Mrs. Smith, after her husband had passed away, has often revised his book "Smith' Sea Fishes" and continued its publication. The book includes most of the fishes seen in markets and their distribution. After reading it, the writer realized that it was not advisable to depend only on the studies of Japanese ichthyologists which centered on Japan. In the field of taxonomy, the specialists have eminent knowledge and opinions in the limited field, but they made many contributions regarding as scholary prejudice or irrelevancy from general view of compilation or generalizations. On these points of view, nonspecialists hardly make such faults due to the absence of entranched lines of thought.

In May, 1979, the letter which the writer had been waiting for arrived. It was from Mr. Robert Sparks, the director of the University Press of Hawaii. It read: In principle the Press is definitely interested in the prospects for serving as the publisher of the English edition of this monumental work, and asked, of course, printing cost.

The publishing had been suggested by Mr. Don Awai, President of Academic Hawaii, a friend of mine. My book had been sent to Mr. Sparks with the recommendation of Drs. Phillip Helfrich and Allison Kay, famous marine biologists of the University of Hawaii.

Unfortunately the publishing was not realized because of its high cost. The publisher was not sure if he could sell the book. The high rate of the dollar then was definitely one of the causes.

Overseas friends were disappointed. One of them was a Mexican fisheries company which wanted to supply each boat with one copy for reference.

I worked for compiling data of poisonous and injurious animals and plants and completed as "Color Encyclopedia of the Poisonous and Injurious Animals and Plants of the Ryukyu Islands" in 1982.

It was when I was having a little rest that this publishing was suggested. I felt heavy responsibility as well as great honor and joy when I heard that the book would not be restricted to the Ryukyu Islands but cover all over Japan and that it would have editions for overseas. It had been easy to compile the studies of the Ryukyu Islands because they were limited in a certain area and because not many data existed. However, the contents of the new book were to be comprehensive for the readers all over the world.

There are about twenty thousands species of fishes in the world. It was impossible to include all of them, but it was desirable to do as many as possible for the readers who would buy the book to know the names of fishes. However, as with all publications, time and cost had to be taken into consideration, for an excellent book might never see the light of day if they were not.

I regarded it the first purpose to be used as many people as possible in the world, so that the Indo-Pacific region was chosen as the range for this encyclopedia to cover. It must have been better to include the Atlantic and the Mediterranean, but limit of the time hindered me to do so I decided to leave it for the future. However, the latter chapter will tell that the Indo-Pacific occupies seventy percent of the whole sea and has sea area where the largest number of marine animals live. Therefore the encyclopedia would be highly useful.

The second purpose was to give edible fishes priority. It might be uninteresting only to take ordinary fishes seen at markets, so the photographs of coral-reef fishes, as many as possible, were added. And fishes of popularity, Coelacanth, Manta, Sunfish, Whale Shark, and Archer Fish were included.

The "Ecological Encyclopedia of the Marine Animals of the Indo-Pacific" is divided into three volumes. Volume 1 introduces Vertebrata including Mammalia, Reptilia, and Pisces such as whales, sea turtles, and sea snakes. Volume 2 contains Mollusca including shells, Echinodermata, and Crustacea. Volume 3 presents Coelenterata centering on Scleractinian corals and some other lower Invertebrata. Therefore the three, in a set, complete a useful book. It will be a great pleasure for the writer, if they are of some help for studying the marine animals of the Indo-Pacific.

I would also like to express my acknowledgement to the people who helped me and allowed me to use references and data.

I would like to extend hearty gratitude and respect to Dr. Tokiharu Abe, the former president of the Ichthyological Society of Japan, for my recommendation of this book and Dr. John E. Randall of the Bishop Museum in Hawaii.

Deep appreciation is extended to the Okinawa Prefectural Fisheries Experimental Station for lending me the precious pictures of deep-sea sharks, and especially to Mr. Kenichi Sakiyama, director, and Mr. Kazunari Kugai, superintendent of the fisheries section.

I also thanks are due the following people who have cooperated in the preparation of this book in some significant way;

Dr. Takeshi Yasumoto, Proffesor of the Tohoku University,

Dr. Osamu Okamura, Proffesor of the Kochi University,

Dr. Masao Katayama, Proffesor Emeritus of the Yamaguchi University,

Dr. Masato Akazaki, Proffesor of the Miyazaki University,

Mr. Takuji Yatou, Kobe-Kita Senior High School,

Mr. Masayoshi Hayashi, Curator of the Yokosuka Municipal Natural Museum,

Mr. Jin Hattori, Tokyo University of Fisheries,

Mr. Hiroshi Kishimoto, Max Diving Ishigaki,

Dr. Kun-Hsiung Chang, Proffesor of the Taiwan University,

Dr. Edgardo D. Gomez, Proffesor of the University of Philippines,

Mr. Jean M. Tapu, Tahiti,

Finally the writer would like to thank Mr. Shuji Fujita, the president of the Shin Nippon Kyoiku Tosho, and Mr. Shunzo Baba, the general manager of the Tokyo branch, who gave form to this encyclopedia.

January, 1986
At Ishigaki Island — a coral island

Shohei Shirai

EXPLANATORY NOTES

1. This encyclopedia has for purpose to give the names of the marine animals living in the Indo-Pacific to the ordinary reader, so that their interpretation is written in easy sentences.
2. It was tried to include as many ecological pictures as possible. Some of them were taken in glass tanks. The pictures of specimens were also added for identification.
3. The order of printing is made from higher animals to lower animals, using general classification systems. The order of some parts is changed owing to the convenience of edition.
4. The encyclopedia is divided into three volumes, Fishes, Shells, and Stony corals. The reason is to carry and use them easily.
5. The encyclopedia includes as many ordinary marine animals as possible, but some were not included, because their photographs were not obtainable. They will be added in revised editions in the future.
6. Special benthos living inside the bay or in sandy mud and lower Invertebrata of deep-sea are deleted because they are not generally known. Micro lives and planktons are excluded, too.
7. It often happens that a species has a few Latin names by authors. In this case the name used prior one is to be adopted. In this encyclopedia synonyms are written for convenience to the comparison with other books.
8. Names in English are commonly called ones, but not the names decided in Taxonomy. Therefore, they differ according to the areas and the people. But they are more known than the Latin names. Here are printed as many common names as possible for ordinary reader.

   The species which have had no common names are newly named in English by the author.
9. The explanation is made simple, because the pictures are sufficient more than descriptions. The specific features are shown on pages 20–23 by the pictures, too.
10. The distribution is taken from the formerly published references. Most of the species are widely distributed in the Indo-Pacific. Those surveyed in restricted areas are included with their definite positions.
11. Most of the fishes, except some venomous, can be eaten. Those written as edible in the diagnosis are generally used fishes for food.

    As far as the venomous fishes are concerned, there are three types; first, being poisonous when eaten, second, by stinging the objects, and third, by biting them. Intoxication caused by ciguatera is frequently found in the Indo-Pacific. Some ciguatera fishes have poison, and others have none within the same species, which depends on the areas and individuals. Therefore the species that have been once taken as poisonous fish and those that have been written so in some areas are called poisonous fish to attract attention to them.
12. The sizes are written in centimeters. However, they differ according to the areas and individuals. This causes troubles in identification. The maximum sizes of standard length are written in order to avoid them as well as to give the standard for the competition of sizes for fishing. The formerly published data were used for them. When records are renewed, the description will be rivised.
13. There may be points, in the contents, with which there could be disagreement, or which could be revised from a narrower scientific viewpoint. Errors which were universally and basically recognized have been mostly corrected. The writer hopes that the reader will realize that the scope of this book covers an area as vast as the Indo-Pacific Ocean.

Vertebrata

Mammals
Reptiles

Dugongidae
Dugong dugon (P. L. Müller) ⇧⇨
Dugong

Diagnosis — A nocturnal shallow-water mammal of the tropics. Often found in small schools. Characterized by clawed flipper-like fore limbs, no hind limbs and a crescent-shaped tail. Incisors of male develop into fangs. Feeds on sea weed. Edible. Internationally protected due to small occurrences. Fabled mermaid. Up to 3.2m in total length.

Distribution — Ryukyu, South East Asia, Arafura Sea, Micronesia, Indian Ocean, Red Sea, African Coast

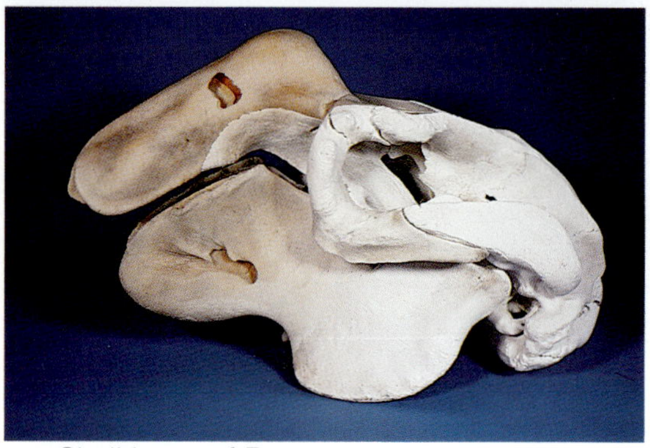

▲ Skull bone of Dugong.

Delphinidae
⇦ *Globicephala macrorhyncha* (Gray)
Short-finned Pilot Whale

Diagnosis — Found in small schools in coastal waters. Head is rounded. Dorsal fin is smaller than that of *G. melaena*. Belly with anchor-shaped white splotches on both sides. Pectoral fins are rather small. Feeds on fish and squid. Edible. Up to 7m in total length.

Distribution — Japan, Ryukyu, Okhotsk Sea

Delphinidae
Pseudorca crassidens (Owen) ⇨
False Killer Whale, Lesser Killer Whale

Diagnosis — Found in schools rather deep waters of the tropics. Head is rounded and protrudes. Mouth and teeth are large. Small pectoral fins situated slightly forward. Dorsal fin is not large. Feeds on fish and squid. Tamable. Up to 6m in total length.

Distribution — Japan, Indo-Pacific-Atlantic

Delphinidae
⇦ *Grampus griseus* (G. Cuvier)
Risso's Dolphin, Gray Grumpus

Diagnosis — Found in small schools near surface of open temperate tropical seas. Head is rounded, no distinct beak. Dorsal fin is tall and pointed. White stripes on body. Feeds on squid. Tamable. Edible. Up to 4m in total length.

Distribution — Japan, Ryukyu, Indo-Pacific-Atlantic

Phocaenidae
⇦⇦ ***Neophocaena phocaenoides***
(G. Cuvier)
[*Neomeris phocaenoides* (G. Cuvier)]
Finless Black Porpoise, Southeast Asiatic Porpoise, Indian Porpoise
Diagnosis — Smallest whales. Found in schools in the temperate coastal waters. Head is rounded, no distinct beak. No dorsal fin. Triangular pectoral fin is pointed. Mouth is small. Feeds on squid and small fish. Tamable. Edible. About 1.8m in total length.
Distribution — Japan, Ryukyu, South China Sea, Indian Ocean, South Africa

Delphinidae
Orcinus orca (Linnaeus) ⇨
Killer Whale
Diagnosis — A large species found in the surface layer of the cold waters. Harmful. Feeds on fish, squid, sea lion and other dolphins. Mouth is large, teeth are sharp. Large, tall dorsal fin. Caudal fin is also large. Distinguished by the large white splotches on the body. Tamable. Up to 9.4m in total length. Female is approximately half size of male.
Distribution — Japan, Formosa, North Pacific-Atlantic-Antarctic, North Sea

Delphinidae
⇦ ***Lagenorhynchus obliquidens*** (Gill)
Pacific White-sided Dolphin, Pacific Striped Dolphin
Diagnosis — Found in schools. Back is black, belly is white. Dorsal fin is tall and falcated. Small and pointed teeth. Feeds on squid and small fish. Not for food. Up to 2.4m in total length.
Distribution — Japan, Korea, North Pacific, Aleutian, Gulf of California

Delphinidae
Tursiops truncatus (Montague) ⇨
[*T. gilli* Dall]
Gill's Bottle-nosed Dolphin, Bottlenose Dolphin, Atlantic Bottlenose Dolphin, Cowfish
Diagnosis — A shallow-water species found in small schools. Head is rounded. Upper jaw resembles a beak, lower jaw is protruded. Dorsal fin is situated approximately in the middle of body. Feeds on squid and fish. Easily tamed. Flesh is unpalatable. Up to 3.7m in total length.
Distribution — Japan, Ryukyu, Indo-Pacific-Atlantic, Mediterranean

Dermochelyidae

⇦ ***Dermochelys coriacea schlegelii*** (Garman)

Seven-banded Turtle, Bat Turtle, Oil Turtle, Trunk Turtle, Hard Turtle, Luth Turtle, Ship Turtle, Boat Turtle, Leatherback Turtle

Diagnosis — A large species. Lacks keratinous shell plate. Body is covered by 1 layer of skin. Carapace has 7 ridges. Carnivorous and dangerous. Lays 90–150 eggs in each spawning. Edible. Up to 1.8m in shell length, 2.8m in standard length, 650kg in weight.

Distribution — Japan, Ryukyu, Indo-Pacific, Mediterranean, England, North America

Cheloniidae

⇦ ***Eretmochelys imbricata squamata*** Agassiz ⇧

[*E. imbricata* (Linnaeus)]

Hawksbill Turtle, Scute Turtle, Shell Turtle, Comb Turtle, Spectacled Turtle, Fowl Turtle

Diagnosis — A tropical species found in coral reefs. Snout is long and pointed. Head is covered with scales. Carapace is oval with 3 ridges. Shell resembles a tile-covered structure, 4 pairs of costal plates. Lays 150–200 eggs in each spawning. Carnivorous. Edible but sometimes causes poisoning. Yields tortoise-shell. Up to 80cm in shell length and 120kg in weight.

Distribution — Japan, Ryukyu, Indo-Pacific

Cheloniidae

⇧ ***Lepidochelys olivacea olivacea*** (Eschscholtz)

[*L. olivacea* (Eschscholtz)]

Pacific Ridley, Ridley

Diagnosis — A small omnivorous species. Carapace is rounded and varies in color from grey, grey-brown to olive. 6 pairs of costal plates. Produces 100–170 eggs in each spawning. Edible. Up to 80cm in shell length.

Distribution — Japan, Ryukyu, Indo-Pacific, Mexico, West Africa

Cheloniidae

Chelonia mydas japonica (Thunberg) ⇩

Meat Turtle, Sand Turtle, Right Turtle, Milk Turtle, Green Turtle, Rock Turtle

Diagnosis — Used for food in all parts of the world. Spawns 120–200 eggs between June and September in the Bonin Islands. Omnivorous. Feeds mostly on seaweed. Carapace is bluish black-brown. 4 pairs of costal plates and 4 one-clawed limbs. Harmless. Body oil greenish. Edible. Up to 1.2m in shell length and 150kg in weight.

Distribution — Japan, Ryukyu, Bonin Islands, Indo-Pacific

Cheloniidae

⇦ ***Caretta caretta gigas*** Deraniyagala

Loggerhead Turtle

Diagnosis — Common in tropical and subtropical waters. Omnivorous and harmful. Spawns 120–150 eggs between June and July on the coast of Japan. Head is large and thick. 5 to 9 pairs of costal plates and 4 two-clawed limbs. Carapace is reddish. Edible. Up to 90cm in shell length, 125kg in weight.

Distribution — Japan, Ryukyu, Indo-Pacific, West Africa

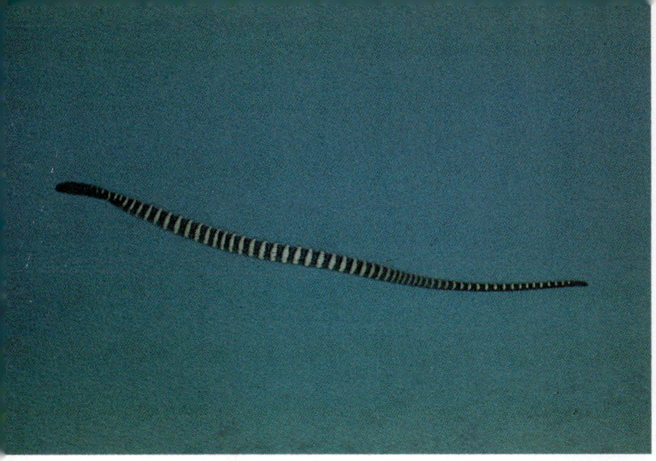

Hydrophiidae

Hydrophis melanocephalus Gray ⇧ ⇨

Black-head Sea Snake

Diagnosis — Inhabits coral reefs, does not move on shore. Ovoviviparous. Brings forth living youngs in October. Trunk has 40−60 and tail 4−9 black bands. Belly is covered with 290−358 ventral scales. Head is thin and black. Highly venomous with records of having caused death to man bitten. Up to 1.3m in total length.

Distribution — Japan, Ryukyu, Formosa

Hydrophiidae

⇧ *Hydrophis cyanocinctus* Daudin

Annulated Sea Snake

Diagnosis — Body is elongated, head is thin. Trunk has 50−70 and tail 2−8 black bands. Belly is covered with 290−390 small ventral scales. Does not move on shore. Dangerous and apt to bite. Highly venomous with records of having caused death to man bitten. Up to 1.9m in total length.

Distribution — Japan, Ryukyu, China, Indo-Pacific, Persian Gulf

Hydrophiidae

⇩ *Pelamis platurus* (Linnaeus)

Yellow-bellied Sea Snake, Pelagic Sea Snake, Hawaiian Yellow-bellied Sea Snake

Diagnosis — Head is elongated; snout is long, dorsal side black and ventral side yellowish. Body is covered with small, spineless scales. Ventral scales are rather small, 265−400 in number. Body color varies widely. A widely distributed species highly adapted to swimming. Ferocious and often attacks man. Highly venomous, sometimes causes death to man bitten. Up to 1.2m in total length.

Distribution — Japan, Ryukyu, China, South East Asia, Indo-Pacific, Persian Gulf, Central and South America

Hydrophiidae
⇦ ***Emydocephalus ijimae*** Stejneger
[*E. annulatus ijimae* (Stejneger)]
Turtle-headed Sea Snake
Diagnosis — Body is cylindrical and somewhat blunt. Body has 30 and tail 4−5 dark brown, serrate-margined bands on yellowish-brown skin. 134−144 broad ventral scales. Also found on shore. Not strongly poisonous. Up to 90cm in total length.
Distribution — Ryukyu, Formosa

Hydrophiidae
⇩ ***Laticauda laticaudata*** (Linnaeus)
[*L. laticaudata affinis* (Anderson)]
Black-banded Sea Snake
Diagnosis — Resembles *L. semifasciata*. Trunk has 30−58, tail has 4−6 black bands. Snout tip is yellow, upper labium is black. Often found on shore. Spawn on shoreline reefs in summer. Highly venomous but not likely to bite. Up to 1m in total length.
Distribution — Ryukyu, Philippines, Australia, Polynesia, India

Hydrophiidae
⇦ ***Laticauda semifasciata*** (Reinwardt)
Erabu Sea Snake
Diagnosis — A common species often found on shore. Body is blunt, tail is flat like a spatula. Ventral scales are broad, 190−215 in number. Trunk has 25−45 and tail 5−7 brown bands. Lays 3−7 eggs in shoreline caves between fall and winter. Not ferocious but highly venomous. Up to 1.8m in total length.
Distribution — Japan, Ryukyu, Formosa, South East Asia, Pacific

Hydrophiidae
⇩ ***Lapemis hardwickii*** Gray
Hardwick's Sea Snake
Diagnosis — Body is blunt. Trunk has 31−41 and tail 5−8 blackish-brown bands. Ventral scales are 120−210 in number, each having 2 spines. Dangerous, often attacks man. Highly venomous with records of having caused death to man bitten. Up to 90cm in total length.
Distribution — Ryukyu, China, South East Asia, North Australia

Hydrophiidae
Laticauda colubrina (Schneider) ⇨
Yellow-lip Sea Snake,
Banded Yellow-lip Sea Snake
Diagnosis — Body is rather thick and rounded. Head is black, snout tip is yellow. Trunk has 30−50 and tail 3−6 black bands. Lays 5−6 eggs on shore. Not ferocious but highly venomous. Up to 1.5m in total length.
Distribution — Southern Japan, Ryukyu, Indo-Pacific

Vertebrata

Cartilaginous Fishes
Bony Fishes

Body is fusiform.

Body is oval or deep.

Body is rounded or remarkably deep.

Measuring and Terms

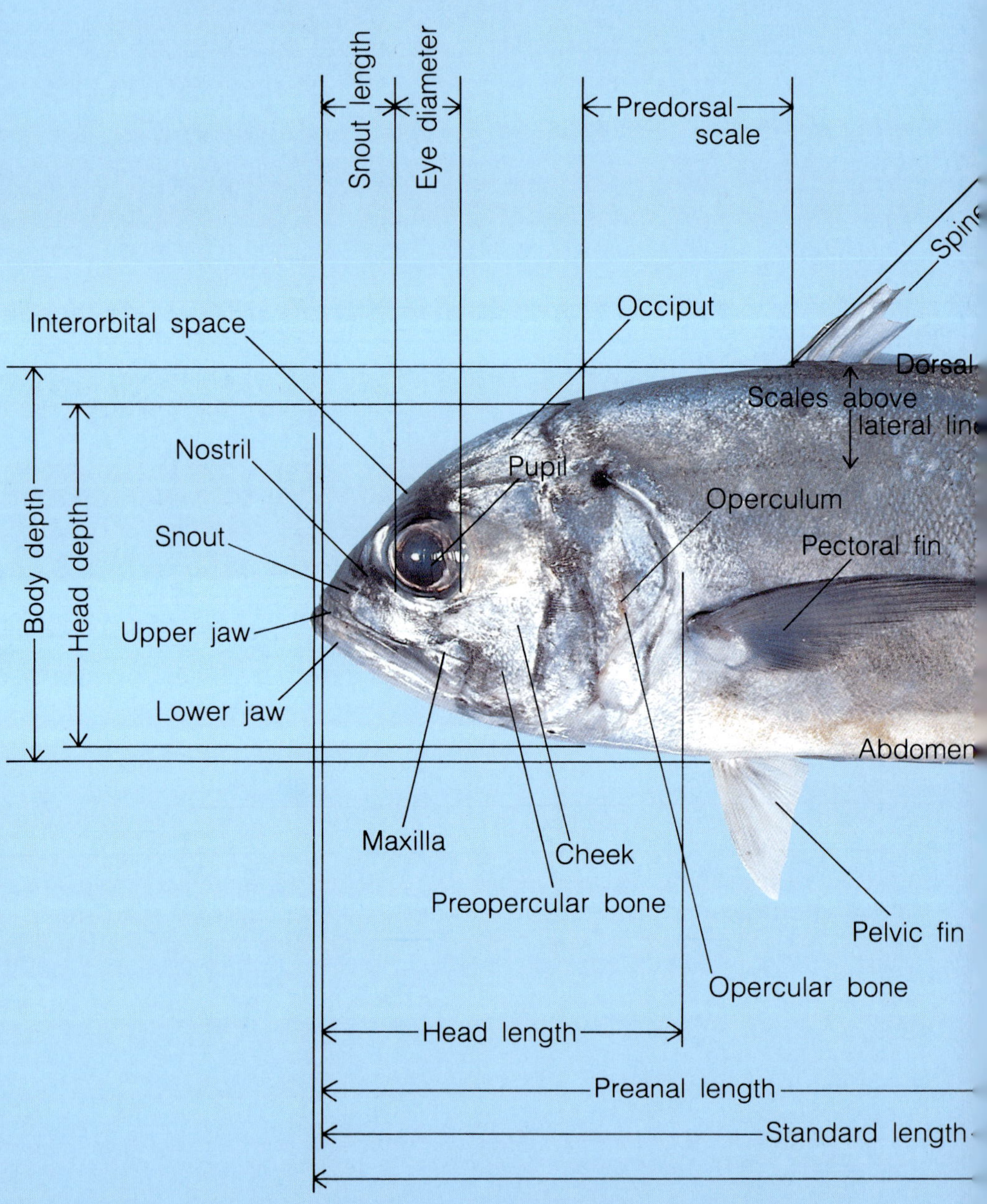

Caudal fin is rounded.

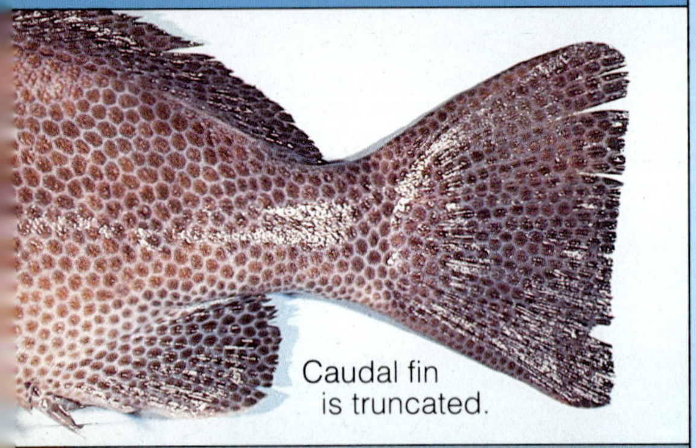
Caudal fin is truncated.

Caudal fin is emarginated.

Caudal lobes are pointed.

Caudal fin is deeply falcated.

Caudal fin is double-emarginated.

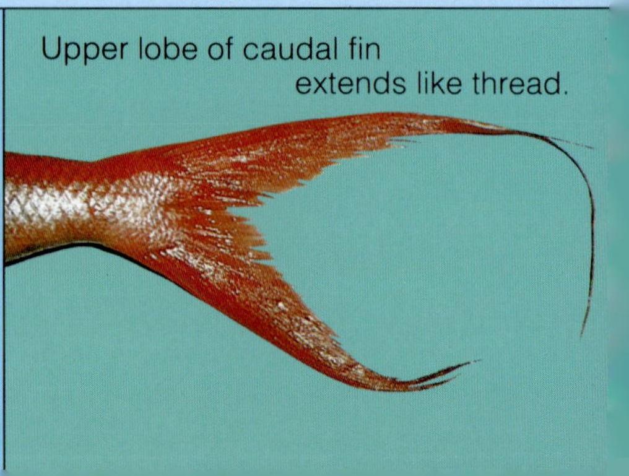
Upper lobe of caudal fin extends like thread.

Body is elongated.

Dorsal fin has a venomous spine.

Pelvic fins form a sucker.

# of Bony Fishes

Dorsal fin forms sucking disk.

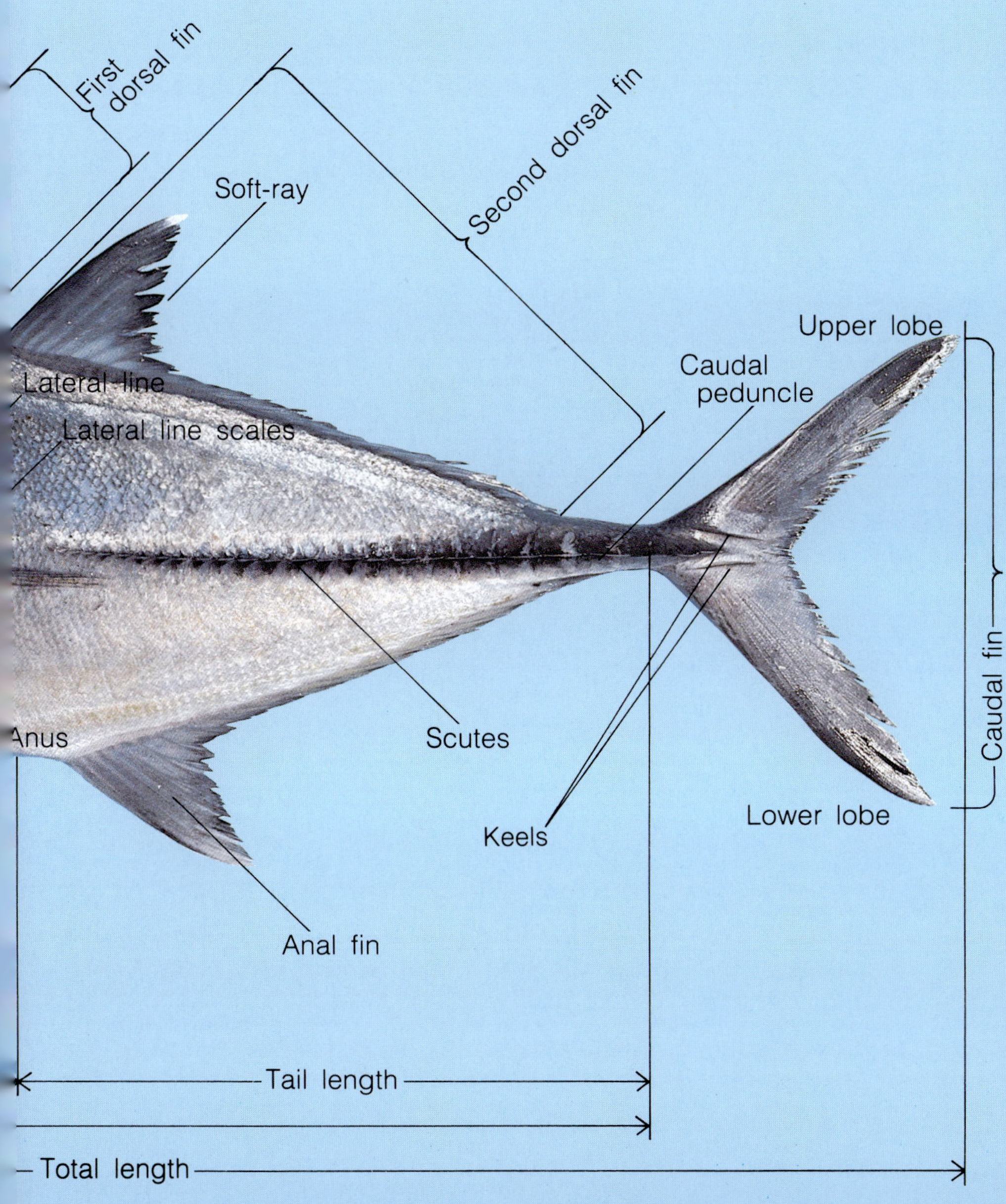

Caudal peduncles have a spine.

First dorsal fin forms a long spine.

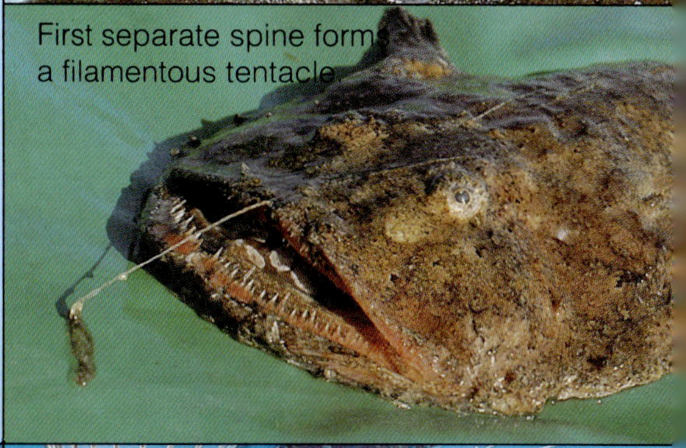
First separate spine forms a filamentous tentacle.

Tail has no caudal fin.

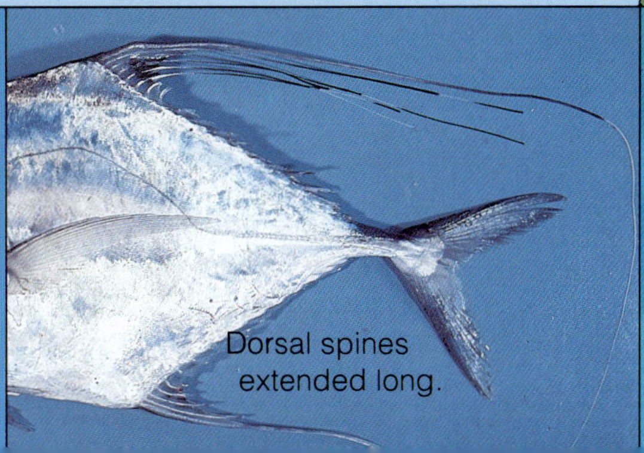
Dorsal spines extended long.

Pectoral fin rays are branched.

Lower jaw has barbels.

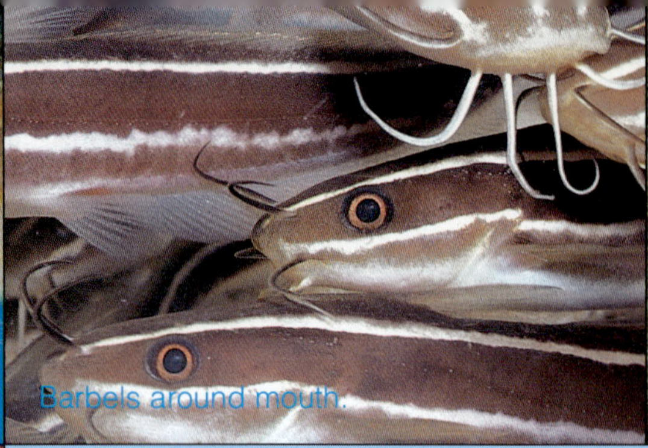

Barbels around mouth.

Mouth has dermal lobes.

Measuring and Terms

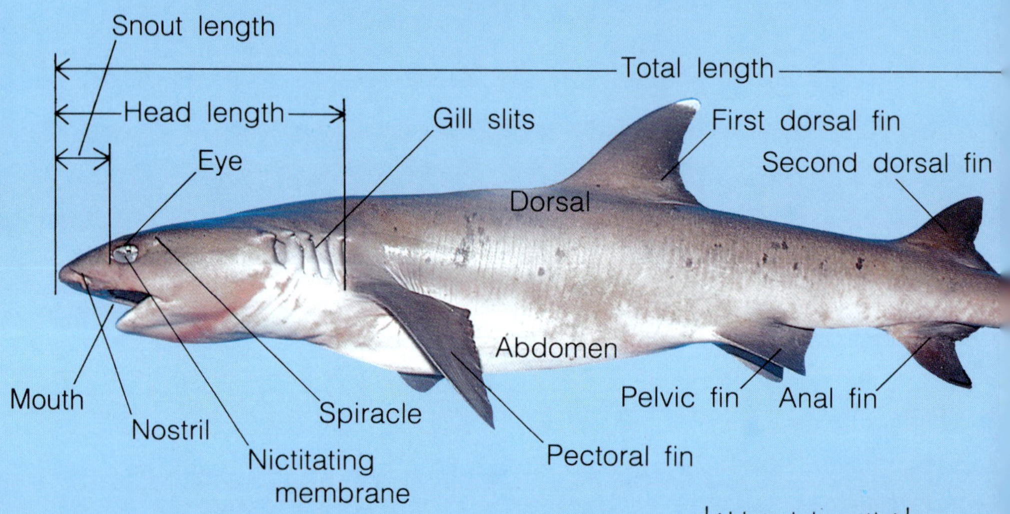

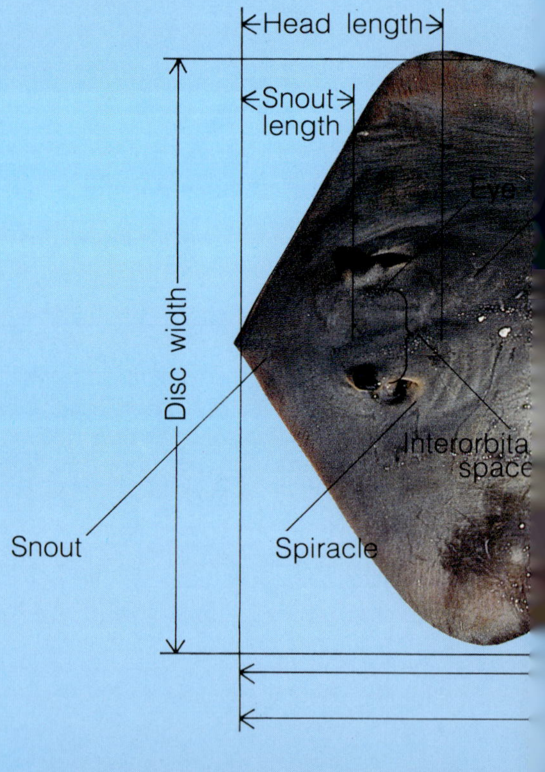

Preoperculum have a strong spine.

Preoperculum have a deep notch.

Eyes have nictitating membrane.

Eyes have adipose eyelids.

Body is flat.

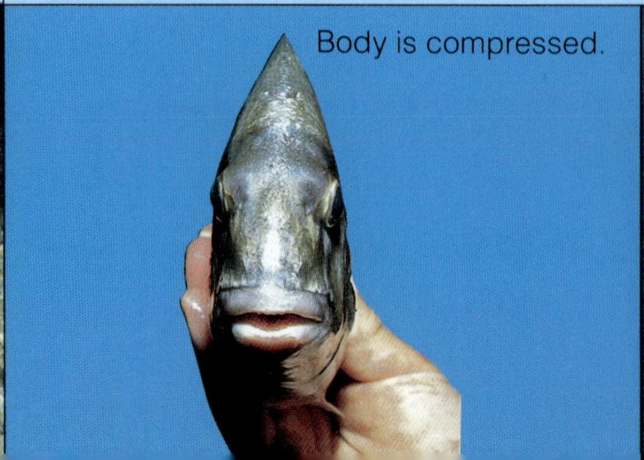

Body is compressed.

Body has oblique stripes.

Lower jaw is elongated.

Snout is projected.

Mouth is extensible.

# of Cartilaginous Fishes

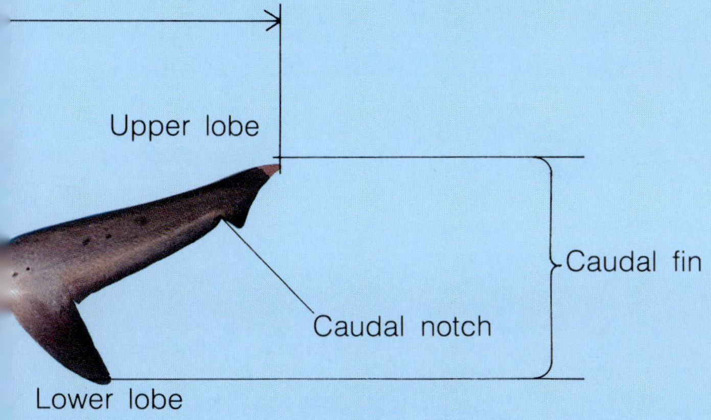

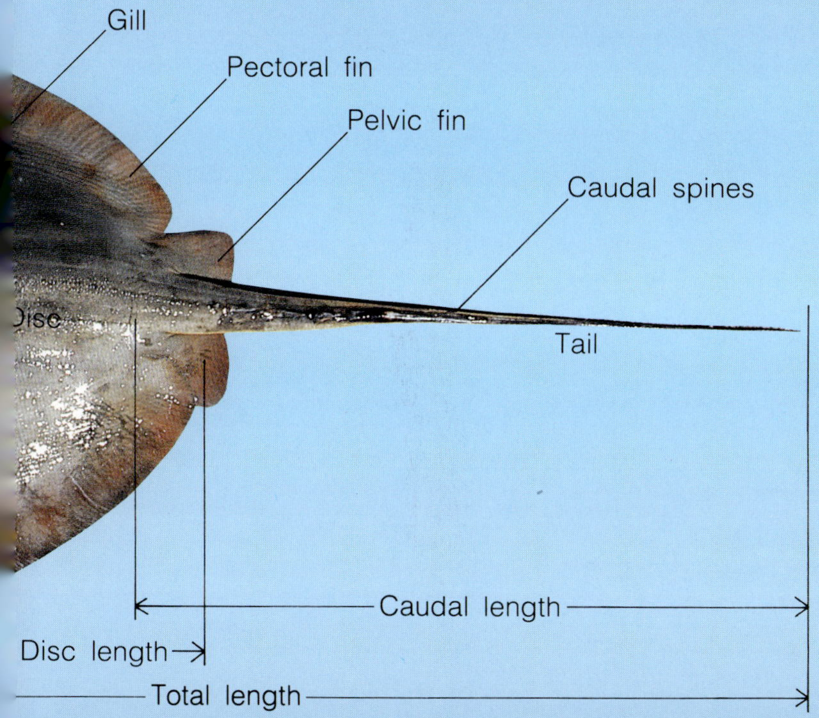

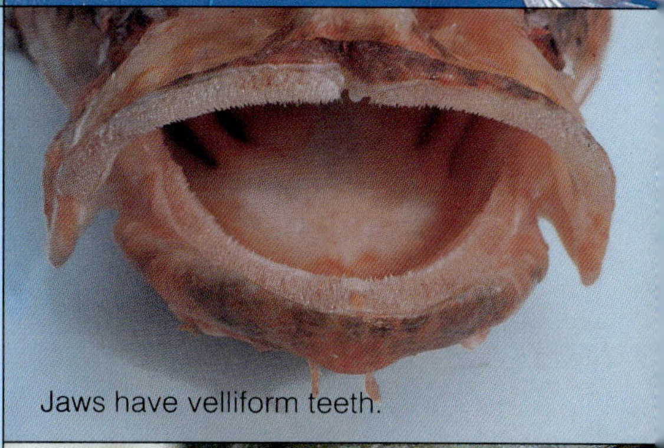

Jaws have velliform teeth.

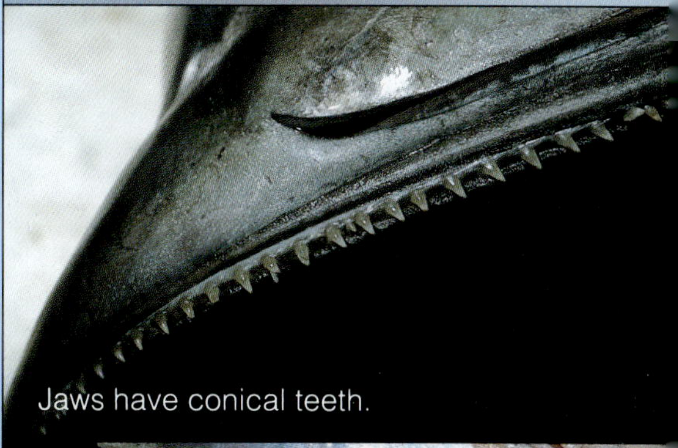

Jaws have conical teeth.

Jaws have canines.

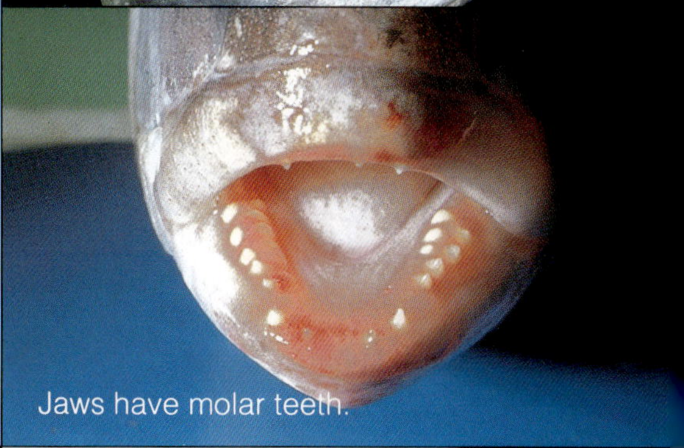

Jaws have molar teeth.

Body has cross bands.

Body has logitudinal stripes.

Jaws have incisor-like teeth.

Heterodontidae

⇦⇩ ***Heterodontus japonicus*** (Duméril)

Port Jackson Shark, Japanese Bullhead Shark, Horned Shark, Cat Shark, Cestracion Shark

Diagnosis — A primitive species found in deeper coastal waters. Head is large, dorsal fins have a strong spine. Feeds on fish and shellfish. Internal fertilization. Large eggs are laid in spring. Not dangerous but apt to bite, sometimes dorsal fin spines cause punctures. Up to 1.5m in total length.

Distribution — Japan, Ryukyu, Korea, East China Sea, East Africa

Hexanchidae

⇩ ***Heptranchias perlo*** (Bonnaterre)

[*H. dakini* Whitley]

Seven-gilled Shark, Sharp-headed, Sharphead Sevengill, Sharp-snouted Seven-gilled Shark, Sharpnose Sevengill Shark

Diagnosis — Inhabits waters more than 300m deep. Dorsal fin is single and small. Caudal fin is long. Snout is elongated with pointed end. Eyes are large, without nictitating membranes. Gills have 7 pairs of slits. Ovoviviparous. Yields oil. Up to 2m in total length.

Distribution — Japan, Formosa, Indo-Pacific-Atlantic, Mediterranean, South Africa

Heterodontidae

⇧ ***Heterodontus francisci*** (Gilard)

Horn Shark, Bullhead Shark, Port Jackson Shark

Diagnosis — Resembles *H. japonicus*, but distinguished by scattered black spots. Dorsal fins have poisonous strong spines. Eyes lack nictitating membranes. Gills have 5 pairs of slits. Up to 1.2m in total length.

Distribution — California, Gulf of California

Carcharhinidae

Carcharhinus obscurus (Le Sueur) ⇨

[*C. brachyurus* (Günther)]

Dusky Shark, Shovel Nose, Bay Shark, Brown Shark, Dusky Ground Shark

Diagnosis — A medium-sized species inhabiting shallow and bottoms of coastal to offshore waters. Claimed to be dangerous to man. Dorsal fins are low, second dorsal fin is smaller than anal fin. Upper teeth are triangular, lower teeth are thin and pointed. Flesh, skin and oil are used industrially. Up to 5m in total length.

Distribution — Japan, Ryukyu, Indo-Pacific-Atlantic

Carcharhinidae

Carcharhinus amblyrhinchos (Bleeker) ⇧⇗

[*C. menisorrah* (Müller et Henle)]

[*C. nesiotes* Snyder]

Gray Reef Shark, Long-nosed Blacktail Shark

Diagnosis — An inhabitant of tropical and subtropical waters. Dangerous to man. Often found in strong currents. Characterized by caudal fin margined with black, fins are black in lower parts. Upper teeth are triangular, lower teeth are pointed. Viviparous. Brings forth some 6 living youngs. Up to 2.5m in total length.

Distribution — Ryukyu, Indo-Pacific, Easter Island

Carcharhinidae

⇧ *Carcharinus galapagensis* (Snodgrass et Heller)

Galapagos Shark

Diagnosis — A widely distributed medium sized tropical and subtropical littoral species. Not ferocious but sometimes bites man. Characterized by broad snout, large and pointed first dorsal fin situated anterior of the body. Viviparous. Brings forth some 10 living youngs. Up to 3m in total length.

Distribution — Ryukyu, Bonin Islands, Indo-Pacific-Atlantic, Galapagos Islands

Carcharhinidae

Carcharhinus falciformis (Bibron) ⇨

[*C. malpeloensis* Fowler]

[*C. floridanus* (Bigelow, Schroeder et Springer)]

Silky Shark, Sickle Shark

Diagnosis — A widely distributed offshore species of tropical waters. Found largely in shallow waters. Body is elongated, surface is silky. Inner corners of second dorsal fin and anal fin are extended. Not ferocious but dangerous to man. Skin and flesh are used industrially. Up to 3m in total length.

Distribution — Southern Japan, Ryukyu, Indo-Pacific-Atlantic, South Africa, California

Carcharhinidae

⇦ *Carcharhinus melanopterus* (Quoy et Gaimard)

Blackfin Reef Shark, Pacific Blacktip, Blacktip Shark, Black-finned Shark, Ground Shark, Blacktip

Diagnosis — A medium-sized species of shallow tropical and subtropical waters. Body is elongated; snout short, head flat and fins black-tipped. Triangular teeth. Not ferocious but sometimes attacks man. Up to 3.5m in total length.

Distribution — Southern Japan, Ryukyu, Indo-Pacific, Red Sea, Mediterranean, South Africa

Carcharhinidae

Prionaca glauca (Linnaeus) ⇦ ⇩

Blue Shark, Blue Whaler, Blue Dog, Great Blue Shark

Diagnosis — A widely distributed species. Body is elongated and dark-blue. Spiracles are absent. Eyes have nictitating membranes. Serrated teeth curved to outside. Ovoviviparous. Brings forth some 50 living youngs. Flesh is unpalatable. Fins and oil are used industrially. Up to 6m in total length.

Distribution — Japan, Ryukyu, Indo-Pacific-Atlantic, Mediterranean

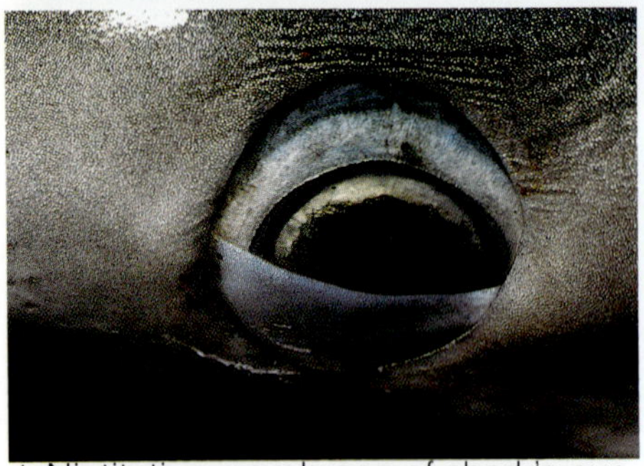

▲ Nictitating membrane of shark's eye.

Carcharhinidae

Negaprion acutidens (Rüppell) ⇨

South Pacific Lemon Shark

Diagnosis — A littoral species. Not ferocious. Characterized by yellowish-brown body; large second dorsal fin; pointed teeth. Ovoviviparous. Brings forth some 10 living youngs. Up to 3m in total length.

Distribution — Ryukyu, Indo-Pacific, Red Sea

Carcharhinidae

Triaenodon obesus (Rüppell) ⇧ ⇨

Reef Whitetip Shark, White-tipped Reef Shark, Whitetip Shark, Blunthead Shark, Hound Shark, Lagoon White-tipped Shark

Diagnosis — A widely distributed littoral species of tropical and subtropical waters. Not ferocious but sometimes bites man. Tips of dorsal and caudal fins are white. Teeth are small. Eyes have nictitating membranes. Spiracles are not distinct. Upper lobe of caudal fin is long. Edible. Up to 2m in total length.

Distribution — Japan, Ryukyu, Bonin Islands, Indo-Pacific, Red Sea

Carcharhinidae
Galeocerdo cuvieri

(Péron et Le Sueur) ⇨

[*G. arcticus* (Faber)] ⇩

Tiger Shark, Leopard Shark, Spotted Shark, Northwestern Tiger Shark, Rayner's Tiger Shark

Diagnosis — A widely distributed large species of tropical and subtropical waters. Ferocious and dangerous to man. Body has black cross bands. Teeth are serrated. Eyes have nictitating membranes. Gills have 5 pairs of slits. Ovoviviparous. Brings forth several living youngs. Flesh is unpalatable. Skin and oil are used industrially. Up to 9m in total length.

Distribution — Japan, Ryukyu, Indo-Pacific-Atlantic

▲ Dangerous man-eater, Tiger Shark.

▲ An accident by shark attack in French Polynesia.

(Courtesy of Mr. J. Tapu)

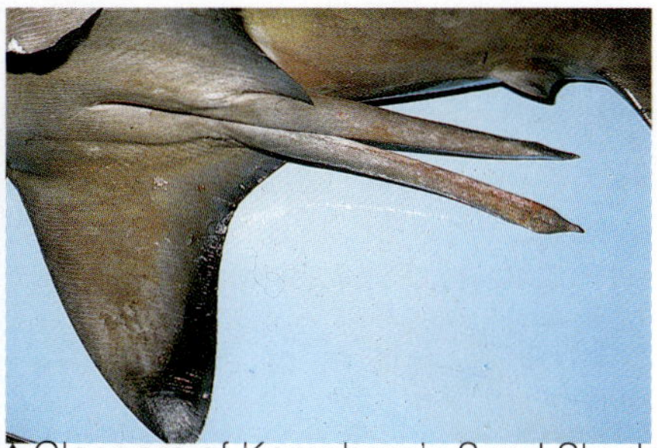

▲Claspers of Kamohara's Sand Shark.

Odontaspididae
⇦ *Pseudocarcharias kamoharai*

(Matsubara) ⇨

Kamohara's Sand Shark

Diagnosis — A deep-sea species. Pelvic fins are larger than either of second dorsal or anal fins. Second dorsal fin is shorter than first dorsal fin. Eyes lack nictitating membranes, spiracles are small. Gills have 5 pairs of long slits. Teeth are curved, forming fangs. Up to 1.5m in total length.

Distribution — Southern Japan, Ryukyu, Korea, Indo-Pacific-East Atlantic

Odontaspididae

⇦ ***Odontaspis ferox*** (Risso)

Fierce Odontaspis, Fierce Shark, Ragged-tooth Shark

Diagnosis — A deep-sea species. Body is blunt, head is depressed. Mouth is large, eyes are small. First dorsal fin is situated close to pectoral fins. Caudal peduncles have no keels. Fins are black margined. Liver yields much squalene. Up to 3m in total length.

Distribution — Ryukyu, Izu Islands, Pacific, Atlantic, Mediteranean

Orectolobidae

⇧ ***Chiloscyllium griseum*** Müller et Henle

Blackbanded Cat Shark, Brown Carpet Shark

Diagnosis — A small species inhabiting bottoms of coastal waters. Body is elongated, tail is long. Back has a single ridge. Dorsal fin is smaller than pelvic fins. Mouth has labial folds. Youngs are black-banded or black-spotted. Edible. Up to 2m in total length.

Distribution — Japan, China, South East Asia, Indo-Pacific, Red Sea, South Africa

▼ Nostrils and mouth of Brown Banded Cat Shark.

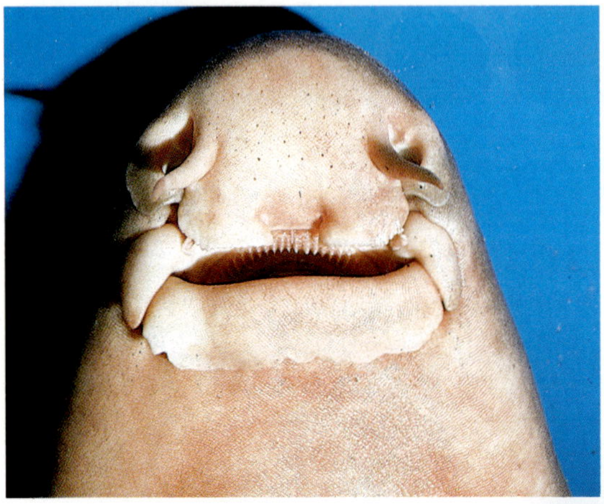

Orectolobidae

Chiloscyllium punctatum Müller et Henle ⇧ ⇨

Brown Banded Cat Shark, Spotted Catshark, Brown-spotted Catshark

Diagnosis — A littoral harmless species. Tail is long, back has a single ridge. Dorsal fin is larger than pelvic fins. Body has more than 10 black bands. Edible. Up to 1m in total length.

Distribution — Formosa, Philippines, Singapore, Indonesia, Australia

Orectolobidae

Chiloscyllium plagiosum (Bennett)

[*Hemiscyllium plagiosum* Bennett]

Catshark

Diagnosis — Body is elongated with 12 transverse bands and white spots. First dorsal fin originates in the middle part of pelvic fin bases. Both dorsal fins are smaller than pelvic fins, have unpointed inner corners. Anal fin base is shorter than lower lobe of caudal fin. Gills have 5 pairs of slits. Mouth has labial folds around. Edible. Up to 1m in total length.

Distribution — Japan, Formosa, China, Indo-Pacific, South Africa

Orectolobidae

Stegostoma fasciatum (Hermann)

[*S. varium* (Seba)]

[*S. tigrinum* Bonnaterre]

Leopard Shark, Monkey-mouth, Pacific Zebra Shark, Tiger Shark, Zebra Shark

Diagnosis — A large species inhabiting bottoms of coastal waters. Tail is long, back has 5 ridges. Second dorsal fin is small. Mouth has labial folds. Ovoviviparous. Feeds on crustacea and shellfish. Up to 3.6m in total length.

Distribution — Japan, Ryukyu, Indo-Pacific, Red Sea, South Africa

Orectolobidae

Orectolobus japonicus Regan

Fringe Shark, Japanese Carpet Shark, Wobbegong

Diagnosis — A small species inhabiting bottom of coastal waters. Flat-bodied. Tail is elongated, snout is broad. Mouth is large, has dermal lobes. Nocturnal. Ovoviviparous. Brings forth some 20 living youngs. Edible. Up to 1m in total length.

Distribution — Japan, Ryukyu, South China Sea, Korea, West Pacific

▼Fringe Shark has dermal lobes around mouth.

Rhincodontidae

Rhincodon typus Smith

Basking Shark, East Indian Basking Shark, Whale Shark

Diagnosis — A widely distributed in tropical and temperate waters. Well known as the largest of all fishes. Head is flat, mouth is broad, has a large number of small teeth. Eyes are small, lack nictitating membranes. Floating in the surface, feeds on small fish, squid and crab. Harmless. Edible. Up to 18m in total length.

Distribution — Japan, Ryukyu, Formosa, Indo-Pacific-Atlantic

▲ Broad mouth has a large numbers of small teeth.

▲ Vertical section of cartilaginous vertebra.

▲ Gill Slits.

▲ Inner epitherial tissue of stomach.

▲ Largest claspers of fishes.

▲ Cross section of tail.

Scyliorhinidae
Halaelurus buergeri
Müller et Henle ⇨

Blackspotted Dogfish, Spotted Cat Shark
Diagnosis — A small species inhabiting bottoms of coastal waters. Body is elongated, tail is thin. Head is small, snout is pointed. First dorsal fin is larger than the second. Mouth has a large number of small three-cusped teeth. Edible. Up to 1m in total length.
Distribution — Japan, Ryukyu, Formosa, South East Asia, India

Scyliorhinidae
Parmaturus pilosus Garman ⇨

Salamander Shark
Diagnosis — A deep-sea species. Head is roundish, snout is short. Enlarged dermal denticles are present on upper margin of caudal fin. Oviparous. Up to 50cm in total length.
Distribution — Japan, Ryukyu, West Pacific

Scyliorhinidae
Galeus eastmani
(Jordan et Snyder) ⇨
[*Pristiurus eastmani* Jordan et Snyder]

Gecko Shark
Diagnosis — A deep-sea species. Body is elongated, snout is long. Spiracles are small. Gill slits are 5-paired. Teeth are small with seven cusps. 2 rows of dermal denticles are present on upper margin of caudal fin. Oviparous. Up to 40cm in total length.
Distribution — Japan, Ryukyu, East China Sea

Scyliorhinidae
Apristurus platyrhinchus
(Tanaka) ⇨

Spatularsnout Cat Shark
Diagnosis — A deep-sea species. Head is flat and broad. Spiracles are small. First dorsal fin is small and situated between pelvic and anal fins. Anal base is long. Oviparous. Up to 70cm in total length.
Distribution — Japan, Ryukyu

▼ Egg-capsules of Dogfish.

Scyliorhinidae
Scyliorhinus torazame (Tanaka) ⇧ ⇨
[*S. rudis* Pietschmann]

Cloudy Dogfish, Dogfish, Tiger Shark
Diagnosis — A cold water species of small size inhabiting bottoms of coastal waters. Head is small, snout is short. Eyes are long and narrow, lack nictitating membranes. Oviparous. Edible. Up to 1.5m in total length.
Distribution — Japan, Korea, Philippines

▼ Swell Shark has small teeth.

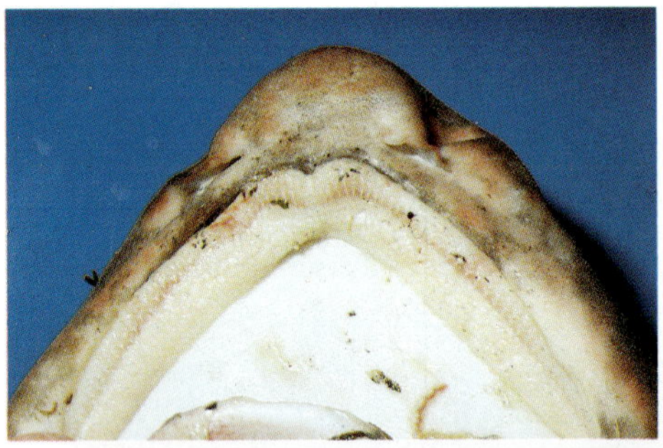

Scyliorhinidae
⇦ *Cephaloscyllium umbratile* Jordan et Fowler
⇩ **Swell Shark, Sevendays Shark, Blotchy Swell Shark**
Diagnosis — A deep-sea species of small size. Head is small, belly is large, inflatable with air. Dorsal fin is situated posterior of the body. Eyes are long and narrow, lack nictitating membranes. Teeth are small, three-cusped. Oviparous. Up to 1m in total length.
Distribution — Japan, Ryukyu, Formosa, East China Sea, Australia

Pseudotriakidae
Pseudotriakis acrales
Jordan et Snyder ⇨

[*Pseudotriacis acrages* Jordan et Snyder]

Dumb Shark

Diagnosis — A deep-sea species. Body is blunt, head is depressed. Mouth is wide. First dorsal fin is short, base is long. Second dorsal and anal fins are large, corresponding with each other. Pelvic fins originate between first and second dorsal fins. Fins are black-margined. Teeth are small and many. Up to 1.3m in total length.

Distribution — Japan, Ryukyu, Hawaii

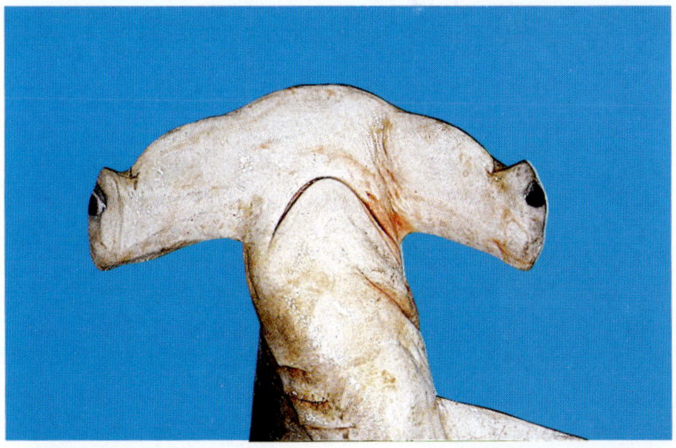

Sphyrnidae
Sphyrna zygaena (Linnaeus)

Common Hammer-head Shark, Smooth Hammer-head Shark, Shovelnose Shark, Cross-staff Shark, Balance Fish, Cornuda, True Hammerhead

Diagnosis — A widely distributed offshore species of warm waters. Ferocious and destructive. Sometimes dangerous to man. Head resembles a double-headed hammer, protruded in the middle, front margin is three-lobed. First dorsal and pectoral fins are approximately equal in size. Flesh is whitish. Viviparous. Brings forth some 35 living youngs. Edible. Up to 5m in total length.

Distribution — Japan, Ryukyu, Indo-Pacific-Atlantic, Mediterranean

◀ Ventral side of Hammer-head Shark.

Sphyrnidae

⇦ ***Sphyrna lewini*** (Griffith et Smith)

[*S. diplana* Spengler]

Scalloped Hammer-head Shark, Kidney-headed Shark, Bronze Hammerhead, Southern Hammerhead Shark, Balance Fish, Cornuda

Diagnosis — A widely distributed large species found in middle layers of warm waters. Dangerous to man. Flesh is reddish. Known as the most abundant of all hammerhead sharks. Characterized by middle-depressed, four-lobed head resembling a double-head hammer. Eyes are large. Edible. Up to 5m in total length.

Distribution — Japan, Ryukyu, Formosa, Indo-Pacific-Atlantic

Triakidae

⇩ ***Triakis semifasciata*** Girard

Leopard Shark, Cat Shark

Diagnosis — A species inhabiting bottoms of coastal waters. Not ferocious but sometimes bites man. Body is elongated, snout is pointed. Back has black spots and bands. Feeds on small fish, crab, shrimp. Edible. Up to 2m in total length.

Distribution — Gulf of California

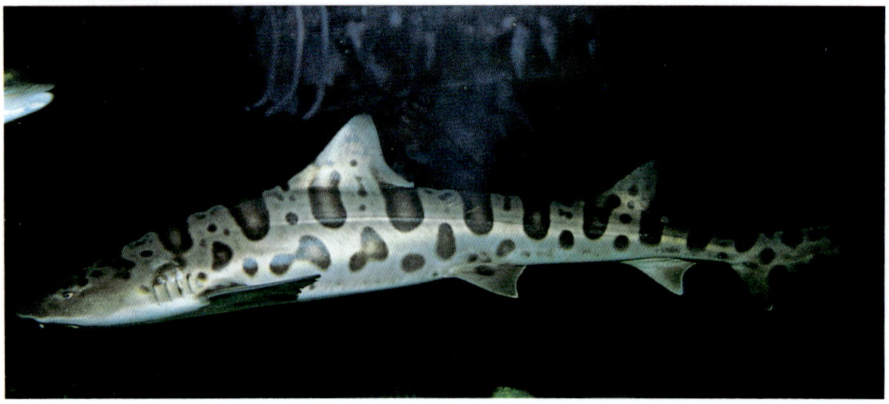

▼ Male (claspers)

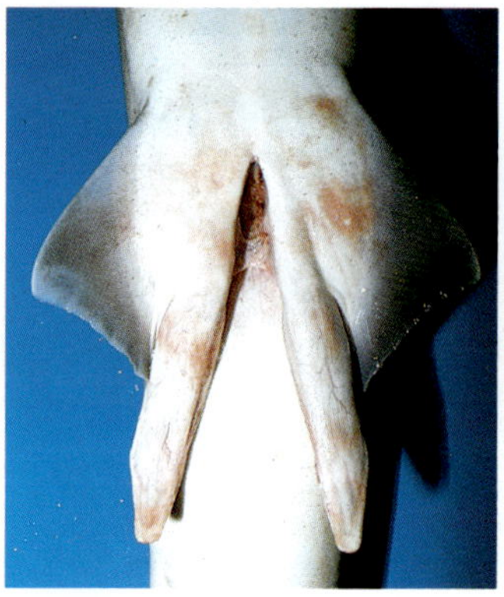

▼ Female

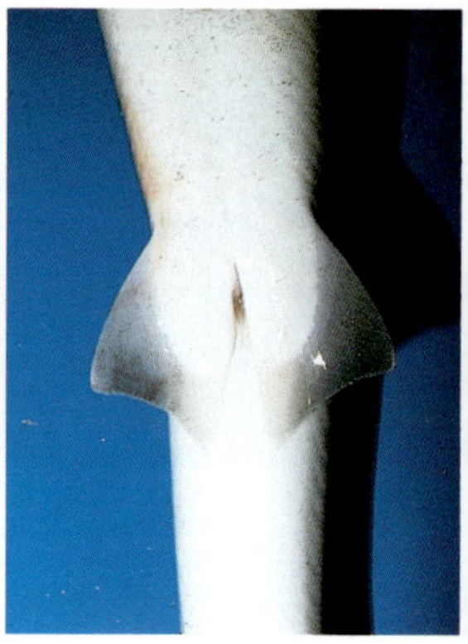

Triakidae

⇦ ***Mustelus kanekonis*** (Tanaka) ⇖⇧

Kaneko's Dog Shark

(New Common Name)

Diagnosis — Body is elongated, tail is thin. Origin of first dorsal fin is closer to snout than to origin of second dorsal fin. Tips of pectoral fins extend beyond first dorsal fin origin. Teeth are flat, not pointed. Edible. Up to 1.5m in total length.

Distribution — Japan, Formosa

Triakidae

Mustelus griseus Pietschmann ⇨

Dog Shark, Japan Gray Smoothhound, Spotless Smoothhound

Diagnosis — A small species inhabiting bottoms of coastal waters. Body is elongated and splotchless. Origin of first dorsal fin is situated approximately between second dorsal fin origin and snout tip. Tips of pectoral fins do not reach first dorsal fin origin. Feeds on small fish and shellfish. Edible. Up to 1.7m in total length.

Distribution — Japan, East China Sea, Indo-West Pacific

Triakidae

Mustelus manazo Bleeker

Smooth Dog Fish, Smooth Hound, Gummy Shark, Japan Smoothhound, Star-spotted Shark, Hound

Diagnosis — A small species inhabiting coastal sandy bottom. Body is elongated, scattered with white spots. Origin of first dorsal fin is closer to snout tip than to second dorsal fin origin. Teeth are flat. Feeds on crustacea and shellfish. Ovoviviparous. Brings forth some 15 living youngs. Edible. Up to 1.5m in total length.

Distribution — Japan, Ryukyu, Formosa, Korea, Indo-West Pacific, Persian Gulf, Mediterranean, South Africa

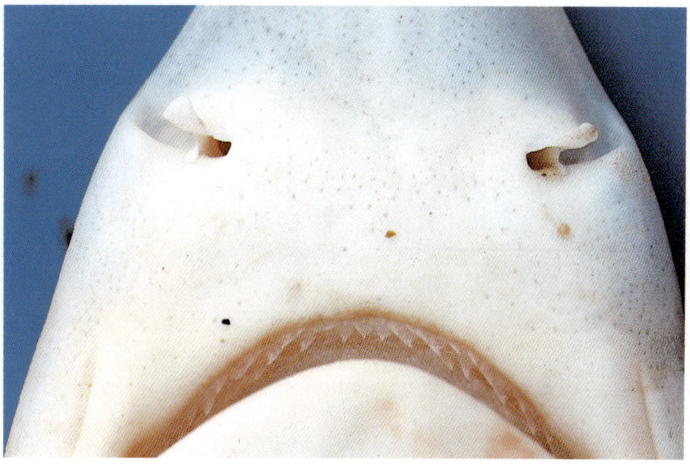

▲ Nostrils and mouth of Smooth Dog Fish.

Triakidae

Triakis scyllia Müller et Henle

Leopard Shark, Koro, Banded Dogfish, Incense Burner

Diagnosis — A small species inhabiting bottoms of coastal waters. Body has dark brown cross bands and small black spots. Eyes have nictitating membranes. Teeth are small and pointed. Nocturnal. Feeds on small fish. Ovoviviparous. Brings forth some 20 living youngs. Edible. Up to 1.5m in total length.

Distribution — Japan, Ryukyu, Formosa, Korea, Indo-West Pacific, Australia

Squalidae
Etomopterus lucifer
Jordan et Snyder ⇨

Deepwater Dogfish, Priest Shark, Spiny Shark

Diagnosis — Inhabits waters not deeper than 1,000m. Body is purplish-black, elongated with black splotches on abdominal side. Caudal fin is slender, upper lobe is extended. Scales are spinelike, form in line on dorsal side. Eyes and spiracles are large. Viviparous. Emit light through a luminous organ. Up to 40cm in total length.

Distribution — Japan, Ryukyu, China, Korea, Indo-West Pacific, South Atlantic

Squalidae
⇦⇩ *Etmopterus princeps* Collett

Crow Shark, Lantern Shark

Diagnosis — A small deep-sea species. Body is thick, snout is rounded and flat. Lower lobe of caudal fin is large. Eyes are black on the upper side. Tooth is three-cusped. Dorsal fins are spined. Liver oil is used for Squalene. Up to 1m in total length.

Distribution — Japan, Kyushu-Palao Ridge, Atlantic

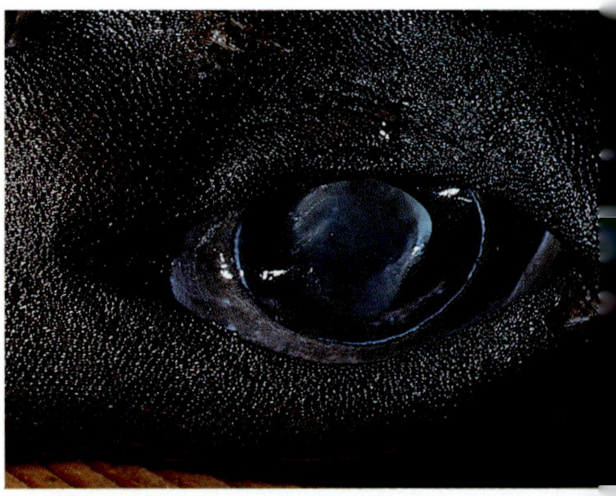

▲ Eyes have no nictitating membrane.

Squalidae
⇦ *Squalus blainvillei* (Risso)

Blainville's Dogfish, Northern Dogfish

Diagnosis — A deep-sea species. Resembles *S. japonicus*, but distinguished by shorter snout and wider mouth. First dorsal fin is large. Ovoviviparous. Liver yields much squalene. Up to 1m in total length.

Distribution — Ryukyu, Indo-Pacific-Atlantic, Mediterranean

Squalidae

⇦⇩ ***Squalus japonicus*** Ishikawa

Japanese Dogfish Shark

(New Common Name)

Diagnosis — First dorsal fin is large, originate in inner corners of pectoral fins. Both dorsal fins have sharp spines. Eyes are large, have no nictitating membranes. Snout is elongated. Dermal denticles are tricusped and broad. Ovoviviparous. Yields oil. Up to 80cm in total length.

Distribution — Southern Japan, Ryukyu

Squalidae

Squalus mitsukurii

Jordan et Fowler ⇨

Mitsukuri's Dogfish

(New Common Name)

Diagnosis — A deep-sea species. Snout is short, mouth is wide. Dorsal fins are short, spines are long. Upper caudal lobe is not lunated. Up to 50cm in total length.

Distribution — Japan, Ryukyu, China, Korea, East China Sea, Hawaii

Squalidae
Centrophorus squamosus (Bonnaterre) ⇨

[*Lepidorhinus squamosus* (Bonnaterre)]

Leafyscale Dogfish

Diagnosis — Inhabits waters 700−1,100m deep. Body is dark-black. First dorsal fin is lower than second, dorsal spines are not long. Inner corner of pelvic fins are not protruded. In both jaws, teeth are curved downward. Ovoviviparous. Liver yields much squalene. Up to 1.5m in total length.

Distribution — Japan, Ryukyu, Indo-Pacific-Atlantic

Squalidae
Cirrhigaleus barbifer Tanaka ⇩

Barbed Spiny Dogfish ⇩

Diagnosis — A small deep-sea species. Dorsal fins have sharp poisonous spines. Nostrils have a pair of long barbels appended to tip. Eyes are large, lack nictitating membranes. Ovoviviparous. Up to 1.5m in total length.

Distribution — Japan, East China Sea, New Zealand

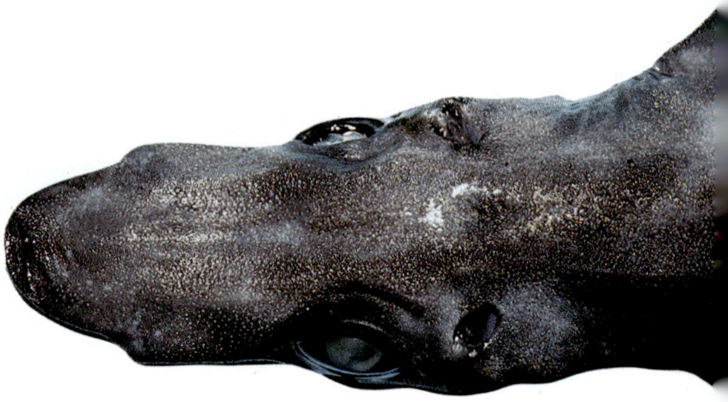

Squalidae
Deania càlcea (Lowe) ⇧

[*D. eglantina* Jordan et Snyder]

Broad Harbor Shark, Beaked Dogfish, Bird's-beak Dogfish, Spiny Shark, Sagami Dogfish

Diagnosis — A small deep-sea species. Body is elongated, snout is protruded, resembling a spatula. Dorsal fins have sharp poisonous spines. Eyes are large, lack nictitating membranes. Ovoviviparous. Liver is large, yields Squalene. Up to 1m in total length.

Distribution — Japan, Ryukyu, Indo-Pacific-North Atlantic, New Zealand, Australia, South Africa

Squalidae

Centrophorus acus Garman ⇨

Taro Shark (New Common Name)

Diagnosis — Inhabits waters more than 600–1,000m deep. Dermal denticles lobe-like with 3–4 longitudinal ridges. First dorsal fin is not high, second is larger than pelvic fins. Inner corners of pectoral fins are short. Eyes are large, have no nictitating membranes. Spiracles are also large. Front of both dorsal fins have sharp spines. Ovoviviparous. Yields oil. Up to 1.6m in total length.

Distribution — Japan, Ryukyu, East China Sea

Squalidae

Centrophorus uyato

(Rafinesque) ⇨

Uyato Dogfish (New Common Name)

Diagnosis — Inhabits waters 400–850m deep. Body is dark-gray, abdominal side is pale. Fins are black-margined. Inner corner of pectoral fins are slightly protruded. Liver yields much squalene. Up to 1.2m in total length.

Distribution — Ryukyu, East Atlantic, Gulf of Mexico

Squalidae

Centrophorus atromarginatus

Garman ⇨

Fin Shark

Diagnosis — A deep-sea species. Upper caudal lobe is lunated. First dorsal fin is longer than second which is smaller than pelvic fins. Inner corner of pectoral fins are slightly extended. Scales are rough. Spiracles are large. Ovoviviparous. Liver yields much squalene. Up to 1.2m in total length.

Distribution — Japan, Ryukyu

Squalidae

⇦ ***Centrophorus niaukang*** Teng

Niaukang Dogfish (New Common Name)

Diagnosis — Inhabits waters 650–1,100m deep. Body is brown, snout is broad and round-tipped. Dorsal spines are short with tips slightly exposed. Second dorsal and pelvic fins are equal in size. Inner corner of pectoral fins are slightly extended. Scales are large conical thorns and scattered. Liver yields much squalene.

Distribution — Ryukyu, Formosa

Squalidae

⇦ ***Centrophorus scalpratus*** McCulloch

[*C. armatus armatus* (Gilchrist)]

Okinawan Dogfish (New Common Name)

Diagnosis — Inhabits waters 300–900m deep. Second dorsal fin is situated close to caudal fin, smaller than first dorsal and pelvic fins, spines are arrowhead-shaped. Inner corner of pectoral fins are extended. Liver yields much squalene. Up to 1m in total length.

Distribution — Ryukyu, South Africa

Squalidae

⇦ ***Centroscymnus owstoni*** Garman

Roughskin Dogfish

Diagnosis — Inhabits waters 800–1,000m deep. Body is dark-brown. Dorsal spines are not stiff, only tips are exposed. Inner corner of pelvic fins are rounded. Scales are rough. Ovoviviparous. Liver yields squalene. Up to 1m in total length.

Distribution — Japan, Ryukyu, New Zealand, West Indian Ocean

Squalidae
Scymnodon squamulosus (Günther) ⇨

[*Centrophorus squamulosus* Günther]
[*C. obscurus* Vaillant]

Velvet Dogfish

Diagnosis — Inhabits waters 450–830m deep, also found in shallow waters. Body is purplish-black. Upper caudal lobe is lunated. Dorsal spines are conspicuously small. Spiracles are large. Scales are tricuspid and rough. Upper teeth are erected upright, lower teeth are directed outward. Ovoviviparous. Up to 1m in total length.

Distribution — Japan, Ryukyu, Atlantic

Pristiophoridae
⇩ *Pristiophorus japonicus* Günther

Japanese Saw Shark, Saw Shark, Halberd Shark

Diagnosis — An inhabitant of shallow shoreline waters and several hundred meters deep waters. Snout is greatly elongated with a sharp-pointed tooth-like structures and a pair of barbels. Mouth is small, teeth are also small and many. Spiracles are large. Dorsal fins are equal in length. Anal fin is absent. Caudal peduncles have a keel. Have a habit of snouting out and eating small benthos in sand. Ovoviviparous. Spawns between spring and summer, producing some 10 youngs in one spawning. Up to 1.5m in total length.

Distribution — Japan, Ryukyu, Formosa, Korea, China

Squatinidae
⇦ *Squatina japonica* Bleeker

Japanese Monk Fish, Canopy Shark, Change Shark, Japanese Angel Shark

Diagnosis — A species inhabiting sandy bottom of shallow waters. Body resembles that of a ray. Pectoral fins are stretched out wide on both sides. Dorsal fins are situated on tail. Back has large scales. Spiracles are large. Skin is rough. Ovoviviparous. Brings forth some 10 living youngs. Skin is used for making files. Up to 2.5m in total length.

Distribution — Japan, Formosa, China, Philippines, East China Sea

Urolophidae

Urotrygon daviesi Wallace ⇨

Flat Round Ray (New Common Name)

Diagnosis — Inhabits waters 300–400m deep. Body is rounded and compressed. Mouth is small. Dorsal fin is absent. Tail has 2 large spines, with a slender fin on tip. Fins are black-margined. Up to 60cm in total length.

Distribution — Kyushu-Palao Ridge, Ryukyu, Hawaii, South Africa

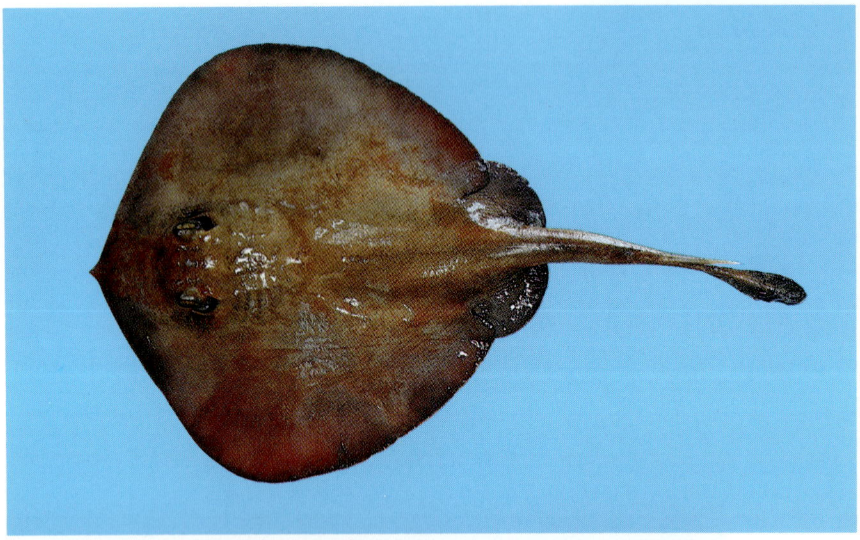

Urolophidae

⇦ *Urolophus aurantiacus* Müller et Henle ⇩

[*U. fuscus* Garman]

Sepia Stingray

Diagnosis — A small species inhabiting sandy bottoms of coastal waters. Resembles *D. akajei*, but distinguished from the latter by shorter tail, also has a caudal fin. Tail has a venom-bearing spine. Ovoviviparous. Brings forth several living youngs. Up to 40cm in total length.

Distribution — Japan, Ryukyu, Formosa, Korea, China,

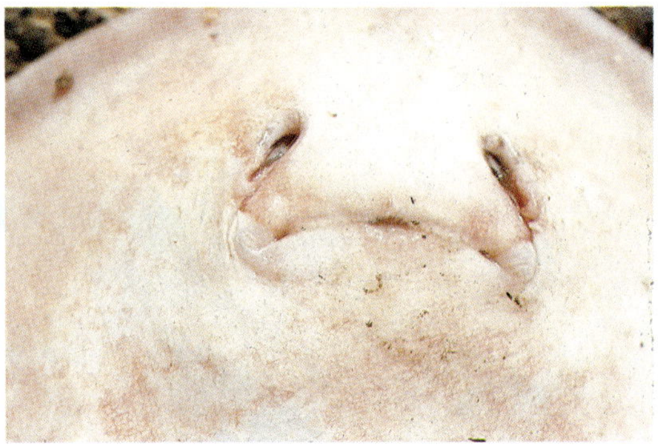

▲ Ventral side of Sepia Stingray.

Dasyatidae

⇦⇩ *Taeniura melanospila* (Bleeker)

[*Dasyatis melanospilos* (Bleeker)]

[*D. maculatus* Miyoshi]

Speckled Stingray, Round Ribbontail Ray, Stingray

Diagnosis — A large species found in shoreline and coral reefs. Disk is rounded, covered with small spines. Tail is longer than disk length, has 1–2 venom-bearing spines. Dorsal spines are large. Up to 2.5m in total length.

Distribution — Japan, Ryukyu, Indo-Pacific, South Africa

▲ Venomous spine of Speckled Stingray.

Dasyatidae

⇦ ***Dasyatis hawaiiensis*** Jenkins

Hawaiian Sting Ray, Brown Sting Ray

Diagnosis — A shallow water species. Pectoral fins are expanded, forming a disk. Spiracles are larger than eyes. Tail is elongated, has a venom-bearing spine, has a single longitudinal fold on top and bottom sides. Ovoviviparous. Feeds on shellfish. Up to 1.5m in total length.

Distribution — Hawaii, California, Peru, East Pacific

Dasyatidae

⇧⇦ ***Dasyatis akajei*** (Müller et Henle)

Red Stingray, Red Skate, Whip Ray, Japanese Sting Ray, Sting Ray

Diagnosis — A species inhabiting sandy-mud bottoms of shallow waters. Disk is broad and rounded. Whip-like tail is long and narrow, has 1−3 venom-bearing spines. Ovoviviparous. Brings forth some 10 living youngs. Sometimes comes to brackish area. Feeds on bivalve, crab, earth worm. Edible. Up to 2m in total length.

Distribution — Japan, East China Sea, Formosa, Korea, China

Dasyatidae

Taeniura lymma (Forskål) ⇨

[*T. halgani* Lesson]

Reef Stingray, Blue-spotted Lagoon Ray, Ribbon-tailed Ray, Fantail

Diagnosis — Inhabits sandy bottoms in shallow coral reefs. Disk is oval in shape, slightly pointed. Tail is longer than disk, has 1−2 venom-bearing spines. Anal fin attached like finlet on underside of tail. Distinguished by bright blue spots, blue-edged tail. Up to 2.5m in total length.

Distribution — Indo-Pacific, Hawaii, Red Sea, South Africa

Dasyatidae

⇦ ***Dasyatis kuhlii*** (Müler et Henle)

Blue-spotted Stingray, Kuhl's Stingray, Spotted Stingray

Diagnosis — A small species inhabiting sandy-mud bottoms of coastal waters. Disk is dotted with blue spots. Snout is short, not protruded. Tail is approximately two times longer than disk length, has 1−2 venom-bearing spines. Up to 1.5m in total length.

Distribution — Japan, Ryukyu, Korea, China, Indo-Pacific, East Africa

▲ Cross Section of disks.

Dasyatidae

⇦ ***Dasyatis zugei*** (Müller et Henle)

Pale-edged Stingray, Stingray

Diagnosis — A large species of shallow waters. Snout is remarkably protruded. Tail is thin, has 1−2 venom-bearing spines; tail length is approximately two times longer than disk. Back has a row of enlarged spines. Edible. Up to 1.5m in total length.

Distribution — Japan, Formosa, China, South East Asia, Indian Ocean

Dasyatidae

Himantura uarnak (Forskål) ⇨

[*H. toshi* Whitley]

[*Dasyatis uarnak* (Forskål)]

Coachwhip Ray, Long-tailed Ray, Stingray, Marbleray, Pylstert, Whipray, Stingaree

Diagnosis — A widely distributed species of large size inhabiting tropical and subtropical waters. Disk is broader than its length. Tail is remarkably long, approximately 3−4 times longer than disk length, has 30−35 dark rings, a venom-bearing spine. Edible. Up to 1.5m at disc width.

Distribution — Ryukyu, Indo-Pacific, Red Sea, South Africa

Dasyatidae

⇦ ***Dasyatis ushiei*** Jordan et Hubbs

Cow Stingray

Diagnosis — A large species of rather deep waters. Disk is broader than its length. Tail is more than two times longer than disk length, has a keel on the upper side; also has a venom-bearing spine. Up to 2m in total length.

Distribution — Japan, East China Sea

Myliobatidae

Aetobatus narinari (Euphrasen) ⇦⇩

[*A. flagellum* Schneider]

Spotted Eagel Ray, Bat Ray, Jumping Ray, Flying Ray, Duckbill Ray, Whipray, Bishop Ray, Bonnet Skate, Spotted Stingray, Lady Ray, Leopard Ray, Spotted Ray

Diagnosis — A widely distributed species of large size occurring in warm waters. Body has no scales. Snout is pointed. Tail resembles a whip, 4 times longer than body length. Mouth has a row of conglutinated teeth. Feeds on bivalve. Edible. Up to 3m in total length.

Distribution — Japan, Ryukyu, Indo-Pacific-Atlantic, Red Sea, South Africa

▼ Ventral side of Kite Ray.

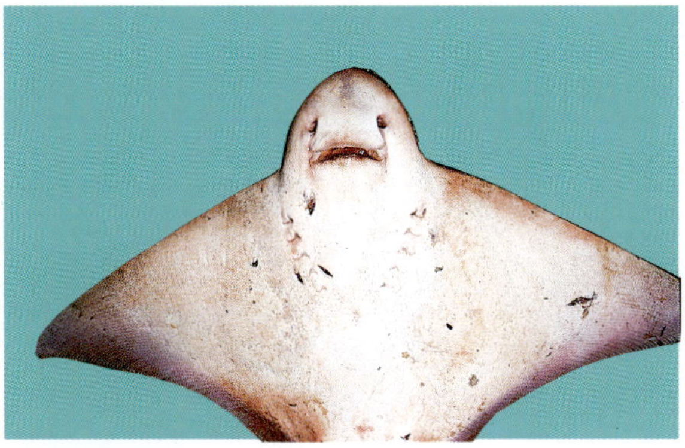

Myliobatidae

Myliobatis tobijei Bleeker ⇧⇨

Kite Ray, Flying Ray, Eagle Ray

Diagnosis — Found in bottom of coastal waters. Disk is broad. Tail is remarkably long like a whip. Dorsal fin has 1–2 venom-bearing spines in the posterior portion. Teeth are hexagonal, strong enough to crush shellfish and crustacea. Ovoviviparous. Brings forth some 8 living youngs. Edible. Up to 2m in total length.

Distribution — Japan, Ryukyu, Formosa, Korea, China

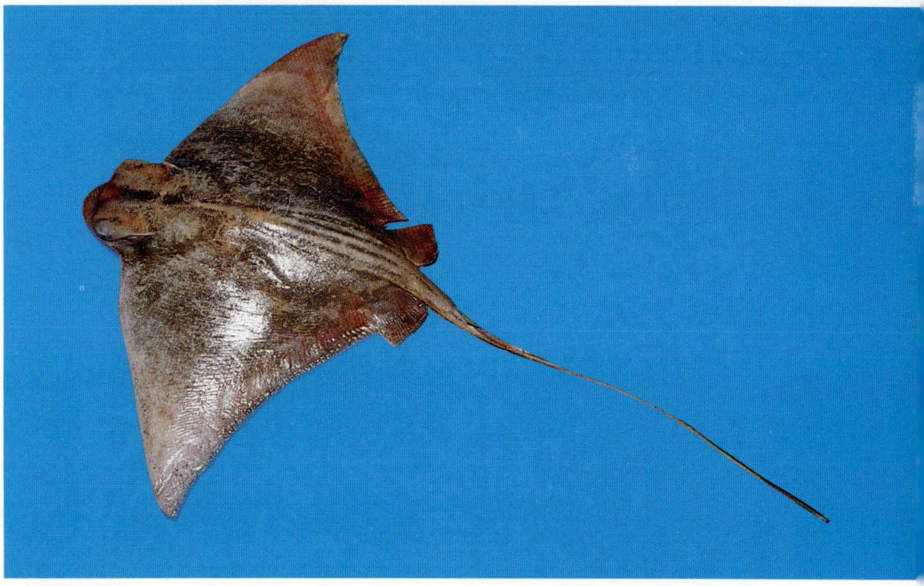

Rhinopteridae

Rhinoptera javanica Müller et Henle ⇩⇗

Japanese Cow Ray, Flapray, Cow-nosed Ray

Diagnosis — A medium-sized species inhabiting middle layer of tropical and subtropical waters. Snout is rounded, slightly V-shaped at tip. Cephalic fins are formed by two separate parts. Mouth has 7 rows of teeth. Tail has a venom-bearing spine. Feeds on shellfish and crustacea. Edible. Up to 1m in total length.

Distribution — Southern Japan, Ryukyu, China, Indo-West Pacific, South Africa

▲ Ventral side of Cow-nosed Ray.

▲ Migrates with Sucking-fishes.

Mobulidae

Manta birostris (Donndorff) ⇧⇨⇩

[*M. ehrenbergi* Edwards et Haime]

[*Deratoptera alfredi* Kreft.]

Manta, Giant Devil Ray, Devilray, Devilfish, Oxray, Eagle Ray

Diagnosis — A large, widely distributed species of shallow tropical waters. Pectoral fins are well-developed, forming a broad disk. Tail is long and whip-like, without a venom-bearing. Mouth is large, opening forward, with cephalic fins. Lower jaw has 130–200 small teeth; upper jaw is toothless. Feeds on small fish. Sometimes attains a disk width of 10m and a weight of 2 tons.

Distribution — Ryukyu, Indo-Pacific-Atlantic, Red Sea, South Africa

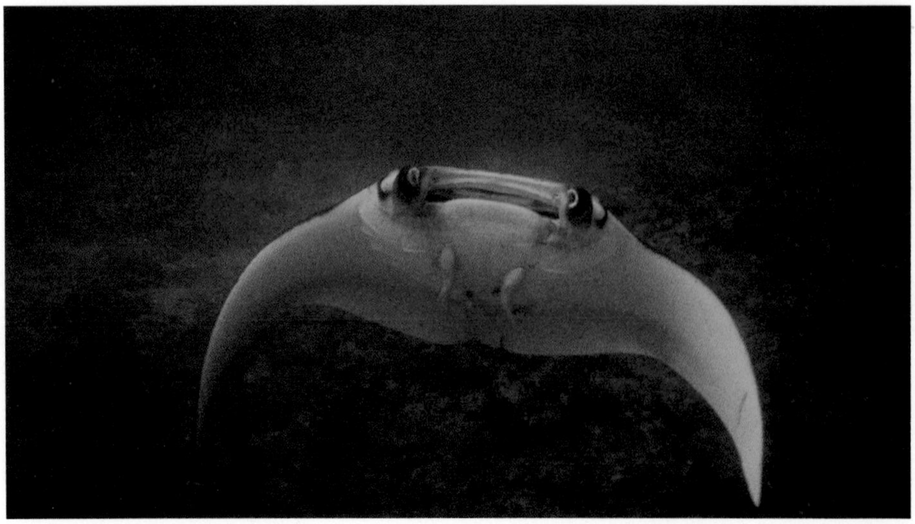

Rhinobatidae

Rhina ancylostoma

Bloch et Schneider ⇨

Short-nosed Mud Skate, Bow-mouthed Angel Fish, Bowmouth Guitarfish, Mud-skate, Shark Ray

Diagnosis — A large shark-like species found in shallow waters. Often swims in the middle layers of the sea. Head is well-developed; snout is broad and short. Mouth is wavy. Spiracles lack skinny projections. Pectoral fins are free from pelvic fins. Ovoviviparous. Up to 2.5m in total length.

Distribution — Japan, Ryukyu, China, Indo-Pacific, Persian Gulf, Red Sea, South Africa

Mobulidae

⇦ ***Mobula japonica*** (Müller et Henle)

Eagle Ray, Spinetail Mobula, Devil Fish, Devil Ray, Spool Ray, Japanese Devil Ray, Japanese Devil Fish

Diagnosis — A large species found in shallow coastal waters. Pectoral fins are well-developed, forming a broad disk. Tail is three times longer than disk length, has a venom-bearing spine. Head has a pair of cephalic fins. Mouth opens downward. Ovoviviparous. Brings forth some 8 living youngs. Feeds on small fish and crustacea. Liver is used for making oil. Up to 3m in total length.

Distribution — Japan, Ryukyu, Formosa, Korea, China, Hawaii

Rajidae

⇩ ***Raja hollandi*** Jordan et Richardson

[*R. porosa* Tortonese]

Holland's Skate

Diagnosis — An inhabitant of shallow waters not deeper than 100m. Disk is square, snout is protruded. Pectoral fins have a pair of ring-shaped splotches in the posterior inner corners. Tail has a caudal fin, 3–5 rows of spines. Spawns between winter and spring. Feeds on crustacea, small fish and ophiuroidea. Up to 60cm in total length.

Distribution — Southern Japan, Formosa, Hong Kong, Yellow Sea, South China Sea

Platyrhinidae

Platyrhina sinensis (Schneider) ⇨

Fanray, Thornback Ray

Diagnosis — An inhabitant of rather deep waters. Disk is rounded, snout is not protruded. Tail is well-developed, has a caudal fin resembling a spatula, lacks venom-bearing spine. Has 2 dorsal fins. Akin to a shark. Body is covered with small scales. Back has large spines. Ovoviviparous. Brings forth several living youngs in spring. Up to 70cm in total length.

Distribution — Japan, Formosa, South China Sea

Rajidae

⇧ ***Raja porosa*** Günther

[*R. tobae* Tanaka]

[*R. meerdervoorti* Jordan et Fowler]

[*R. katsukii* Tanaka]

Ocellate Skate

Diagnosis — A medium-sized species inhabiting shallow waters not deeper than 100m. Disk is square; snout is pointed; tail is short, has 5 rows of spines, also has 2 small dorsal fins in the posterior portion. Varies widely in color and splotch. Spawns between summer and fall. Up to 50cm in total length.

Distribution — Japan, Korea, China, East China Sea

Rhinobatidae
⇩ *Rhinobatos schlegeli* Müller et Henle
[*Rhinobatus holcorhynchus* Norman]
[*R. natalensis* Fowler]

Brown Guitarfish, Sandkruper, Sand Shark, Shovelnose, Skate Shark

Diagnosis — A shark-like species inhabiting sandy-mud bottom not deeper than 200m. Gill slits are situated on the ventral side. Body is silky; snout is long and pointed. Spiracles are large, have two skinny projections. Interval between outer margin of nostril and outer margin of snout is longer than nostril's length. Ovoviviparous. Brings forth some 10 living youngs in summer. Edible. Fins are dried, used for making shark's fin soup. Up to 1m in total length.

Distribution — Japan, Ryukyu, Formosa, Indo-Pacific, Persian Gulf, Red Sea, South Africa

Rhinobatidae
Rhynchobatus djiddensis (Forskål) ⇩

Fiddle Shark, Giant Guitarfish, Sand Shark, Shovel-nose, Spotted Guitarfish, White-spotted Ray, Shovel-nose Sandy

Diagnosis — A large species found at bottom of tropical and subtropical waters. Snout is long and pointed. Spiracles have 2 skinny projections. Caudal fin is large. First dorsal fin is situated above pelvic fins. Ovoviviparous. Edible. Up to 3m in total length.

Distribution — Japan, Ryukyu, Formosa, Korea, Indo-West Pacific, Red Sea, South Africa

Rhinobatidae
Rhinobatos formosensis Norman ⇧

Taiwan Guitarfish (New Common Name)

Diagnosis — Resembles *R. schlegeli*, but interval between outer margin of nostril and outer margin of snout is shorter than nostril's length. Up to 1m in total length.

Distribution — Formosa

Rhinobatidae
Rhinobatos hynnicephalus Richardson ⇨

Ringstraked Guitarfish

Diagnosis — Inhabits sandy-mud bottom of shallow waters. Closely resembles *R. schlegeli*, but distinguished from the latter by body covered with small ring-shaped black spots, pelvic fins free of pectoral fins. Ovoviviparous. Brings forth some 8 living youngs in summer. Feeds on shellfish and crustacea. Up to 1m in total length.

Distribution — Southern Japan, Formosa, East China Sea

Pristidae

Pristis cuspidatus Latham ⇨

Pointed Sawfish

Diagnosis — Inhabits coastal sandy-mud bottoms. Sometimes moves into freshwater areas. Body is elongated like a shark. Gill slits open on the ventral side. Snout is long and protruded, has regularly set strong dentations on both sides. Has 2 dorsal fins and 1 caudal fin. Anal fin is absent. Spiracles are large. Eyes lack nictitating membranes. Mouth has a large number of small teeth. Viviparous. Feeds on shellfish, sea urchin, crustacea. Snout was once used for ornament or weapon. Up to 5m in total length.

Distribution — Southern Japan, Ryukyu, Formosa, China, Indo-West Pacific-Atlantic, Red Sea

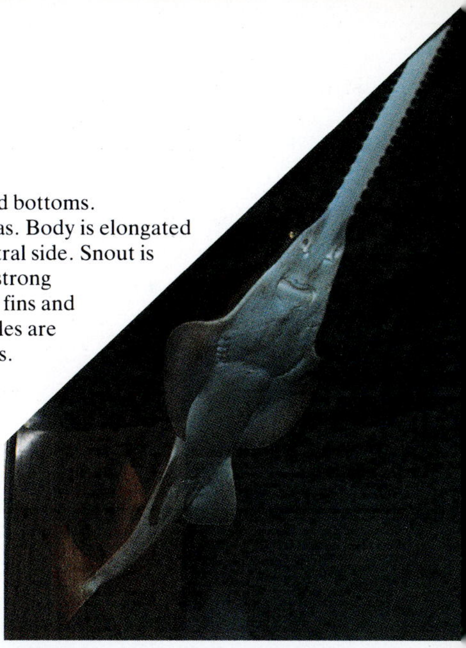

Torpedinidae

Narcine timlei (Bloch et Schneider) ⇧

Speckled Numbray (New Common Name)

Diagnosis — Disk is rounded; snout is not protruded. Back has 2 dorsal fins. Caudal fin is well-developed. Origin of first dorsal fin is situated posterior of the rear end of pelvic fin base. Back is dotted with splotches larger than spiracle. Disk has electric organs on both sides, giving strong electric discharges. Ovoviviparous. Up to 50cm in total length.

Distribution — Southern Japan, Formosa, China, Indonesia, Indian Ocean

Torpedinidae

Narke japonica Temminck et Schlegel ⇨

Electric Ray, Numb-ray, Torpedo, Japanese Spotted Torpedo, Shocked Ray, Numbfish

Diagnosis — A small species of rather deep waters. Body is soft and scaleless. Back has a single dorsal fin. Caudal fin is well-developed, lacks venom-bearing spine. Eyes are small; spiracles are large. Disk has electric organs on both sides, giving strong electric discharges of 30–80 volts, 60 amperes over the body surface. Ovoviviparous. Brings forth some 5 living youngs in early summer. Up to 40cm in total length.

Distribution — Japan, Ryukyu, Formosa, China, Philippines, South China Sea

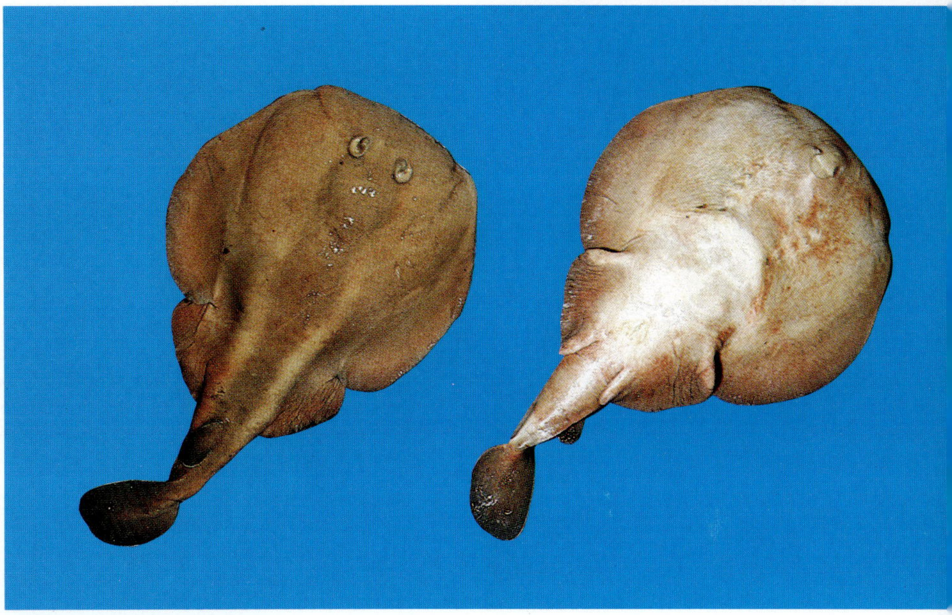

Chimaeridae

⇦ ***Chimaera phantasma*** Jordan et Snyder

Chimaera, Ghost Shark, Silver Chimaera, Silver Shark

Diagnosis — Inhabits waters not deeper than 500m. Body is compressed, head is large. Snout is short, teeth are fused to form dental plates. Scales are present on lateral lines. First dorsal fin has a depressible venom-bearing spine, caudal fin is extended like thread. Oviparous. Feeds on small benthos. Up to 1m in total length.

Distribution — Japan, Ryukyu, Korea, China, East China Sea

Chimaeridae

⇦ ***Chimaera jordani*** Tanaka

Jordan's Ghost Shark (New Common Name)

Diagnosis — A deep-sea species. Body is dark-brown, snout is short. Surface between dorsal and caudal fins, anal and caudal fins are lunated. Upper caudal lobe is extended like thread. Clasper is trifid. Up to 1m in total length.

Distribution — Japan, Ryukyu

Clupeidae

⇦ ***Konosirus punctatus*** (Temminck et Schlegel)

[*Dorosoma thrissa* (Linnaeus)]

[*Clupanodon punctatus* (Temminck et Schlegel)]

Dotted Gizzard Shad, Gizzard Shad, Spotted Sardine

Diagnosis — Inhabits coastal waters, often comes to bays for spawning between March and June. Dorsal fin has 16–20 rays, final ray is slender and thready. Anal fin has 19–27 rays. Scutes on abdominal margin are sharp. Pectoral fins have a black splotch. Toothless. Feeds on plankton. Edible. Up to 30cm in total length.

Distribution — Japan, Korea, China, New Guinea, Polynesia, India, Persian Gulf

Clupeidae

⇦ ***Nematalosa come*** (Richardson)

Threadfin Shad, Gizzard-shad, Hairback Herring

Diagnosis — Found in schools in muddy-sand bottoms of bays. Body is compressed and deep. Lower jaw is small. Dorsal fin has 17–18 rays with extending last fin-ray, anal fin has 23–25 rays. Operculum has a black spot on rear corner. Edible. Up to 25cm in total length.

Distribution — Ryukyu, Indonesia, Australia

Clupeidae

Sardinella zunasi (Bleeker) ⇨

Japanese Scaled Sardine, Japanese Shad, Scaled Sardine

Diagnosis — Inhabit sandy-mud bottom of bays. Occasionally found in brackish areas. Flat-bodied. Abdominal margin is pointed. Dorsal fin has 17–19 rays, anal fin has 18–20 rays. Lateral lines are absent. Spawn between April and September. Edible. Up to 20cm in total length.

Distribution — Japan, Ryukyu, Korea, China, South East Asia

Clupeidae

⇦ ***Hilsa ilisha*** (Buchanan-Hamilton)

River Shad

Diagnosis — A gregarious fish of coastal waters. Dorsal fin is closer to snout than to caudal fin. Dorsal fin has 18–20 rays, anal fin has 18–21 rays. 45–50 lateral scales. Toothless. Edible. Up to 50cm in total length.

Distribution — Indian Ocean, Persian Gulf, Red Sea

Clupeidae
⇦ ***Herklotsichthys quadrimaculatus*** (Rüppell)

[*H. punctatus* (Rüppell)]

[*H. ovalis* (Bennett)]

Gold-spot Herring, Spotted Herring, Murray Island Sardine, Torres Strait Herring

Diagnosis — A migratory fish of coastal waters. Occasionally found in brackish areas. Abdominal margin is pointed, has scutes. Scales are large and strong. Pelvic fins are situated posterior of dorsal fin. Dorsal fin has 17–20 rays, anal fin has 16–19 rays. Feeds on plankton. Spawns in spring. Edible. Up to 15cm in total length.

Distribution — Ryukyu, Indo-Pacific, Red Sea, South Africa

Clupeidae
Amblygaster leiogaster (Valenciennes) ⇨

[*A. clupeoides* (Bleeker)]

[*Clupea okinawensis* Kishinouye]

Sharpnose Trenched Sardine

Diagnosis — A migratory fish of coastal waters. Resembles *Sardinops melanostictus*, but has no black splotches, pelvic fins are formed from the origin of dorsal fin. Abdominal scutes are not distinct. Dorsal fin has 16–18 rays, anal fin has 17–21 rays. Edible and also used for bait. Up to 25cm in total length.

Distribution — Southern Japan, Ryukyu, South East Asia, Indo-West Pacific, Red Sea

Clupeidae
Amblygaster sirm (Walbaum) ⇨

Spotted Trench Sardine
(New Common Name)

Diagnosis — Resembles to *A. leiogaster*, but distinguished by dorsal fin origin closer to snout tip than to middle part of dorsal margin. Back has rows of dark spots. Dorsal fin has 17–19 rays, anal fin has 15–19 rays. Edible. Up to 25cm in total length.

Distribution — Ryukyu, Indo-West Pacific

Clupeidae
⇦ ***Spratelloides gracilis*** (Temminck et Schlegel)

[*S. japonicus* (Houttuyn)]

Banded Blue Sprat, Blue Sprat, Whitestriped Blue Sprat

Diagnosis — A pelagic migratory fish often found in large schools. Body is elongated, rounded on the ventral side. Scales are easily removable. Pelvic fins are formed from the posterior end of dorsal fin base. Dorsal fin has 12–13 rays, anal fin has 12–13 rays. Toothless. Lateral lines are absent. Comes to bays for spawning between spring and summer. Edible. Up to 10cm in total length.

Distribution — Japan, Ryukyu, Korea, China, Indo-Pacific, Red Sea, East Africa

Clupeidae

Clupea pallasii Valenciennes

California Herring, Herring, Eastern Herring, North Pacific Herring, Oriental Herring

Diagnosis — Found in surface of offshore waters. Pelvic fins correspond with dorsal fin. Dorsal fin has 15–19 rays, anal fin has 14–19 rays. Abdominal margin is formed by frail scutes. Feeds on zooplankton. Spawns in coastal waters in spring. Youngs grow fully in 4 years. Edible. Up to 35cm in total length.

Distribution — Northern Japan, Korea, North China, Okhotsk Sea, Bering Sea, North America

Clupeidae

Sardinops melanostictus (Temminck et Schlegel)

[*S. sagax melanostictus* (Temminck et Schlegel)]

Japanese Sardine, True Sardine, Spotlined Sardine, Japanese Pilchard, Pilchard

Diagnosis — A pelagic and gregarious fish of temperate offshore waters, lives in lower layers during the day, surfaces at night. Move north between spring and summer, south between fall and winter. Scales are easily removable. Body has 1–3 rows of black spots. Dorsal fin has 16–21 rays, anal fin has 16–20 rays. Opercles have radial stripes. Feeds on phytoplankton. Edible and also used for bait. Up to 25cm in total length.

Distribution — Japan, Ryukyu, Korea, East China Sea, China, Okhotsk Sea, Kamchatka, Mediterranean

Clupeidae

Sardinella jussieu (Lacépède)

[*S. gibbosa* (Bleeker)]

[*Clupea exile* Kishinouye]

Gold-striped Sardine, Tembang

Diagnosis — A migratory fish of coastal waters. Body is slightly compressed. Abdominal margin has 15–16 scutes. Dorsal fin has 17–20 rays, anal fin has 16–19 rays. Dorsal fin base has black splotches. Toothless. Edible. Up to 18cm in total length.

Distribution — Bonin Islands, China, Indo-Pacific, Persian Gulf, South Africa

Clupeidae

⇦ ***Dussumieria hasseltii*** Bleeker

[*D. elopsoides* Jordan et Evermann]

Van Hasselt's Sprat, Hasselt's Sprat, Round Herring

Diagnosis — Dorsal fin origin is closer to caudal fin than to snout. Dorsal fin has 18−19 rays, anal fin has 16 rays. Abdomen is rounded. Margin of caudal fin is black. Dorsal and pectoral fin bases have a black spot. 52−56 lateral scales. Edible. Up to 20cm in total length.

Distribution — Formosa, China, Indo-West Pacific

Clupeidae

⇦ ***Etrumeus teres*** (Dekay)

[*E. micropus* (Temminck et Schlegel)]

Big-eye Sardine, Herring, Market Sardine, Pacific Round Herring, Round Herring

Diagnosis — A pelagic migratory fish. Comes to bays for spawning between April and June. Body is rounded. Pelvic fins are situated posterior of dorsal fin. No lateral lines. Dorsal fin has 19−22 rays, anal fin has 10−12 rays. Teeth are fairly small. Feeds on zooplankton. Edible. Up to 30cm in total length.

Distribution — Japan, Ryukyu, Korea, China, Indo-Pacific-Atlantic, Red Sea, South Africa

Engraulidae

Engraulis japonicus (Houttuyn) ⇩

[*E. capensis* Gilbert]

Japanese Anchovy, Anchovy, Halfmouthed Sardine, Half Mouth Sardine

Diagnosis — A small pelagic fish found in schools. Body is elongated and cylindrical. Mouth is large, lower jaw is short. Toothed. Maxilla does not extend. Abdomen is rounded, has no scutes. Dorsal fin has 14−16 rays, anal fin has 15−18 rays. Scales are easily removable. Edible and also used for bait. Up to 18cm in total length.

Distribution — Japan, Ryukyu, Formosa, Korea, China, Sakhalin, South Africa

Engraulidae

Thrissa dussumieri (Valenciennes) ⇧

Dussumier's Anchovy

Diagnosis — Found in coastal waters. Maxilla extends down as far as pelvic fins. Ventral side is covered with 22−24 scutes. Dorsal fin has 15−16 rays, anal fin has 34−38 rays. 40−42 lateral scales. Edges of caudal fin are black. Edible. Up to 20cm in total length.

Distribution — Bonin Islands, Formosa, China, South East Asia, India

Engraulidae

Stolephorus indicus (Van Hasselt) ⇨

[*Engraulis baliensis* Bleeker]

Indian Anchovy

Diagnosis — Found in coastal waters, occasionally found in brackish areas. Maxilla extends down as far as preopercle. Dorsal fin has 16 rays, anal fin has 18−21 rays. 37−42 lateral scales. Edible. Up to 15cm in total length.

Distribution — Ryukyu, Formosa, China, Indo-Pacific, Red Sea, South Africa

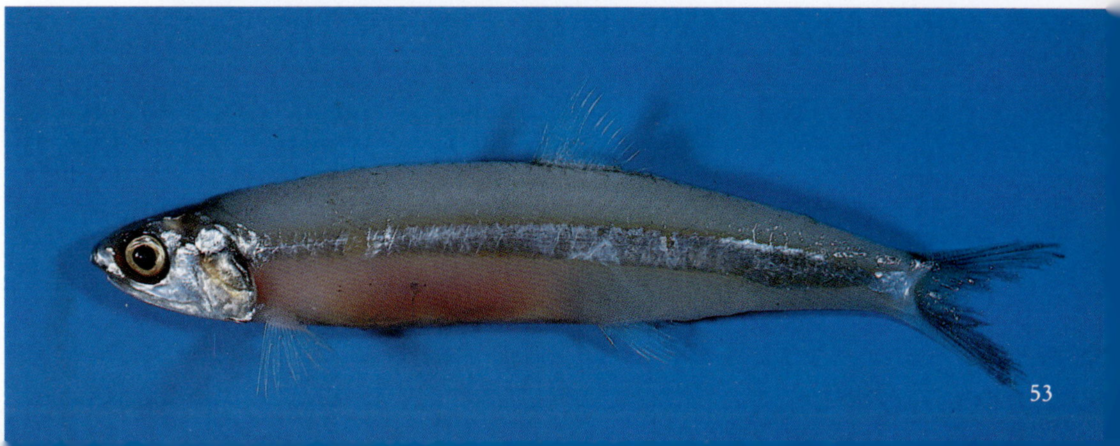

Elopidae

Megalops cyprinoides (Broussonet) ⇨

[*M. filamentosus* Lacépède]

[*Brisbania staigeri* Castle]

Bastard Mullet, Big-eyed Herring, Bony Mullet, Broussonet's Tarpon, Lady Fish, Ox Eye Herring, Ox Eye Tarpon, Pacific Tarpon, Small Tarpon

Diagnosis — Found in both fresh and salt waters. Bred in Formosa. Juveniles pass the "leptocephalus" stage before they grow. Dorsal fin has 16–20 rays, last ray is slender and thready. Anal fin has 23–31 rays. 36–40 lateral scales. Lower jaw is protruded, mouth is directed sharply upward. Edible. Up to 1m in total length.

Distribution — Japan, Ryukyu, Formosa, Indo-Pacific, Red Sea

Elopidae

⇦ ***Elops hawaiensis*** Regan

[*E. machnata* (Forskål)]

Big-eye Herring, Bony Fish, Giant Herring, Jackmariddle, Ladyfish, Pacific Ladyfish, Ten Pounder, Whole Sardine

Diagnosis — Body is elongated. Caudal fin is falcated. Dorsal fin has 23–27 rays, is situated in the middle of dorsal margin, corresponding with pelvic fins. Anal fin has 15–17 rays. Scales are small cycloid. Eyes have no adipose eyelids. Teeth are small and many. Juveniles pass the "leptocephalus" stage. Edible. Up to 1m in total length.

Distribution — Japan, Ryukyu, China, Indo-Pacific, Hawaii

Albulidae

⇦ ***Albula vulpes*** (Linnaeus)

[*A. conorhynchus* Valenciennes]

Bone Fish, Lady Fish, Banana Fish, Tarpon, Grubber

Diagnosis — Found in coastal sandy bottoms or mangrove swamps. Juveniles pass the "leptocephalus" stage before they grow. Snout is pointed, lower jaw is short. Has lateral lines. Dorsal fin has 15–19 rays, anal fin has 8–9 rays. Caudal fin is falcated. Teeth are fairly small. Edible and also used for bait. Up to 90cm in total length.

Distribution — Japan, Ryukyu, Indo-Pacific-Atlantic, Red Sea

Albulidae

Pterothrissus gissu Hilgendorf ⇨

Deepsea Bonefish

Diagnosis — Inhabits waters 200–1,000m deep. Body is elongated, snout is projected. Mouth is situated on lower surface. Caudal fin is deeply falcated. Dorsal fin has 55–65 rays, base is long. Anal fin has 12 rays. 107–113 lateral scales. Spawns in winter, larvae experience leptocephalus. Edible. Up to 60cm in total length.

Distribution — Japan, Ryukyu, China

Anguillidae
Anguilla japonica Temminck et Schlegel

Eel, Japanese Eel

Diagnosis — An elongated freshwater fish, down to the sea for spawning. Juveniles pass the "leptocephalus" stage before they grow. Dorsal, caudal and anal fins are connected. Pelvic fins are absent. Scales are remarkably small, sink under skin. Edible. Up to 1m in total length.

Distribution — Japan, Ryukyu, Bonin Islands, Korea, Formosa, China, South China Sea, Philippines, Vietnam

▼ Bred eels in Japan.

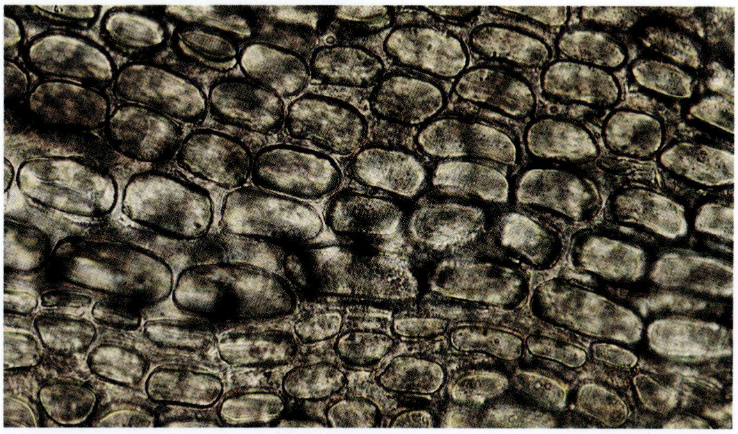

▲ Small scales sink under skin.

Anguillidae
Anguilla marmorata Quoy et Gaimard

[*A. mauritiana* Bennett]

Giant Long-finned Eel, Marbled Eel

Diagnosis — Found in both fresh and salt waters. Resembles *A. japonica*, but body is thicker and larger with dark brown splotches. Has a pair of gill slits. Scales sink under skin. Feeds on small fish, crustacea and shellfish. Edible. Up to 2m in total lenght.

Distribution — Japan, Ryukyu, Bonin Islands, Indo-Pacific, East Africa

Eurypharyngidae
Eurypharynx pelicanoides Vaillant

Pelican Gulper Fish

Diagnosis — Queer in appearance. Found in depths of 1,500–3,000m of temperate and tropical waters. Head is large; tail is elongated. Lower jaw is large and expansible. No pelvic fins. Pectoral fins are remarkably small. Have 6 gill slits. Eyes and teeth are very small. Juveniles pass the "leptocephalus" stage before they grow. Up to 70cm in total length.

Distribution — Japan, Bonin Islands, Indo-Pacific-Atlantic, Hawaii

Congridae

⇩ ***Conger myriaster*** (Brevoort)

⇩ **Common Japanese Conger, Whitespotted Conger**

Diagnosis — Inhabits sandy-mud bottoms of bays. Nocturnal. Body is elongated. Dorsal, caudal and anal fins are connected. Pelvic fins are absent. Has a pair of gill slits. Lateral line pores are white. Rows of white spots show behind eyes, along lateral lines. Juveniles pass the "leptocephalus" stage before they grow. Edible. Up to 1m in total length.

Distribution — Japan, Korea, East China Sea

▲ Whitespotted Conger in aquarium.

Congridae

⇧ ***Conger cinereus*** Rüppell

[*C. noordzicki* Bleeker]

[*C. marginatus* Valenciennes]

[*Leptocephalus riukiuanus* Jordan et Snyder]

Ashen Conger-eel, Black Conger, Blackedged Conger, Gray Conger, Indian Conger, Mustache Conger

Diagnosis — A large species inhabiting in coral reefs of tropical waters. Snout is pointed. Dorsal fin formed from the middle of pectoral fins. Pectoral fins have black splotches. Edges of dorsal, anal and caudal fins are black. Juveniles have 16−18 cross stripes. Edible. Up to 1m in total length.

Distribution — Southern Japan, Ryukyu, Indo-Pacific, Red Sea, South Africa

Congridae

⇩ ***Anago anago*** (Temminck et Schlegel)

Sea Conger, Silvery Conger, Conger

Diagnosis — Body is fairly blunt; head is swollen. Mouth is small; jaws are equal in length. Dorsal fin formed from the origin of pectoral fins. Caudal margin is white. Eyes have 2 brown spots behind. Edible. Up to 1m in total length.

Distribution — Japan, Formosa, Indo-Pacific, Souh Africa

Congridae

⇧ ***Ariosoma shiroanago major*** (Asano)

Black-fin Conger Eel (New Common Name)

Diagnosis — Inhabits muddy-sand bottoms of coastal waters. Body is pale and blunt. Fins are black. Edible. Up to 55cm in total lenth.

Distribution — Japan, Formosa, East China Sea

Congridae

Ariosoma anagoides (Bleeker) ⇨

[*Alloconger anagoides* (Bleeker)]

Blackchin Conger

Diagnosis — Inhabits muddy-sand bottoms of coastal waters. Dorsal fin originates anterior to pectoral fins. Upper lip is inverted, lower jaw is black. Anus is situated anterior to the middle of body. Tail end is rounded and hard. Edible. Up to 55cm in total length.

Distribution — Japan, Ryukyu, Formosa, China, East China Sea, Indonesia

Congridae

⇦ ***Conger japonicus*** Bleeker

[*C. vulgaris* Temminck et Schlegel]

[*Leptocephalus erebennus* Jordan et Snyder]

[*L. kiusiuanus* Jordan et Snyder]

Beach Conger, Swarthy Conger, Conger

Diagnosis — Resembles *C. myriaster*, but body is larger, more blackish. Dorsal fin formed from the posterior end of pectoral fins or from a point further backward. Has a pair of gill slits. Jaws are equal in length. Skin is hard. Up to 2.4m in total length.

Distribution — Japan, Ryukyu, Korea

Muraenesocidae

Muraenesox bagio (Hamilton) ⇨

[*M. yamaguchiensis* Katayama et Takai]

Brown Pike Conger

Diagnosis — Resembles *M. cinereus*, but distinguished by smaller number of lateral line pores (33−39), smaller eyes, thinner posterior part of opercular bone. Edible. Up to 2m in total length.

Distribution — Japan, Formosa, East China Sea, Indo-Pacific

Muraenesocidae

Oxyconger leptognathus (Bleeker) ⇨

Longbill Pike Conger

Diagnosis — A deep-sea inhabitant. Body is elongated; snout is long and pointed. Tail is short and finned. Pectoral fins are slender. Mouth is toothed in 3 rows with canines. Up to 60cm in total length.

Distribution — Japan, Formosa, East China Sea, South China Sea

Muraenesocidae

⇦ ***Muraenesox cinereus*** (Forskål)

[*M. arabica* Bleeker]

[*M. longirostris* Bennett]

Silver Conger Eel, Purple Pike Conger, Pike Eel, Conger Pike, Arabian Pike-eel, Sharp-toothed Eel

Diagnosis — Inhabits sandy-mud bottoms of inland seas. Nocturnal. Snout is pointed. Mouth is large, has strong canines. No pelvic fins and scaleless. Dorsal fin formed from a point above gill slit. Juveniles pass the "leptocephalus" stage before they grow. 40−47 lateral line pores. Edible. Up to 2m in total length.

Distribution — Japan, Ryukyu, Formosa, Korea, Indo-Pacific, Red Sea, South Africa

Large mouth has strong canines. ▶

Nettastomatidae

Saurenchelys fierasfer (Jordan et Snyder) ⇨

Blacktail Duckbill

Diagnosis — A deep-sea species. Slender-bodied. Tail is elongated and thready. Caudal margin is black. No pectoral fins and scaleless. Upper jaw is long; mouth is pointed. Nostrils open toward both sides of head. Cheek is silver-banded. Up to 50cm in total length.

Distribution — Japan, Formosa, East China Sea, South China Sea

Dysommidae

Dysomma anguillare Barnard ⇨

Arrowtooth Eel

Diagnosis — Body is elongated. Eyes sink under skin, are rather degenerated. Dorsal fin formed from a point above gill slit. Pectoral fins are small. Snout is long and narrow. Canineless. Up to 50cm in total length.

Distribution — Japan, Formosa, China, Indo-Pacific, South Africa

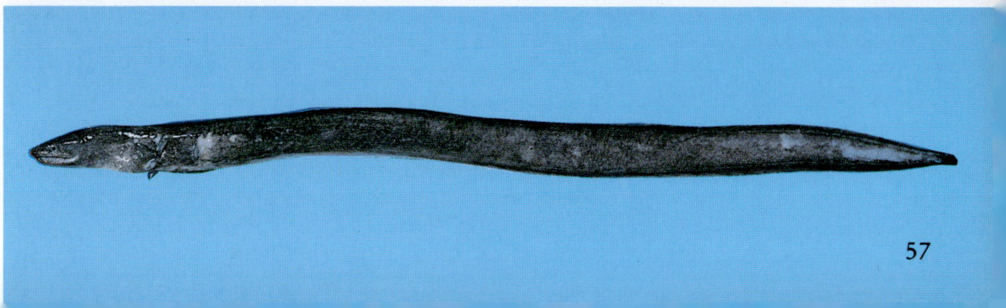

Muraenidae

⇧ *Gymnothorax melanospilus* ⇗
(Bleeker) ⇨

Spotted Purply Moray,
Black-spotted Moray

Diagnosis — Head is thick. Gill slits are not black. Tail is elongated. Resembles *G. pescadores*, but black splotches over body are oval or paired. Upper jaw is toothed in 3 rows. Causes ciguatera poisoning. Up to 1.8m in total length.

Distribution — Japan, Ryukyu, Formosa, Indonesia, West Pacific

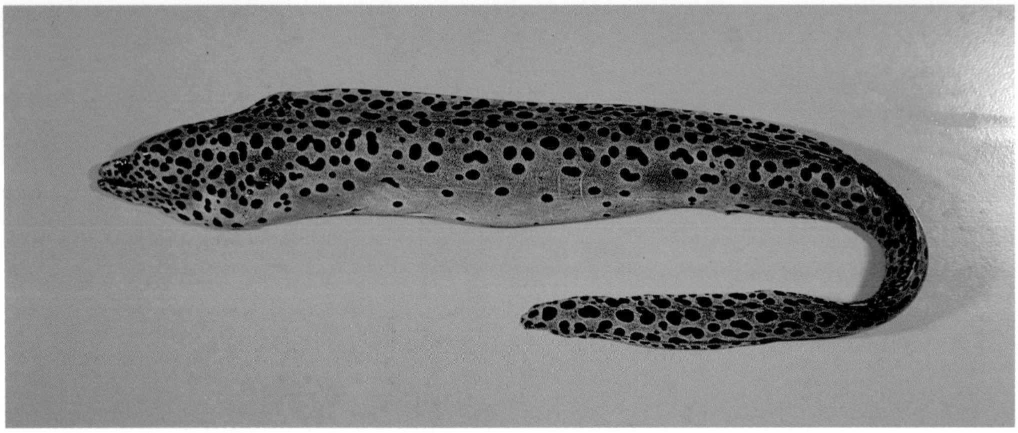

Muraenidae

⇦ *Gymnothorax pseudothyrsoideus* (Bleeker)

Yellowlined Moray

Diagnosis — Body is brown with narrow, reticulate network. Snout is short. Dorsal fin is slightly high. Mouth is reddish. Up to 80cm in total length.

Distribution — Ryukyu, Indo-Pacific

Muraenidae

⇦ *Gymnothorax favagineus* Bloch et Schneider

[*Muraena tessellata* Day]

Tessellated Reef Eel, Leopard Moray, Black-blotched Reef-eel, Black-lined Moray, Coral Eel, Honeycomb Moray

Diagnosis — A large species of moray inhabiting shallow coral reefs. Body is covered with large dark brown splotches. Jaws are toothed in single row, upper jaw is large-fanged. Up to 2m in total length.

Distribution — Ryukyu, Indo-Pacific, South Africa

Muraenidae

Enchelycore schismatorhynchus (Bleeker) ⇨

[*Lycodontis schismatorhynchus* (Bleeker)]

[*Gymnothorax schismatorhynchus* (Bleeker)]

White-edged Yellow Moray

Diagnosis — Found in coral reefs. Body is elongated. Dorsal and anal margins are white. Mouth is large, jaws are so curved that they can not be closed completely. Snout is pointed. Body is grayish-brown. Up to 1.2m in total length.

Distribution — Ryukyu, Pacific

Muraenidae

Gymnothorax javanicus (Bleeker)

Javanese Moray Eel, Giant Moray

Diagnosis — Largest in moray family. Body is thick; head is large. Snout is thick and short. Upper jaw is toothed in single row. Gill slits are large and black. Body is covered with 3 rows of dark brown splotches. Tail is reticulated. Often causes ciguatera poisoning. Up to 2.5m in total length.

Distribution — Ryukyu, Indo-Pacific, Red Sea, Hawaii

Muraenidae

Gymnothorax undulatus (Lacépède)

Leopard Moray, Mottled Reef-eel, Common Hawaiian Moray, Leopard Reef-eel

Diagnosis — Blunt-bodied. Snout is black and short. Body has a reticulated pale pattern. Head is spotted. Gill slits and caudal margin are white. Jaws are toothed in single row. Dorsal fin formed from a point above gill slit or anterior of gill slit. Causes ciguatera poisoning. Up to 1.5m in total length.

Distribution — Southern Japan, Ryukyu, Formosa, Indo-Pacific-Atlantic, South and West Africa

Muraenidae

Gymnothorax berndti Snyder

Berndt's Moray Eel, Spotted-lip Moray

Diagnosis — Resembles Y-lined Moray, but distinguished by slightly longer lower jaw. Head is covered with small black spots forming reticulate pattern. Body has irregular stripes. Margin of anal fin is white. Up to 1m in total length.

Distribution — Japan, Ryukyu, Formosa, Pacific, Hawaii

Muraenidae

Gymnothorax sp.

Y-lined Moray

Diagnosis — Resembles *G. berndti*, but distinguished by slightly longer upper jaw. Head has no black spots. Body has some 30 straight or Y-shaped bars. Up to 1m in total length.

Distribution — Japan, Formosa, Hawaii

Muraenidae
Gymnothorax kidako
(Temminck et Schlegel)

Brutal Moray, Moray Eel

Diagnosis — Inhabits shallow shoreline reefs. Harmful, ready to bite at objects. Has no pectoral and pelvic fins. Thick-skinned. Has a black gill slit. Nocturnal. Spawns in summer. Edible. Skin can be tanned. Up to 80cm in total length.

Distribution — Japan, Formosa, Korea

Muraenidae
Gymnothorax eurostus (Abbott)
[*G. wakanourae* Tanaka]

Wakanoura Moray, Abbott's Moray Eel

Diagnosis — Snout is short. Upper jaw is toothed in 2 or more rows on both sides. Body has densely distributed white spots smaller than eyes, sparsely distributed black spots larger than eyes. Up to 80cm in total length.

Distribution — Japan, Ryukyu, Indo-Pacific, Hawaii, Easter Island, Costa Rica, South Africa

Muraenidae
Enchelycore lichenosa
(Jordan et Snyder)

Moss Moray

Diagnosis — Body is dark-brown, covered with dark moss-like irregular blotches. No pectoral fins. Characterized by 2 rows of over 30 tiltable canine-like teeth in jaws and vomerines. Up to 90cm in total length.

Distribution — Japan, Ryukyu, Galapagos Islands

Muraenidae
Gymnothorax leucostigma
Jordan et Richardson

Whitespotted Brown Moray

Diagnosis — Inhabits rocky reefs of shallow waters. Body is rather thick. Gill slits are not black. Jaws are toothed in single row. Body is profusely dotted with white larger than eyes. Up to 1m in total length.

Distribution — Japan, Ryukyu, Formosa, South China Sea

Muraenidae
Gymnothorax thyrsoideus
(Richardson)
[*Muraena griseo-badia* Richardson]

Slender Moray

Diagnosis — Found in coral reefs of tropical and subtropical waters. Head is black. Snout and lower jaw are pale. Upper jaw is toothed in 2 rows. Up to 60cm in total length.

Distribution — Japan, Ryukyu, China, Pacific

Muraenidae

⇦ ***Gymnothorax fimbriatus*** (Bennett)

[*G. undulatus* var. *isingteenus* Richardson]

[*G. undulatus* var. *fimbriatus* (Bennett)]

[*G. stellatus* (Snyder)]

Dark-spotted Moray

Diagnosis — A small species inhabiting coral reefs. Body is covered with irregular black splotches. Margins of dorsal and anal fins are yellow. Gill slits are not black. Up to 80cm in total length.

Distribution — Southern Japan, Ryukyu, Bonin Islands, Formosa, China, Indo-Pacific, Red Sea, East Africa

Muraenidae

⇧⇦ ***Gymnothorax reticularis*** Bloch

Net Moray

Diagnosis — A small species found in deep waters. Body has 14–22 dark cross bands. Up to 60cm in total length.

Distribution — Japan, Ryukyu, Formosa, Indonesia, Indian Ocean, East Africa

Muraenidae

Scuticaria tigrina (Lesson) ⇨

Tiger Moray Eel

Diagnosis — Body is elongated and cylindrical. No fins. Anus is situated in posterior part of body. 2 rows of teeth in jaws. Nostrils are tubular. Eyes are remarkably small. Up to 1.2m in total length.

Distribution — Indo-Pacific, Hawaii, South Africa

Muraenidae

⇩ ***Echidna nebulosa*** (Ahl)

Nebulous Moray Eel, Floral Moray, Clouded Reef Eel, Starry Moray, Starry Reef-eel, Snowflake Moray

Diagnosis — A small species inhabiting shoreline reefs and other coral reefs of shallow waters. Has 2 rows of black ring-shaped blotches on body. Jaws have 1–2 rows of blunt conical teeth. Up to 75cm in total length.

Distribution — Southern Japan, Ryukyu, Indo-Pacific, Red Sea, South Africa

Muraenidae

Gymnothorax meleagris (Shaw et Nodder) ⇩⬂

[*G. laysanus* (Steindachner)]

[*G. chlorostigma* Bleeker]

Guineafowl Moray, Speckled Moray

Diagnosis — A large species inhabiting coral reefs. Body is profusely dotted with white spots smaller than eyes. Caudal margin is white. Upper jaw is toothed in 2 rows, lower jaw has single row. Gill slits are black. Causes ciguatera poisoning. Up to 1m in total length.

Distribution — Ryukyu, Formosa, Indo-Pacific, South Africa

Muraenidae

⇦ ***Muraena pardalis*** (Temminck et Schlegel) ⇩

Dragon Moray Eel, Leopard Moray Eel, Hawaiian Dragon Eel

Diagnosis — Inhabits shoreline reefs of shallow waters. Ferocious, often bites man. Jaws are hooked, snout is pointed. Canines are present. Nostrils are tubular, posterior nostrils are long and movable. Spawns in spring. Up to 1m in total lengh.

Distribution — Japan, Ryukyu, Korea, Indo-Pacific, Hawaii

Ophichthyidae

⇦ ***Ophichthus erabo*** (Jordan et Snyder)

Buffoon Snake Eel, Serpent Eel

Diagnosis — Inhabits muddy bottoms of coastal waters. Tip of tail is stick-like. Caudal fin is absent. Snout is pointed. Eyes and gill slits are small. Lateral lines are distinct. Ferocious, readily bites man. Up to 70cm in total length.

Distribution — Japan, Ryukyu, Formosa, Indo-Pacific

Ophichthyidae

⇩ ***Pisodonophis cancrivorus*** (Richardson)

Burrowing Snake Eel, Longfin Snake-eel, Snake Eel

Diagnosis — Inhabits sandy-mud bottoms of bays. Caudal fin is absent; tail is long. Anterior part of dorsal fin is dark brown. Dorsal fin originates above pectoral fins. Dorsal and anal fins elevated near tip of tail. No blotches on body. Molar teeth in jaws. Up to 1m in total length.

Distribution — Japan, Ryukyu, China, Indo-Pacific

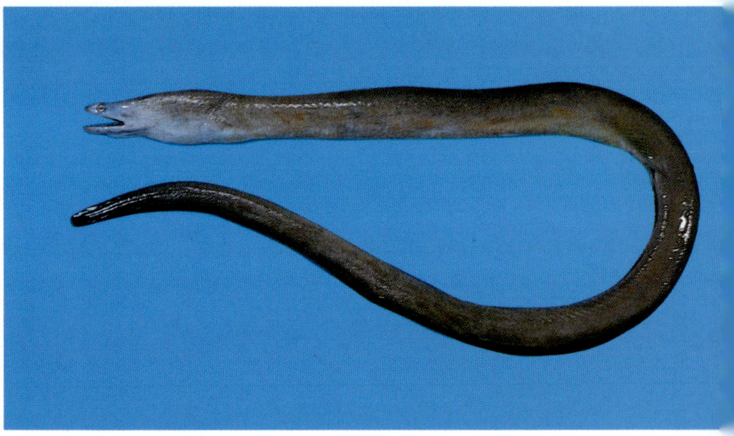

Muraenidae

⇧ ***Evenchelys macrurus*** (Bleeker)

Giant Moray, Longtailed Eel, Slender Giant Moray

Diagnosis — A large species inhabiting muddy shallow waters. Body has no blotches; tail comprises more than 60% of body length. Up to 3m in total length.

Distribution — Ryukyu, Formosa, Indo-Pacific, South Africa

Ophichthyidae

Myrichthys maculosus (Cuvier) ⇨

[*M. rupestris* Snyder]

[*Ophichthus miyamotonis* Tanaka]

Clown Snake Eel, Spotted Snake Eel, Ocellated Snake Eel

Diagnosis — A species inhabiting sandy bottoms of shallow waters. Caudal fin is absent. Pectoral fins are short. Dorsal fin originates at a point anterior to gill slit. Teeth granulated. 3 rows of black oval blotches on body. Nocturnal. Up to 1m in total length.

Distribution — Japan, Izu Islands, Ryukyu, Indo-Pacific, Red Sea, East Africa

Ophichthyidae

Brachysomophis cirrhocheilus (Bleeker) ⇩

Yellow-fin Snake Eel

Diagnosis — A large species inhabiting muddy bottoms of deep waters. Upper lip is margined with barbels. Head length is 4–5 times longer than pectoral fin length. Dark irregular cross bands on body. Fins are yellow with brown margin. 2 rows of teeth in upper jaw, single row in lower jaw. Up to 1.5m in total length.

Distribution — Southern Japan, Formosa, Indo-West Pacific, Persian Gulf

Ophichthyidae

Ophichthus evermanni Jordan et Richardson ⇧

[*O. cephalozonus* Bleeker]

Blacknape Snake Eel

Diagnosis — A large species inhabiting muddy bottoms of deep waters. Body is lined with brown irregular cross bands. Dorsal fin originates above pectoral fins. A single row of teeth in jaws. Ferocious, often bites man. Up to 1m in total length.

Distribution — Southern Japan, Ryukyu, Formosa, Pacific

Ophichthyidae

⇧ *Ophichthus urolophus* (Temminck et Schlegel) ⇧

Greattail Snake Eel

Diagnosis — Inhabits bays approximately 200m deep. Has no blotches on body. Caudal fin is absent. A single row of teeth on jaws. Dorsal fin originates from a point posterior to pectoral fins. Feeds on benthos. Up to 60cm in total length.

Distribution — Japan, Indo-Pacific

Ophichthyidae

⇧ *Myrichthys colubrinus* (Boddaert)

Barred Snake Eel

Diagnosis — Inhabits coral reefs. Body is remarkably elongated and has 24–34 black cross bands. Pectoral fins are extremely small. Teeth are molar. Tips of snout and tail are white. Up to 75cm in total length.

Distribution — Ryukyu, Indo-West Pacific

Salangidae

Salangichthys microdon Bleeder ⇨

Vermilion Fish, Common Ice-fish, Japanese Icefish, White Fish, Glassfish, Salanx, Shirauo Icefish

Diagnosis — Inhabits brackish areas of bays. Body is transparent, turns white when the fish dies. Female is scaleless. Male has 16–18 scales on base of anal fin. Adipose fin is situated posterior to dorsal fin. Pelvic fins are situated further posterior. Attains full growth in one year. Ascends rivers, spawns in muddy bottoms. Eggs are adhesive, hatched in 10–20 days, juveniles descend rivers into the sea. Edible. Up to 10cm in total length.

Distribution — Sakhalin, Japan, Korea

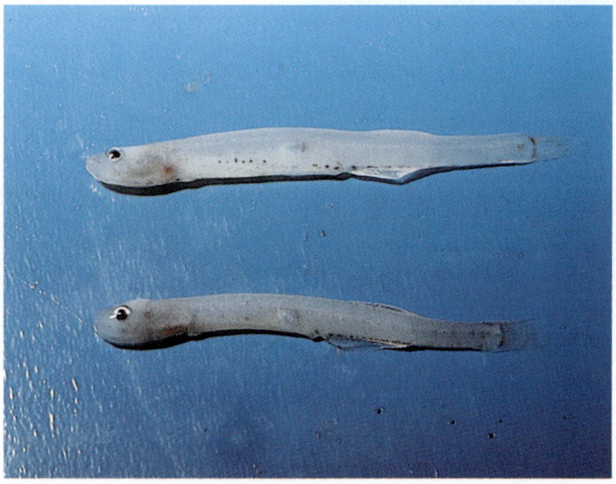

Ophichthyidae

⇧ *Ophisurus serpens* (Linnaeus)

[*O. macrorhynchus* Bleeker]

Longbill Snake Eel, Off-shore Snake, Sand Snake

Diagnosis — Largest species widely distributed in deep waters. Mouth is large. Snout is pointed. Canines are sharp. Caudal fin is absent. Some 200 lateral line pores. Dorsal fin originates from a point posterior to tip of pectoral fins. Edible. Up to 3m in total length.

Distribution — Japan, Ryukyu, Korea, Indo-Pacific-Atlantic, Mediterranean, South Africa

Chanidae

Chanos chanos Forskål

[*C. salmoneus* Forster]

Milk-fish, Bony Salmon, Salmon Herring, Bangos, Giant Herring, Morton Bay Salmon, White Mullet

Diagnosis — A littoral species, often moves into brackish areas of bays or river mouths. Bred in Formosa. Mouth is small and toothless. Scales are small. Dorsal fin has 14–16 rays and anal fin 9–11 rays. Lateral lines present. Caudal fin is falcated. Edible but causes ciguatera poisoning on rare occasions. Up to 1.8m in total length.

Distribution — Southern Japan, Ryukyu, Formosa, Indo-Pacific, Red Sea

Ariidae

Arius maculatus (Thunberg)

[*Tachysurus sinensis* Lacépède]

[*T. falcarius*]

Spotted Catfish, Sea Barbel, Sea Catfish

Diagnosis — A littoral species resembling a catfish. Head is compressed, skeleton structure is hard. Caudal fins falcated. Adipose fin is present. Dorsal fin has 1 spine and 6 rays, anal fin has 15 rays. Spines of dorsal and pectoral fins are venom-bearing. Mouth is situated close to ventral side, 2 barbels in upper, 4 barbels in lower jaw. Has a group of villiform teeth in palatine. Edible. Up to 40cm in total length.

Distribution — Southern Japan, Ryukyu, Formosa, South China Sea, South East Asia, Indian Ocean

Ariidae

Arius thalassimus (Rüppell)

[*Netuma osakae* Jordan et Hubbs]

[*N. nasuta* Bleeker]

Giant Catfish, Giant Salmon Catfish

Diagnosis — Resembles *A. maculatus*, but distinguished by 1 spine and 7 rays on dorsal fin. Barbels of upper jaw are long enough to reach base of pectoral fins. 3 groups of villiform teeth on palatine. Spines of dorsal and pectoral fins are strong, venom-bearing, dangerous to man. Scaleless. Edible. Up to 60cm in total length.

Distribution — Southern Japan, Formosa, China, Indo-Pacific, Persian Gulf, Red Sea, East Africa

Plotosidae

Plotosus lineatus (Thunberg)

[*P. anguillaris* (Lacépède)]

Sea Catfish, Striped Catfish, Stinging Catfish, Striped Catfish Eel, Plotosid Catfish, Marine Catfish, Barbel Eel, Catfish-eel, Barber

Diagnosis — A gregarious species found in coastal shoreline reefs and coral reefs. Has a venom-bearing spine on first dorsal and pectoral fins. Second dorsal fin is connected with caudal fin. 8 barbels on margin of mouth. Feeds on small benthos. Spawns on ledges. Nocturnal. Aquarium fish. Up to 30cm in total length.

Distribution — Japan, Ryukyu, Korea, Indo-Pacific, Red Sea, South Africa

Synodontidae

Trachinocephalus myops (Schneider)

[*Saurus trachinus* Temminck et Schlegel]

[*S. brevirostris* Poey]

Ground Spearing, Offshore Lizardfish, Painted Grinner, Painted Lizardfish, Painted Saury, Short-headed Lizardfish, Short-nosed Lizardfish, Snakefish

Diagnosis — A widely distributed species inhabiting shallow temperate waters. Snout is shorter than eye's diameter. Head is large. Eyes are close to each other, situated on top of head. Inner soft rays of pelvic fins are longer than outer rays. Gill rakers are absent. Feeds on small fish. Nocturnal. Edible. Up to 30cm in total length.

Distribution — Japan, Ryukyu, Formosa, Indo-Pacific-Atlantic, Persian Gulf, South Africa

Synodontidae

Saurida gracilis (Quoy et Gaimard)

[*S. nebulosa* (Cuvier et Valenciennes)]

Graceful Lizardfish, Slender Grinner, Slender Saury, Slender Lizardfish

Diagnosis — Inhabits coastal coral reefs. Dorsal fin has 10−12 rays, anal fin has 9−10 rays, pectoral fins have 12−13 rays. Body is dotted with irregualr blotches. Up to 30cm in total length.

Distribution — Japan, Ryukyu, Indo-Pacific, Red Sea, South Africa

Synodontidae

Synodus hoshinonis Tanaka

Hoshino's Lizardfish

Diagnosis — A littoral small species. Base of dorsal fin is longer than half of head length. Body is dotted with 8−10 T-shaped blotches. Caudal fin is pale. A black blotch on upper side of opercles. Up to 25cm in total length.

Distribution — Japan, Ryukyu, West Pacific

Synodontidae

Saurida undosquamis (Richardson)

[*S. argyrophanes* (Richardson)]

[*S. macrolepis* Tanaka]

[*S. grandisquamis* Günther]

Large-scaled Grinner, Large-scaled Saury, Largescale Lizardfish, Lizardfish, True Lizardfish, Brushtooth Lizardfish

Diagnosis — Inhabits sandy bottoms of shallow waters. Body is elongated. Mouth is large, teeth are sharp. Upper margin of caudal fin is dotted with black spots. Dorsal fin has 11−12 rays, anal fin has 10−12 rays. 48−54 lateral scales. Nocturnal. Feeds on small fish, shrimp, octopus and squid. Spawns between spring and summer. Edible. Up to 70cm in total length.

Distribution — Japan, Ryukyu, Formosa, Korea, Indo-Pacific, Red Sea, South Africa

Synodontidae

Saurida elongata (Temminck et Schlegel)

[*S. eso* Jordan et Herre]

[*S. microlepis* Wu et Wang]

Lizard Fish, Shortfin Lizardfish

Diagnosis — An inhabitant of sandy-mud bottoms of shallow waters. Nocturnal. Feeds on small fish and crustacea. Pectoral fins are short, do not reach origin of pelvic fins. No black spot on upper margin of caudal fin. Soft rays of pelvic fins are uniform in length. Scales are small, not easily removable. 62−66 lateral scales. 11−12 rays on both dorsal and anal fins. Up to 50cm in total length.

Distribution — Japan, Ryukyu, Formosa, Korea, Indo-Pacific

Aulopodidae
Aulopus damasi Tanaka ⇨
[*Hime damasi* (Tanaka)]

Deepwater Aulopus

Diagnosis — Inhabits sandy-mud bottoms of deep waters. Single dorsal fin has 14 rays. Adipose fin present. Eye's diameter is shorter than snout length. 35−37 lateral scales. Up to 30cm in total length.

Distribution — Japan, Ryukyu

Synodontidae
⇦ ***Saurida tumbil*** (Bloch)
[*Saurus badi* Cuvier et Valenciennes]

[*S. argyrophanes* Richardson]

Greater Lizardfish

Diagnosis — Body is elongated and cylindrical. Mouth is large, lower jaw is rather short. Dorsal fin has 11−13 rays, anal fin has 10−11 rays. Pectoral fins are too short to reach origin of pelvic fins. 47−53 lateral scales. Insides of opercles have a large black blotch. Edible. Up to 45cm in total length.

Distribution — Ryukyu, Formosa, China, Indo-Pacific

Synodontidae
Synodus englemani Schultz ⇨
Engleman's Lizardfish (New Common Name)

Diagnosis — Body is cylindrical. Resembles *S. ulae*. Dorsal fin has 12−13 rays, anal fin has 9−10 rays. 62−63 lateral scales. Body has reddish-brown cross bands. Up to 20cm in total length.

Distribution — Ryukyu, Formosa, Pacific

Synodontidae
Synodus ulae Schultz ⇧ ⇨
[*Saurus japonicus* Houttuyn]

[*Synodus dermatogenys* Fowler]

[*S. variegatus* (Lacépède)]

Redbrown Lizardfish, Red Lizard Fish, Variegated Lizard Fish

Diagnosis — Inhabits shoreline reefs of shallow coastal waters. Snout is pointed. Eyes and mouth are large. Inner soft rays of pelvic fins are longer than outer rays. Dorsal fin has 13−14 rays, anal fin has 8−10 rays. 60−66 lateral scales. Body varies widely in color. Up to 40cm in total length.

Distribution — Japan, Ryukyu, China, Indo-Pacific, Hawaii, Red Sea, South Africa

Synodontidae
Synodus variegatus (Lacépède) ⇨
Variegated Lizardfish

Diagnosis — Inhabits rocky reefs and in sandy bottoms. Resembles *S. ulae*, but distinguished by somewhat longitudinally compressed body with 8−9 blotches. Dorsal fin has 10−13 rays, anal fin has 8−10 rays. 56−61 lateral scales. Up to 30cm in total length.

Distribution — Ryukyu, Indo-Pacific

Chlorophthalmidae
Chlorophthalmus nigromarginatus (Kamohara) ⇨
[*C. acutifrons nigromarginatus* Kamohara]

Blacktipped Greeneye

Diagnosis — A deep-sea species. Large eyes are bright green, approximately equal to snout in diameter. Margins of dorsal and anal fins are black. Dorsal fin has 11 rays, anal fin has 9 rays. 49–53 lateral scales. A single black cross band on pelvic fins. Up to 30cm in total length.

Distribution — Japan, Formosa, East China Sea

Myctophidae
Myctophum asperum Richardson ⇨

Prickly Lanternfish, Rough Lanternfish

Diagnosis — A widely distributed small deep-sea species. Body is shallow, length is 4.7 times longer than depth. Snout is rounded. Eyes are large. Margin of opercles are rounded. Scales are ctenoid. Body is equipped with a number of luminous organs which emit light. Up to 10cm in total length.

Distribution — Japan, Ryukyu, Indo-Pacific-Atlantic

Belonidae
Strongylura anastomella (Valenciennes) ⇧ ⇨
[*Ablennes anastomella* (Valenciennes)]

Green Gar, Houndfish, Needlefish, Pacific Needlefish, Round-bodied Garfish

Diagnosis — An inhabitant of shallow waters and bays. Body is compressed. Jaws curved inward, lower jaw of juvenile is long, teeth are thin. Nostrils are triangular. No ridges on caudal peduncles. Dorsal fin has 17–21 rays, anal fin has 21–25 ray. Spawns between spring and early summer. Edible. Up to 1m in total length.

Distribution — Japan, Korea, China, South China Sea

Belonidae
Platybelone argalus platyura (Bennett) ⇨

Gill-rakered Needlefish (New Common Name)

Diagnosis — A small pelagic species. Body is elongated, caudal peduncles are depressed. Gill rakers are present. Dorsal fin has 13–15 rays, anal fin has 17–19 rays. Edible. Up to 50cm in total length.

Distribution — Ryukyu, Bonin Islands, Indo-Pacific

Belonidae
Strongylura incisa (Valenciennes) ⇨

Garfish, Squaretail Alligator Gar

Diagnosis — Body is elongated, somewhat compressed laterally. Head is large. Caudal margin is slightly emarginated, caudal peduncles have no ridges. Pectoral fins are large. Scales are also large. Jaws are long. Dorsal fin has 18–20 rays, anal fin has 21–23 rays. 103–113 predorsal scales. Edible but may cause ciguatera poisoning. Up to 1m in total length.

Distribution — Ryukyu, Indo-Pacific

▲ Angling Needlefishes in Tuamotu Atoll.

Belonidae

Tylosurus crocodilus crocodilus (Le Sueur)

[*T. giganteus* (Temminck et Schlegel)]

[*T. indicus* (Le Sueur)]

[*T. choram* Rüppell]

[*Belone annulata* Cuvier et Valenciennes]

Offshore Needlefish, Common Gar, Bilfish, Choram Long-tom, Crocodile Needlefish, Fork-tail Alligator Gar, Houndfish, Needlefish

Diagnosis — A widely distributed pelagic species of warm waters. Body is cylindrical. Jaws are strong, short, sharp canines forward directed. Dorsal fin has 21–24 rays, anal fin has 19–22 rays. Base of anal fin is long. Caudal peduncles have a ridge. Edible. Attracted to light. Harmful. Up to 2m in total length.

Distribution — Japan, Ryukyu, Bonin Islands, Indo-Pacific-Atlantic, South Africa

▼ Dangerous to man caused by jumping to light.

Belonidae

⇦ ***Tylosurus acus melanotus*** (Bleeker)

Alligator Gar, Blackfin Longtom, Blackfin Needlefish, Garfish

Diagnosis — Resembles *T. crocodilus crocodilus*, but distinguished by elongated jaws. In some individuals, lower jaw has projections at end. Caudal peduncles have a ridge. Dorsal fin has 24–27 rays, anal fin has 22–24 rays. 280–310 predorsal scales. Edible. Up to 1m in total length.

Distribution — Japan, Ryukyu, Korea, China, Indo-West Pacific

Scomberesocidae

Cololabis saira (Brevoort) ⇩

Pacific Saury, Saury, Mackerel Pike, Skipper

Diagnosis — A migratory fish of surface of cold waters. Body is elongated. Snout is pointed. Teeth are not strong. Stomach is absent. Feeds on zooplankton. A single dorsal fin situated close to tail, pelvic fins situated in the middle of body. Has 6–7 finlets. Edible. Up to 40cm in total length.

Distribution — Kuril Islands, Japan, Ryukyu, Korea, North Pacific, Mexico

Belonidae

Ablennes hians (Valenciennes) ⇧

[*Belone schismatorhynchus* Bleeker]

[*B. melanostigma* Cuvier et Valenciennes]

Flat Needlefish, Flat Billfish, Barred Long-tom, Gaping Gar, Gaping Needlefish, Silver Gar, Barred Needlefish, Compressed Body Gar

Diagnosis — A widely distributed pelagic species inhabiting warm waters.Body is slightly compressed. Jaws are curved slightly inward. Scales are remakably small. Body is lined with 6–10 cross bands. Dorsal fin has 23–27 rays, posterior rays are extended. Anal fin has 24–27 rays. Bones are blue. Edible. Up to 1.2m in total length.

Distribution — Southern Japan, Ryukyu, Indo-Pacific-Atlantic, Persian Gulf, Red Sea, South Africa

Exocoetidae

⇦⇦ ***Exocoetus volitans*** Linnaeus

[*E. evolans* Day]

Common Flyingfish, Cosmopolitan Flyingfish, Tropical Two Wing Flyingfish, Sharp-finned Flyingfish

Diagnosis — A small species of offshore waters. Snout is short; eyes are large. Pectoral fins are long and gray, have 14–16 rays. Dorsal fin is not elevated, has 13–15 rays. Pelvic fins are short, situated anterior to dorsal fin. Anal fin has 12–15 rays, corresponds with dorsal fin. 18–21 predorsal scales. Edible. Up to 25cm in total length.

Distribution — Japan, Ryukyu, Indo-Pacific-Atlantic

Exocoetidae

⇦⇦ ***Parexocoetus brachypterus brachypterus*** (Richardson)

Sailfin Flyingfish, Short-finned Flyingfish, Slender Two-winged Flyingfish

Diagnosis — A small species. Snout is short. Juveniles have short fleshy tentacles. Pectoral fins are pale, too short to reach the posterior end of dorsal fin. Pelvic fins are also short. In male, both pelvic and caudal fins are light-pink. Dorsal fin is elevated, upper side blackish, has 11–14 rays. 19–25 predorsal scales. Edible. Up to 18cm in total length.

Distribution — Japan, Ryukyu, China, Indo-Pacific, Hawaii

Exocoetidae
Hirundichthys oxycephalus
(Bleeker) ⇨⇨

Mirror-finned Flyingfish, Sharphead Flyingfish

Diagnosis — Body is elongated. Dorsal fin is not black, corresponds with anal fin. Pectoral fins are dark, fin rays are forked except first one. Dorsal fin has 10–12 rays, anal fin has 11–12 rays, pectoral fins have 15–17 rays. 28–37 predorsal scales. Edible. Up to 26cm in total length.

Distribution — Japan, Ryukyu, Pacific-Atlantic

Exocoetidae
Prognichthys agoo
(Temminck et Schlegel) ⇨⇨

[*Cypselurus agoo agoo* (Temminck et Schlegel)]

Flying Fish, Opaquewing Flyingfish

Diagnosis — A medium-sized species abundant in waters of Japan. Distinguished from other flying fishes by non-forked first and second fin rays of pectoral fins, anal fin originated at posterior of dorsal fin base, black-spotted pectoral and pelvic fins of juvenile. Dorsal fin has 10–12 rays, anal fin has 9–11 rays. 33–37 predorsal scales. Edible. Up to 36cm in total length.

Distribution — Japan, Ryukyu, Formosa, Korea

Exocoetidae
Cypselurus katoptron
(Bleeker) ⇨⇨

Bicolorwing Flyingfish

Diagnosis — A species of tropical waters. Pectoral fins are lined with yellowish-white bands. Lower lobe of caudal fin is dark, upper lobe is pale. Dorsal fin is large, has 12–15 rays. Anal fin has 8–11 rays. 22–29 predorsal scales. Teeth are large. Edible. Up to 30cm in total length.

Distribution — Japan, Ryukyu, Pacific

Exocoetidae
Cypselurus cyanopterus
(Valenciennes) ⇨⇨

Margined Flyingfish

Diagnosis — Dorsal fin is elevated, has a black blotch. Pectoral fins are black, have 12–15 rays. Anal fin has 9–11 rays. Teeth are large. A pair of long fleshy barbels on lower jaw of juvenile. 34–40 predorsal scales. Edible. Up to 45cm in total length.

Distribution — Japan, Ryukyu, Indo-Pacific-Atlantic

Exocoetidae
Cypselurus starksi Abe ⇨⇨

Snubnose Flyingfish

Diagnosis — Blunt-bodied. Eyes are large; head is rounded. Pectoral fins are bluish-black, lower sides are pale. Dorsal fin has 12–14 rays, anal fin has 8–10 rays. No fleshy barbels on juvenile. 30–35 predorsal scales. Edible. Up to 30cm in total length.

Distribution — Japan, Izu Islands

Exocoetidae
Cypselurus antoncichi
Wood et Schultsz ⇨⇨

[*C. unicolor* (Valenciennes)]

Bigeye Summer Flyingfish

Diagnosis — A large species of tropical waters. Eyes are large; lower jaw is protruded. Pectoral fins are transparent, have 14−15 soft rays. Anal fin has 9−11 rays. Teeth are extremely small. 30−37 predorsal scales. Edible. Up to 40cm in total length.

Distribution — Japan, Ryukyu, Pacific

Exocoetidae
Cypselurus spilonotopterus
(Bleeker) ⇨⇨

Brown-winged Flyingfish

(New Common Name)

Diagnosis — A large species of tropical waters. Dorsal fin has a large black blotch. Pectoral fins are reddish-brown, no black blotch. A pair of fleshy barbels on juvenile. 34 or less predorsal scales. Edible. Up to 40cm in total length.

Distribution — Ryukyu, Indo-Pacific

Exocoetidae
Cypselurus angusticeps
Nichols et Breder ⇨⇨

Small Brown Flyingfish

(New Common Name)

Diagnosis — A medium-sized species of reddish flying fish. Fleshy barbels on juvenile. Dorsal fin has 13−14 rays and anal fin 8−9 rays. Upper side of pectoral fins are dark reddish-brown, lower side are transparent. 25−28 predorsal scales. Up to 25cm in total length.

Distribution — Ryukyu, West Pacific

Exocoetidae
Cypselurus suttoni
(Whitley et Colefax) ⇨⇨

Great Brown Flyingfish

(New Common Name)

Diagnosis — A large species. Pectoral fins are reddish, dotted profusely with black spots. Dorsal fin has 12−14 rays and one black blotch, anal fin has 9−14 rays. Fleshy barbels on lower jaw of juvenile. 38−42 predorsal scales. Edible. Up to 38cm in total length.

Distribution — Ryukyu, West Pacific

Exocoetidae
Cypselurus poecilopterus
(Valenciennes) ⇨⇨

Spotted Flyingfish, Yellow-wing Flyingfish

Diagnosis — Body is blunt. No fleshy barbels on juvenile. Dorsal fin has 11−13 rays, no black blotch. Pectoral fins are yellowish, dotted with oval blotches. Anal fin has 7−9 rays. Teeth are large. 25−31 predorsal scales. Edible. Up to 30cm in total length.

Distribution — Japan, Ryukyu, Formosa, Korea, China, Pacific

Hemiramphidae
⇦ *Hyporhamphus dussumieri*
(Valenciennes)

Halfbeak, Needlefish, Dussumier's Garfish, Dussumier's Halfbeak

Diagnosis — Body is cylindrical, middle of body is square in cross-section. Lower jaw is long; snout has scales. Dorsal fin has 14–17 rays, anal fin has 13–17 rays. 52–56 lateral scales. Edible. Up to 40cm in total length.

Distribution — Ryukyu, Indo-West Pacific, Red Sea, Mediterranean, South Africa

Hemiramphidae
Euleptorhamphus viridis
(Van Hasselt) ⇨

[*E. longirostris* (Cuvier)]

Long-beaked Garfish, Long-finned Garfish, Ribbon Halfbeak, Slender Longfin Halfbeak

Diagnosis — A pelagic species inhabits among schools of flying fish. Body is elongated and laterally compressed. Lower jaw and pectoral fins are remarkably long. Dorsal and anal fins are elevated. Pelvic fins are extremely small. Dorsal fin has 21–25 rays, anal fin has 20–24 rays. 110 lateral scales. Capable of gliding in the air. Edible. Up to 50cm in total length.

Distribution — Japan, Ryukyu, Indo-Pacific, South Africa

Hemiramphidae
⇧ *Hemiramphus far* (Forskål)

[*H. commersoni* Cuvier]

[*H. fasciatus* Bleeker]

Barred Garfish, Black-barred Garfish, Candle Fish, Half-beak, Needle Fish, Spotted Halfbeak

Diagnosis — Inhabits coastal waters and bays. Body is deep and compressed. Lower jaw is extended, tip is orange, upper jaw is scaleless. 4–9 black blotches on body. Dorsal fin has 12–14 rays, anal fin has 10–13 rays. 50–55 lateral scales. Edible. Up to 65cm in total length.

Distribution — Ryukyu, Indo-Pacific, Red Sea, Mediterranean, South Africa

Hemiramphidae
Hyporhamphus yuri
Collette et Parin ⇨

Yuri's Halfbeak (New Common Name)

Diagnosis — Resembles *H. dussmmieri*, but distinguished by slightly longer, regular, triangular upper jaw. Lower jaw is long. Dorsal fin has 16–17 rays, anal fin has 17–19 rays. 43–46 predorsal scales. Edible. Up to 30cm in total length.

Distribution — Ryukyu

Hemiramphidae
Hyporhamphus quoyi
(Valenciennes) ⇧

[*Hemiramphus micropterus* Jordan et Dickerson]

[*H. quoyi* Valenciennes]

Smallfin Halfbeak

Diagnosis — An inhabitant of bays. Body is cylindrical. Lower jaw is short. Dorsal fin has 15−17 rays, anal fin has 13−16 rays. 55 lateral scales. Edible. Up to 35cm in total length.

Distribution — Southern Japan, Ryukyu, Korea, Indo-West Pacific

Hemiramphidae
⇦⇧ *Hyporhamphus sajori* (Temminck et Schlegel)

[*Hemiramphus sajori* (Temminck et Schlegel)]

Japanese Halfbeak, Halfbeak Fish, Japanese Needlefish, Half Beak

Diagnosis — Inhabits coastal waters, bays and river mouths. Body is elongated. Lower jaw is extended. Dorsal fin corresponds with anal fin. Feeds on zooplankton. Spawns in seaweed beds between spring and summer. Male grows to full size in 1−2 years, female in 2 years. Edible. Up to 45cm in total length.

Distribution — Sakhalin, Japan, Formosa, Korea, Japan Sea

Hemiramphidae
Hyporhamphus acutus acutus
(Günther) ⇨

[*Hemiramphus acutus* Günther]

Polynesian Half-beak

Diagnosis — Body is slightly compressed. Middle of body is square in cross-section. Dorsal and anal fins approximately equal in size. Lower lobe of caudal fin is long. Dorsal fin has 13−14 rays, anal fin has 17 rays. Upper side of snout is scaled. Scales are large. Edible. Up to 30cm in total length.

Distribution — Wake Island, Polynesia, Easter Island

▲ Yellow variation of Trumpet-fish.

Aulostomidae
Aulostomus chinensis

(Linnaeus)

Trumpet-fish, Painted Flute-mouth, Spiny-back Trumpet-fish, Stick Fish

Diagnosis — Singly inhabits coral reefs of shallow waters. Body is elongated and compressed. Snout is long, tubular in shape. Dorsal fins corresponds with anal fin. A barbel on lower jaw. Teeth and gill rakers are degenerated. Feeds on zooplankton. Two different body clolors; yellow and brown. Edible. Up to 1m in total length.

Distribution — Japan, Ryukyu, Indo-Pacific-Atlantic, East Africa

Fistulariidae
Fistularia petimba Lacépède

[*F. villosa* Klunzinger]

[*F. serrata* Cuvier]

Cornet Fish, Flute Mouth, Red Cornetfish, Smooth Flute Mouth

Diagnosis — A littoral species. Body is elongated and bluish. Snout is long and tubular. Small spines on body surface. Dorsal fin has 13−15 rays, anal fin has 13−15 rays corresponding with dorsal fin. A single row of scales on anterior and posterior ends of dorsal and anal fins. Edible. Up to 1.5m in total length.

Distribution — Japan, Ryukyu, Indo-Pacific, Red Sea, South Africa

▼ Tubular snout of Cornet Fish.

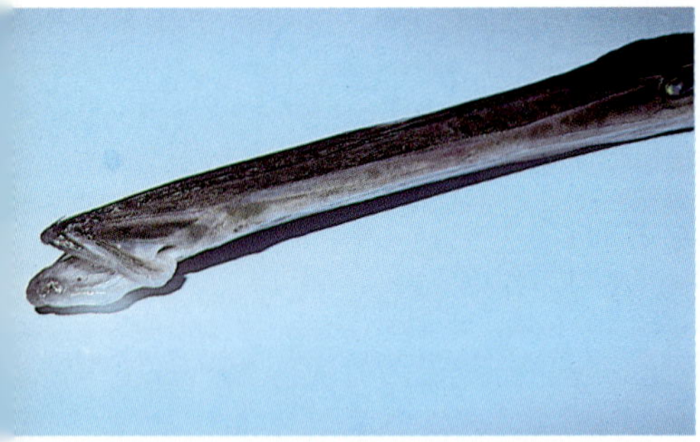

Fistulariidae
Fistularia commersonii

Rüppell

[*F. depressa* Günther]

Cornet Fish, Flute Mouth, Smooth Flutemouth

Diagnosis — Inhabits deep offshore waters. Body is elongated; snout is long and tubular. Dorsal fin has 16−17 rays, anal fin has 14−15 rays corresponding with dorsal fin. 2 soft rays in the middle of caudal fin are filiform. Scaleless. Spawns in winter. Edible. Up to 1.5m in total length.

Distribution — Japan, Ryukyu, Indo-Pacific, Red Sea, South Africa

Centriscidae

Aeoliscus strigatus (Günther) ⇧⇗

Jointed Razor-fish, Razorfish, Shrimp Fish

Diagnosis — A gregarious species inhabiting coral reefs of shallow waters. Has a strange habit of swimming with head down. Body is remarkably flat like a blade, covered with 3 rows of horny plates. Mouth resembles a fountain-pen ink filler. Feeds on small zooplankton. Fins are concentrated on tail. First spine of dorsal fin is large, extending posteriorly. Aquarium fish. Up to 16cm in total length.

Distribution — Japan, Ryukyu, Indo-Pacific, Persian Gulf

Macrorhamphosidae

Macrorhamphosus japonicus (Günther) ⇨

[*M. gracilis* (Lowe)]

Shortspine Snipefish, Egret-piper

Diagnosis — Resembles *M. scolopax*, but distinguished by less deep, slightly smaller blackish body. Second spine of dorsal fin is not long enough to reach caudal fin. Dorsal fin has 4–5 spines and 11 rays, anal fin has 18–19 rays. Up to 15cm in total length.

Distribution — Japan, China, Indo-Pacific, South Africa

Aulorhynchidae

⇧ *Aulichthys japonicus* Brevoort

Japanese Tubesnout

Diagnosis — A small species inhabiting seaweed beds of bays. Body is elongated; snout is long and tubular in shape. Dorsal fin has 14–16 separate spines and 9–10 rays, anal fin has 1 spine and 9–10 rays. 55 lateral plates. Feeds on zooplankton. Spawns in sea squirt, *Halocynthia roretzi* in spring. Male has outer reproductive organs. Up to 15cm in total length.

Distribution — Japan, Korea

Macrorhamphosidae

⇦ *Macrorhamphosus scolopax* (Linnaeus)

[*M. sagifue* Jordan et Starks]

[*M. japonicus* (Günther)]

Japanese Snipefish, Bird Flute, Trumpet Fish, Woodcock-fish, Longspine Snipefish, Japanese Spinefish, Egret Piper, Bellows Fish

Diagnosis — A small gregarious species inhabiting rather deep waters. Snout is long. No teeth and lateral lines. Dorsal fin has 5 spines and 12 rays, second spine well-developed, reaching caudal fin. Anal fin has 18 rays. Scales are minute, arranged irregularly. Up to 20cm in total length.

Distribution — Japan, Indo-Pacific, South Africa

Syngnathidae
⇩ *Doryrhamphus japonicus*
Araga et Yoshino

Japanese Blue-striped Pipefish
Diagnosis — Found in pairs in small caves of shoreline reefs of shallow waters. Dorsal fin has 21–23 rays, anal fin has 3–4 rays. Resembles *D. melanopleura melanopleura*, but distinguished by different splotches on body and caudal fin. Up to 8cm in total length.
Distribution — Japan

Syngnathidae
Syngnathoides biaculeatus
(Bloch) ⇧

Alligator Pipefish, Double-ended Pipefish, Horned Pipefish, Two-barbel Pipefish
Diagnosis — Inhabits bays. Body is compressed and straight. Caudal and pelvic fins are absent. Tail prehensile like sea-horse. Dorsal fin has 35–50 rays, anal fin has 4–6 rays. Eggs are exposed on abdomen. Up to 25cm in total length.
Distribution — Japan, Ryukyu, Indo-Pacific, South Africa

Syngnathidae
⇧ *Corythoichthys haematopterus*
(Bleeker)

Banded Pipe Fish, Messmate Pipe Fish
Diagnosis — Inhabits shallow coral reefs. Body is elongated; snout is remarkably short. A ridge is present on opercular surface. Dorsal fin has 23–33 rays, anal fin has 4 rays. Posterior head is ridged in front of eyes. Eggs are exposed. Up to 18cm in total length.
Distribution — Japan, Ryukyu, Indo-Pacific, Red Sea, South Africa

Syngnathidae
Corythoichthys amplexus
Dawson et Randall ⇨
[*C. flavofasciatus* (Rüppell)]

Banded Pipefish
Diagnosis — Inhabits coral reefs. Snout is short. Distinct cross bands on body. Dorsal fin has 23–30 rays, anal fin has 4 rays. Eggs are exposed on abdomen. Up to 10cm in total length.
Distribution — Ryukyu, Indo-West Pacific

Syngnathidae
⇦⇩ *Corythoichthys schultzi* Herald

Gilded Pipefish
Diagnosis — Inhabits coral reefs. Snout is long. 5 rows of orange spots on body. Caudal fin is reddish and white margin. Dorsal fin has 25–31 rays, anal fin has 4 rays. Up to 16cm in total length.
Distribution — Ryukyu, Indo-Pacific, Red Sea

Syngnathidae
⇩ *Dunckerocampus dactyliophorus* (Bleeker)

Banded Pipefish

Diagnosis — Inhabits coral reefs of shallow waters. Body is elongated, covered with minute spines. Snout is remarkably long, dorsally with 3 low ridges. Dorsal fin has 21–23 rays, anal fin has 4 rays. Up to 18cm in total length.

Distribution — Ryukyu, South East Asia, West Pacific

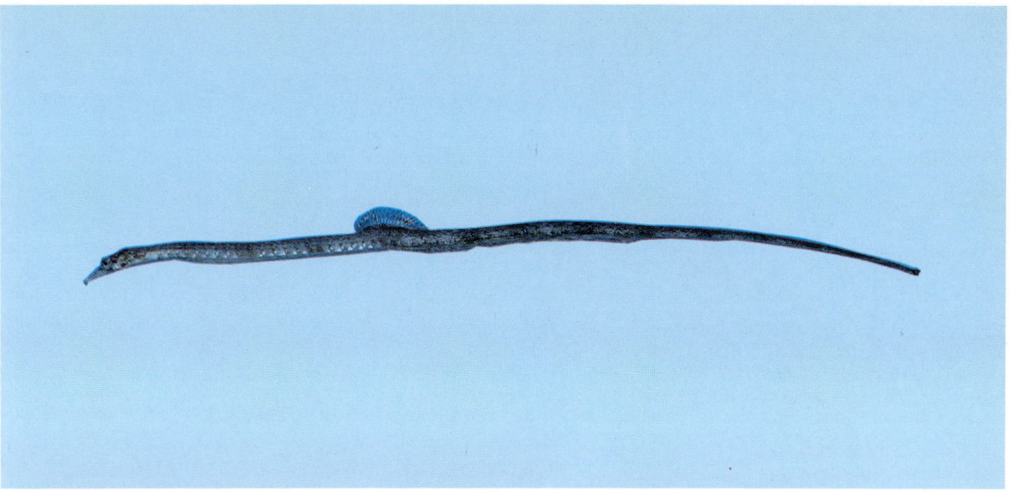

Syngnathidae
Trachyrhamphus serratus (Temminck et Schlegel) ⇧

Crested Pipe Fish, Goggle-eye Pipefish

Diagnosis — A large species of pipe fish inhabiting shoreline reefs in shallow waters. Snout is short, dorsally with serrated ridge. Dorsal fin has 25–29 rays, fin base is elevated. 9–12 dark cross bands on body. Up to 30cm in total length.

Distribution — Japan, Ryukyu, South East Asia, Indo-West Pacific

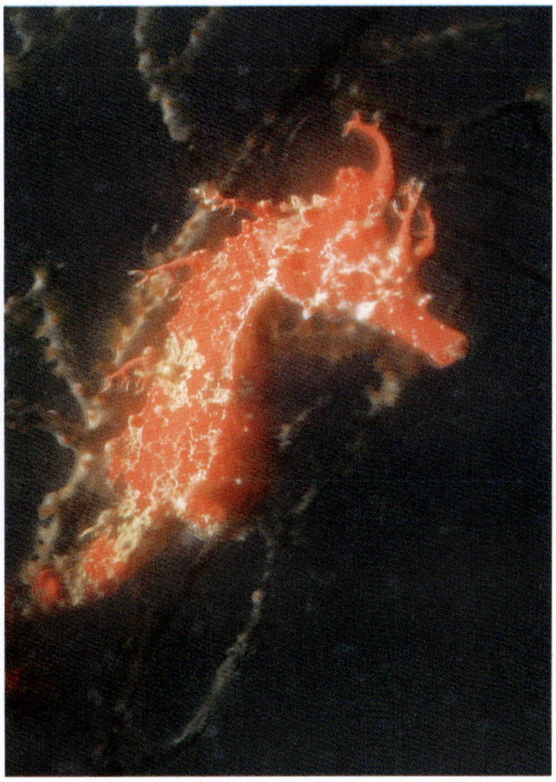

Syngnathidae
⇩ *Hippocampus takakurae* Tanaka

Threespot Seahorse

Diagnosis — Inhabits coastal waters. Snout resembling a fountain-pen ink filler is capable of sucking in zooplankton and juveniles. Body is covered with short spines. Back has 3 black splotches. Has 11 body rings and 41 tail rings. Dorsal fin has 21 rays. Male has a broad pouch in the belly. Aquarium fish. Up to 18cm in total length.

Distribution — Japan, Ryukyu, China

Syngnathidae
⇧ *Hippocampus coronatus* Temminck et Schlegel

[*H. mohnikei* Bleeker]

Japanese Sea-horse, Seahorse, Red-and-black Seahorse, Horned Seahorse

Diagnosis — Inhabits seaweed bed in shallow waters. Spines or fleshy tentacles extending from body rings. Black bands on body and tail. Dorsal fin is short, has 13–14 rays. 10 body rings and 38–40 tail rings. Snout is short. Aquarium fish. Up to 9cm in total length.

Distribution — Japan, Ryukyu, Korea, China

Syngnathidae
Hippocampus kuda Bleeker ⇩

[*H. kelloggi* Jordan et Snyder]

[*H. polytaenia* Bleeker]

Kellogg's Seahorse, Seeperd

Diagnosis — A large species of seahorse, varying widely in color. Caudal fin is absent, tail is prehensile. Dorsal fin has 16–18 rays. Has 11 body rings and 36–39 tail rings. Spawns in all seasons. Aquarium fish. Up to 30cm in total length.

Distribution — Japan, Ryukyu, Formosa, Indo-Pacific, South Africa

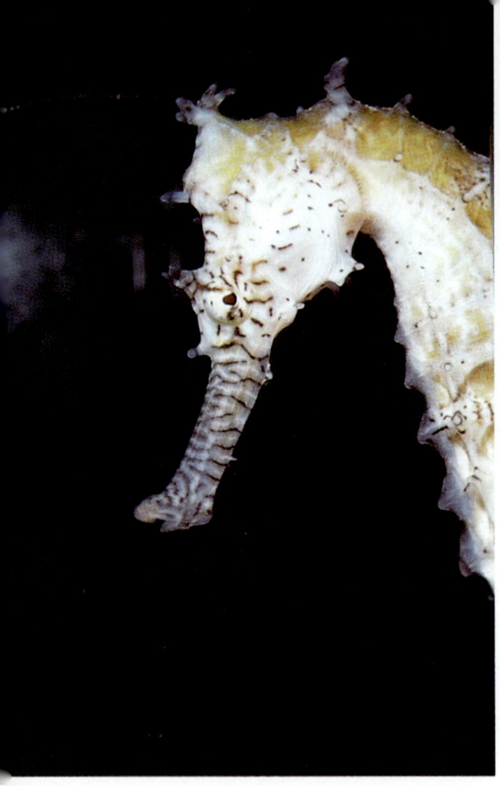

Syngnathidae

⇦⇧ ***Hippocampus histrix*** Kaup

[*H. hystrix* Kaup]

Seahorse, Seeperd, Spiny Seahorse, Thorny Seahorse

Diagnosis — Inhabits shallow coral reefs. Has 11 body rings with sharp spines; 33–34 tail rings. Snout is long and striped. Aquarium fish. Up to 17cm in total length.

Distribution — Japan, Ryukyu, Indo-Pacific, Red Sea, South Africa

Pegasidae

⇧ ***Pegasus volitans*** Linnaeus

Shorttail Dragonfish, Winged Dragonfish

Diagnosis — Inhabits sandy-mud bottoms of deep waters. Body is broad, longitudinally compressed. Snout is protruded. Pectoral fins are well-developed, fifth ray is swollen, forming a spine. Has 11 movable tail rings. Up to 8cm in total length.

Distribution — Japan, Formosa, China, Indo-West Pacific, South Africa

Gadidae

⇧ ***Gadus macrocephalus*** Tilesius

Alaska Codfish, Cod, Codfish, Gray Cod, Grayfish, Pacific Cod

Diagnosis — Inhabits waters some 300m deep. Feeds on small fish, squid, octopus, shrimp and crab. Lower jaw is long, has long barbels. 3 dorsal fins, 13–15 rays on first dorsal, 15–18 on second, 15–17 on third. 2 anal fins, 16–17 rays on first anal, 16–17 on second. Grows fully in 5 years. Spawns in coastal waters in winter. Edible, also used for making oil. Up to 1m in total length.

Distribution — Japan, Korea, Okhotk Sea, Bering Sea, North Pacific, North America

Gadidae

⇙⇩ ***Theragra chalcogramma*** (Pallas)

Walleye Pollock, Whiting, Pollock, Pacific Pollock, Alaska Pollock

Diagnosis — A swimmer, inhabits waters several-hundred meters deep. Body is elongated. Head is small. Lower jaw is long, has short barbels. 3 dorsal fins, 10–13 rays on first dorsal, 14–18 on second, 15–20 on third. 2 anal fins, 17–22 rays on first anal, 16–21 on second. Spawns in coastal waters between winter and spring. Edible. Up to 90cm in total length.

Distribution — Japan, Korea, Okhotk Sea, Bering Sea, North Pacific, North America

▲ Ovary of Alaska Pollock for food.

Macrouridae

Coelorhynchus multispinulosus Katayama ⇨

[*C. vermicularis* Matsubara]

Spearnose Grenadier

Diagnosis — Inhabits waters some 300m deep. Head is large. Snout is remarkably protruded. Tail is extended. Caudal fin is absent. Ventral luminous organ is blackish. A short barbel on lower jaw. Spinules of scales are 5-pointed. Head is 2.2 times longer than snout. Up to 40cm in total length.

Distribution — Japan, Formosa, East China Sea

▼ Luminous organ found from anus to surface between pelvic fins.

▼ Mouth opens underside.

▼ Spinous scale of Hardhead Grenadier.

Macrouridae

⇦ *Coelorhynchus japonicus* (Temminck et Schlegel) ⇧

Hardhead Grenadier

Diagnosis — A deep-sea species. Snout is protruded. A single barbel on lower jaw. Lateral margins of snout are straight. A single spinous carinae on top of head scales. 3–5 divergent spinous ridges on scales of body. Lower side of head is scaled. Fins are blackish. Edible. Up to 75cm in total length.

Distribution — Japan, Indo-West Pacific

Macrouridae

⇦ *Coelorhinchus hige* Matsubara

Lantern Grenadir

Diagnosis — A deep-sea species. Body is long, head is developed. Snout tip is sharp-pointed. Scales above lateral lines are in 5 rows. Lower surface of head is scaleless. A broad luminous organ is present anterior to anus. Up to 50cm in total length.

Distribution — Japan, Ryukyu

Moridae

⇦ ***Physiculus japonicus*** Hilgendorf

Japanese Hakeling

Diagnosis — Inhabits sandy-mud bottoms of deep waters. Feeds on small fish, squid and octopus. 9–10 rays on first dorsal fin, 60–68 on second, 60–71 rays on anal fin. A short barbel on lower jaw. Luminous organ, resembling a black spot, situated between pelvic fins. Edible. Up to 40cm in total length.

Distribution — Japan

Moridae

Laemonema filodorsale Okamura ⇨

Filamentous Morid Cod (New Common Name)

Diagnosis — A deep-sea species. Body is somewhat thick, tail is thin. Lower jaw is short, lips are thick. Pelvic fins are thread-like, long enough to reach anus. Upper surface of snout is covered by scales. First dorsal fin has 4–5 spines, first spine is extended like thread. Second dorsal fin has 52–56 rays, anal fin has 49–52 rays. 124–139 lateral scales. Edible. Up to 35cm in total length.

Distribution — Kyushu-Palao Ridge, Ryukyu

Moridae

⇦ ***Physiculus maximowiczi*** (Herzenstein)

Brown Hakeling

Diagnosis — A deep-sea species. 9–10 rays on first dorsal fin, 67–69 on second, 63–76 rays on anal fin. Second dorsal fin corresponds with anal fin, but not connected with caudal fin. A barbel on lower jaw. Feeds on small benthonic fish. Edible. Up to 30cm in total length.

Distribution — Japan

Antennariidae

⇧ ***Phrynelox tridens*** (Temminck et Schlegel)

Trilobate Frogfish, Toadfish, Fishing Frog

Diagnosis — Inhabits shoreline reefs of rather deep waters. Body is covered with stream-shaped splotches of protective color. Attracts preys by moving filamentous tentacle above snout. This tentacle is originally the first separate spine of dorsal fin. Skin is rough. Crawls on bottom with pelvic fins. Black splotches on caudal fin larger than eyes. Up to 12cm in total length.

Distribution — Japan, Ryukyu, Korea, Formosa, China, Indonesia, Indian Ocean

Antennariidae

⇧ ***Antennarius chironectes*** (Lacépède) ⇧

[*A. chironemus* (Lacépède)]

[*A. commersonii* Bleeker]

[*A. laysanius* Jordan et Evermann]

Fishing Frog, Variable Frog Fish, Black Angler

Diagnosis — Inhabits rocky reefs in rather deep waters. Body is dotted with small black spots. Filamentous tentacle on snout is elongated. Body colors vary widely. Dorsal fin has 3 separate spines and 13–14 rays, anal fin has 7–8 rays. Crawls on bottoms by moving pectoral fins. Has venom-bearing spines. Up to 25cm in total length.

Distribution — Ryukyu, Indo-Pacific, Hawaii, South Africa

Antennariidae

Antennarius hispidus Bloch et Schneider ⇧

[*A. scriptissimus* Jordan]

Fishing Frog, Shaggy Angler, Shaggy Fishing-frog, Toadfish, Zebra Angler-fish

Diagnosis — Body is oval, dark stripe-patterned. Skin is rough. Pelvic fins are short. Filamentous tentacle on snout is equal in length to second separate spine. Dark round blotches on second dorsal fin and caudal fin. Gill slits are small, opened posterior to pelvic fins. Up to 40cm in total length.

Distribution — Japan, Ryukyu, Indo-Pacific, South Africa

Antennariidae

Antennarius moluccensis Bleeker ⇧ ⇩

Black Angler, Pacific Yellow Angler, Molucca Islands Angler Fish

Diagnosis — Inhabits rocky reefs. Body is rounded. Body and blotches vary widely in color. Filamentous tentacle on snout is elongated, has some 15 cross stripes. Separate spines of dorsal fin contain venomous glands, cause painful punctures. Up to 30cm in total length.

Distribution — Japan, Ryukyu, Pacific

Lophiidae

⇩ *Lophius litulon* (Jordan)

Fishing-frog, Yellow Goosefish

Diagnosis — Inhabits sandy-mud bottoms of waters not deeper than 500m. Body is rounded, longitudinally compressed. Mouth is large, directed upward. 2 rows of teeth in both jaws. Body is light-brown, covered with reticulated blotches. Many short fleshy flaps on lower jaw and body. Dorsal fin has 6 separate spines and 10 rays. First separate spine serves as a filamentous tentacle, attracts small fish. Anal fin has 9 rays. Mouth and peritoneum are white. Edible. Up to 1.5m in total length.

Distribution — Japan, Formosa, Korea, Philippines, East China Sea

Ogcocephalidae

Halieutopsis vermicularis Smith et Radcliffe ⇩

Vermiculated Butfish

Diagnosis — Body is oval, longitudinally compressed. Gill slits are small, opened posterior to pectoral fins. Toothless. Dorsal fin has 4 rays, anal fin has 3 rays. Pectoral fins, caudal fin, snout and lower side of body are reddish. Up to 15cm in total length.

Distribution — Southern Japan, Formosa, Philippines

Lophiidae

Lophiomus setigerus (Vahl) ⇨

Angler, Fishing Frog, Blackmouth Goosefish

Diagnosis — Inhabits in rocky reefs more than 30m deep. Body is flat, scaleless, has flaps. Tail is narrow and long. Lower jaw has 1–3 rows of teeth. Mouth is black inside. First dorsal spine extends long, is used to attract small fish. Creeps about the sea bottom on pectoral fins. Spawns in spring. Edible. Up to 1.5m in total lengh.

Distribution — Japan, Ryukyu, Formosa, Indo-Pacific, Mexico, South Africa

Berycidae
Beryx splendens Lowe
Alfonsino, Berycoid Fish

Diagnosis — Inhabits warm and deep waters. Mouth is large, opened diagonally upward, lower jaw is protruded. Has slender teeth and small hard scales. Dorsal fin has 4 spines and 13–15 rays, anal fin has 4 spines and 26–30 rays, base is long. 71–76 lateral scales. Preorbital spines are small. Feeds on zooplankton. Edible. Up to 50cm in total length.

Distribution — Japan, Ryukyu, Indo-Pacific-Atlantic, Mediterranean, Gulf of Mexico, South Africa

Berycidae
Beryx decadactylus Cuvier
[*B. longipinnis* Barnard]
[*Actinoberyx jugeati* Boule]

Broad Alfonsino

Diagnosis — A deep-sea species. Body is deep, covered with large and hard scales. Eyes are remarkably large, preorbital spines are strong. Dorsal fin has 4 spines and 16–20 rays, anal fin has 3–4 spines and 25–30 rays. 52–66 lateral scales. Edible. Up to 45cm in total length.

Distribution — Japan, Ryukyu, Indo-Pacific-Atlantic, Mediterranean

Berycidae
Centroberyx druzhinni (Busakhin)
Flathead Alfonsino

Diagnosis — Inhabits rather deep waters. Resembles *Beryx decadactylus* , but distinguished by extended, red-tipped lobes of caudal fin. Mouth is large, lower jaw is protruded. Abdominal scales have spines. Dorsal fin has 5–7 spines and 12–15 rays, anal fin has 4 spines and 15–17 rays. 53–62 lateral scales. Edible. Up to 22cm in total length.

Distribution — Southern Japan, Ryukyu, Bonin Islands, South China Sea, Indo-West Pacific

Holocentridae

Sargocentron ruber (Forskål) ⇨⇨

[*Adioryx ruber* (Forskål)]

Red Soldierfish, Red Squirrelfish

Diagnosis — Inhabits shoreline reefs and coral reefs of shallow waters. 9 reddish stripes on body. Upper and lower sides of caudal fin are red-margined. Preopercular bones are spined. Dorsal fin has 11 spines and 13–14 rays, anal fin has 4 spines and 8–10 rays. 34–37 lateral scales. Edible but sometimes causes ciguatera poisoning. Up to 25cm in total length.

Distribution — Japan, Ryukyu, Indo-Pacific, Red Sea, Mediterranean, South Africa

Holocentridae

Sargocentron cornutus (Bleeker)

[*Adioryx cornutus* (Bleeker)] ⇨⇨

Horned Squirrelfish, Red and White Soldierfish

Diagnosis — Large black blotches on dorsal fin, anal fin and caudal peduncles. Dorsal fin has 11 spines and 13 rays, anal fin has 4 spines and 9 rays. Preopercular bones are spined. Edible. Up to 27cm in total length.

Distribution — Southern Japan, Ryukyu, Indo-West Pacific

Holocentridae

Sargocentron ittodai

(Jordan et Fowler) ⇨⇨

[*Adioryx ittodai* (Jordan et Fowler)]

Japanese Soldier Fish

Diagnosis — Dorsal fin has 11 spines and 12–14 rays, a dark blotch on anterior part of dorsal fin. Anal fin has 4 spines and 9–11 rays. 46–50 lateral scales. White stripes on body. Spawns in early summer. Edible. Up to 20cm in total length.

Distribution — Japan, Ryukyu, Formosa, Indian Ocean

Holocentridae

Sargocentron furcatus (Günther)

[*Adioryx furcatus* (Günther)] ⇨⇨

Yellowstriped Squirrelfish

Diagnosis — Inhabits rocky reefs of rather deep waters. Yellow stripes on body. Dorsal fin has 11 spines and 14–15 rays, margins of fin membrane are slightly falcated. Anal fin has 4 spines and 10–11 rays. 43–47 lateral scales. Edible. Up to 30cm in total length.

Distribution — Japan, Ryukyu, Indo-Pacific, Hawaii

Holocentridae

Flammeo sammara (Forskål) ⇨⇨

Blood-spot Squirrelfish, Longchin Soldierfish, Sam Soldier, Lion Male

Diagnosis — Inhabits shoreline reefs and coral reefs. Body is elongated. Mouth is large, lower jaw is protruded. Strong spines on preopercular bone. 10 spines on first dorsal fin, a black blotch on anterior part of it. 1 spine and 12–13 rays on second. Anal fin has 4 spines and 7–9 rays. Edible. Up to 28cm in total length.

Distribution — Japan, Ryukyu, Indo-Pacific, Red Sea, South Africa

Holocentridae

⇦⇧ ***Sargocentron spinifer*** (Forskål) ⇧

[*Adioryx spinifer* (Forskål)]

Scarlet-fin Soldierfish, Spiny Squirrelfish, Spine-bearing Squirrel Fish, Sabre Squirrelfish

Diagnosis — A large species of soldier fish inhabiting shoreline reefs and coral reefs of shallow waters. Ferocious, has its own territory. Nocturnal. Snout is pointed. A red blotch on upper side of opercles. Dorsal fin has 11 spines and 14–16 rays, fin membrane is red. Anal fin has 4 spines and 9–11 rays. 40–47 lateral scales. Edible but sometimes causes ciguatera poisoning. Up to 45cm in total length.

Distribution — Japan, Ryukyu, Indo-Pacific, Hawaii, Red Sea

Holocentridae

⇩ ***Myripristis chryseres*** Jordan et Evermann

Yellow-fin Squirrelfish

Diagnosis — Inhabits waters deeper than 100m. Second dorsal, anal and caudal fins are yellowish. Preopercular bone has no spines. 10 spines on first dorsal fin, 1 spine and 13–14 rays on second. Anal fin has 4 spines and 11–13 rays. 33–37 lateral scales. 3 rows of scales above lateral line. Edible. Aquarium fish. Up to 20cm in total length.

Distribution — Southern Japan, Ryukyu, South China Sea, Hawaii, Indo-Pacific

Holocentridae

Pristilepis oligolepis (Whitley) ⇧

[*Ostichthys pillwaxii* (Steindachner)]

Slender Soldierfish

Diagnosis — Inhabits shoreline reefs of rather deep coastal waters. Body is not deep. Preopercular bone has no spines. White spots on scales. 12 spines on first dorsal fin, 14 rays on second. Anal fin has 4 spines and 11 rays. Edible. Aquarium fish. Up to 30cm in total length.

Distribution — Japan, Ryukyu, Pacific, Hawaii, Easter Island

Holocentridae

Sargocentron caudimaculatus (Rüppell) ⇨

[*Adioryx caudimaculatus* (Rüppell)]

[*Holocentrus andermanense* Day]

Silverspot Soldier

Diagnosis — Resembles *S. spinifer*, but distinguished by white tail, silver blotches on upper side of opercles and bases of pectoral fins. Dorsal fin has 11 spines and 14–15 rays, anal fin has 4 spines and 9–10 rays. 40–43 lateral scales. Edible. Aquarium fish. Up to 25cm in total length.

Distribution — Southern Japan, Ryukyu, Indo-Pacific, South Africa

Holocentridae

⇦ ***Ostichthys kaianus*** (Günther)

Twelve-spined Soldierfish (New Common Name)

Diagnosis — Resembles *O. japonicus*, but distinguished by 12 spines on first dorsal fin, 2.5 rows of scales above lateral line. Anal fin has 4 spines and 11 rays. 28–30 lateral scales. Edible. Aquarium fish. Up to 30cm in total length.

Distribution — Japan, Ryukyu, South China Sea, Indo-West Pacific

Holocentridae

Sargocentron lacteoguttatus (Cuvier) ⇨

[*Adioryx lacteoguttatus* (Cuvier)]

[*Holocentrus argenteus* Cuvier]

[*H. gracilispinus* Fowler]

White-spotted Soldier Fish

Diagnosis — Inhabits shoreline reefs and coral reefs of shallow waters. Body is pinkish-silver. Margin of dorsal fin is red. Silver blotches on margins of opercles. Dorsal fin has 11 spines and 12–14 rays, anal fin has 4 spines and 9–10 rays. 42–45 lateral scales. Edible. Aquarium fish. Up to 15cm in total length.

Distribution — Southern Japan, Ryukyu, Indo-Pacific, South Africa

Holocentridae

⇦ ***Sargocentron spinosissimus*** (Temminck et Schlegel)

[*Adioryx spinosissimus* (Temminck et Schlegel)]

Soldierfish, Squirrelfish, Spiny Squirrelfish

Diagnosis — Inhabits rocky reefs of rather deep waters. Active at sunrise and sunset. Body is slightly elongated, has 8–9 white stripes. Strong spines on preopercular bones. Dorsal fin has 11 spines and 13 rays, anal fin has 4 spines and 8–9 rays. 35–38 lateral scales. Spawns in early summer. Edible. Up to 25cm in total length.

Distribution — Japan, Ryukyu, Formosa

Holocentridae

⇦ ***Sargocentron tiere*** (Cuvier)

[*Adioryx tiere* (Cuvier)]

Tahitian Squirrel Fish

Diagnosis — Inhabits shoreline reefs and coral reefs of shallow waters. Body is large; snout is pointed. Strong spines on preopercular bones. 8–9 shiny blue stripes on body. Dorsal fin has 11 spines and 14–15 rays, anal fin has 4 spines and 9–10 rays. 47–52 lateral scales. Caudal fin is red. Edible but sometimes causes ciguatera poisoning. Up to 30cm in total length.

Distribution — Ryukyu, Indo-Pacific, Hawaii

Holocentridae
⇦ *Flammeo opercularis* (Valenciennes)

Black-finned Squirrelfish, Blackfin Soldier

Diagnosis — Inhabits coral reefs of rather deep waters. 10 spines on first dorsal fin, fin membrane is black. 1 spine and 12–13 rays on second. Anal fin has 4 spines and 9 rays. 38–40 lateral scales. Edible. Up to 30cm in total length.

Distribution — Ryukyu, Indo-Pacific

Holocentridae
Sargocentron diadema (Lacépède) ⇨

[*Adioryx diadema* (Lacépède)]

Crowned Soldier Fish, Crowned Squirrel Fish, Soldier Fish

Diagnosis — Inhabits coral reefs of shallow waters. Dorsal fin has 11 spines and 13–14 rays, anterior membrane of dorsal fin is black. Anal fin has 4 spines and 8–10 rays. 46–50 lateral scales. A white band between snout and lower side of opercles. Edible but sometimes causes ciguatera poisoning. Up to 23cm in total length.

Distribution — Ryukyu, Indo-Pacific, Red Sea, South Africa

Holocentridae
Myripristis adustus Bleeker ⇧

[*M. murdjan* var. *adustus* Günther]

Blacktip Soldierfish, Blue Squirrelfish, Shadowfin Soldier

Diagnosis — Inhabits shoreline reefs and coral reefs of coastal waters. Margins of second dorsal, anal and caudal fins are black, 10 spines on first dorsal fin, 1 spine and 14–16 rays on second. Anal fin has 4 spines and 12–14 rays. 27–29 lateral scales. Preopercular bone has no spines. Edible. Aquarium fish. Up to 20cm in total length.

Distribution — Japan, Ryukyu, Indo-Pacific, East Africa

Holocentridae
⇧ *Myripristis randalli* Greenfield

Randall's Soldier Fish (New Common Name)

Diagnosis — Resembles *M. berndti*, but distinguished by shorter and rounded snout. No small scales on axil of pectoral fins. Lower jaw has 1 pair of tooth patche. First dorsal fin has 10 spines, second has 1 spine and 13–14 rays. Anal fin has 4 spines and 12–13 rays. 28 lateral scales. Edible. Aquarium fish. Up to 15cm in total length.

Distribution — Southern Japan, Ryukyu, Indo-Pacific

Holocentridae
Myripristis kuntee Cuvier ⇨

Cuvier's Squirrel Fish, Shoulderbar Soldierfish

Diagnosis — Resembles *M. berndti*, but distinguished by a black band running from over operculum down to pectoral fins. First dorsal fin has 10 spines, has yellow membranes, second has 1 spine and 15–17 rays. Anal fin has 4 spines and 14–16 rays. 38–43 lateral scales. Edible. Aquarium fish. Up to 15cm in total length.

Distribution — Southern Japan, Ryukyu, Indo-Pacific, Hawaii

Holocentridae
Myripristis berndti
Jordan et Evermann ⇨⇨

[*M. parvidens* Cuvier]

Big-eyed Soldier Fish, Blacktipped Soldierfish, Blotch-eye, Blotcheye Soldier, Crimson Squirrel Fish, Soldier Fish

Diagnosis — Inhabits shoreline reefs. Active at sunrise and sunset. Ferocious, has its own territory. 10 spines on first dorsal fin, 1 spine and 13–15 rays on second. Anal fin has 4 spines and 11–14 rays. 28–30 lateral scales. Margin of opercles are black. Scales are rough and hard. Dorsal, anal and caudal fins are black-tipped. Edible. Aquarium fish. Up to 30cm in total length.

Distribution — Japan, Ryukyu, Indo-Pacific, Red Sea, South Africa

Holocentridae
Myripristis violaceus Bleeker
Violet Soldierfish ⇨⇨

Diagnosis — Inhabits coral reefs of rather deep waters. 10 spines on first dorsal fin, 1 spine and 13–16 rays on second. Anal fin has 4 spines and 12–14 rays. 27–29 lateral scales. Preopercular bone has no spines. Back is black. Edible. Aquarium fish. Up to 20cm in total length.

Distribution — Ryukyu, Indo-Pacific

Holocentridae
Ostichthys japonicus (Cuvier) ⇦⇦
Hardscale Soldierfish, Deep-water Squirrelfish, Golden Perch, Brocade Perch, Sea Goldfish, Ebisu Perch

Diagnosis — A beautiful species inhabits rocky reefs of rather deep waters. Body is deep, covered with large and hard scales. 12–13 spines on first dorsal fin, 1 spine and 13 rays on second. Anal fin has 4 spines and 10–12 rays. 28–30 lateral scales. Preopercular bone has no spines. Edible. Aquarium fish. Up to 45cm in total length.

Distribution — Japan, Ryukyu, Formosa, Korea, Indo-Pacific

▼ Soldier Fishes taken by net from Tuamotu lagoon.

Holocentridae
Myripristis hexagonus
(Lacépède) ⇨

Oval Squirrelfish (New Common Name)

Diagnosis — Resembles *M. berndti*, but distinguished by unprojected lower jaw, light-colored body. Lower jaw has 2 pairs of tooth patches. First dorsal fin has 10 spines, second has 1 spine and 14–15 rays. Anal fin has 4 spines and 11–13 rays. 27–28 lateral scales. Edible. Aquarium fish. Up to 20cm in total length.

Distribution — Ryukyu, Indo-West Pacific

Holocentridae
Myripristis melanostictus
Bleeker ⇨

Bigscale Soldierfish, Chubby Soldier

Diagnosis — Resembles *M. berndti*, but distinguished by roundish body, unprojected lower jaw. No small scales on axil of pectoral fins. Lower jaw has 2 pairs of tooth patches. First dorsal fin has 10 spines, second has 1 spine and 13–14 rays. Anal fin has 4 spines and 12–13 rays. 28 lateral scales. Edible. Aquarium fish. Up to 20cm in total length.

Distribution — Southern Japan, Ryukyu, Indo-Pacific

Holocentridae
Myripristis murdjan (Forskål) ⇨

Big-eye Soldier Fish

Diagnosis — Resembles *M. berndti*, but distinguished by slightly shorter head. Body has small black dots. First dorsal fin has 10 spines, second has 1 spine and 13–15 rays. Anal fin has 4 spines and 12–14 rays. 27–29 lateral scales. Edible. Aquarium fish. Up to 20cm in total length.

Distribution — Ryukyu, Indo-Pacific

Regalecidae
⇩ *Regalecus russellii* (Shaw)

Slender Oarfish

Diagnosis — A deep-sea species, but rarely found ashore. Body is silver-white, laterally compressed and long, resembling a ribbon. Gray round blotches on back. No scales and teeth. Dorsal fin originates anterior to eyes, continues to posterior end of body. 6 soft rays in anterior part of dorsal fin are extended like filiforms. Pelvic fins also long and filiform with membraneous appendix. Anal fin is absent. Caudal fin is filiform. Fins are red. Mouth is extensible. Up to 10m in total length.

Distribution — Japan, Indo-Pacific-Atlantic, California

Monocentridae

Monocentris japonica (Houttuyn)

Knight Fish, Pineapple Fish, Pine-cone Fish

Diagnosis — Inhabits rather deep coastal waters. Body is covered with hard carapace. 5–7 spines changeable into sharp spines on first dorsal fin, 10–12 rays on second. Strong spines on pelvic fins. A pair of luminous organs on lower jaw, emit light. Spawns between spring and fall. Nocturnal. Edible. Aquarium fish. Up to 16cm in total length.

Distribution — Japan, Ryukyu, Korea, Indo-Pacific, South Africa

▲ Scale of Pineapple Fish.

Lampridae

Lampris regius (Bonnaterre)

[*L. guttatus* (Brünnich)]

[*L. luna* (Gmelin)]

[*L. immaculata* Gilchrist]

[*L. pelagicus* Smith]

Opah, Moonfish, Kingfish, Jerusalem Haddock, Mariposa, San Pedro Fish, Glance Fish, Gudlax, Cravo, Sunfish

Diagnosis — Inhabits upper layers of open sea. Body is disk-shaped. Swims with pectoral fins. Lips, snout and fins are scarlet. Mouth is toothless. Feeds on jellyfish. Dorsal fin has 50–55 rays, anal fin has 34–41 rays. Flesh is pink, rich in fat. Edible. Up to 2m in total length.

Distribution — Southern Japan, Ryukyu, Indo-Pacific-Atlantic, South Africa

Polymixiidae

Polymixia japonica Günther

[*P. nobilis* Lowe]

[*P. berndti* Gilbert]

Silver Eyes, Beardfish, Japanese Beardfish, Japanese Salmon de Alto, Salmon de Alto

Diagnosis — Inhabits sandy-mud bottoms of deep waters. Body is thick and rounded. Head is small. Eyes are large and silvery white. A pair of barbels on lower jaw. Scales are hard. Dorsal fin has 4–6 spines and 31–35 rays, anal fin has 3–5 spines and 15–16 rays. 32–35 lateral scales. Spawns in spring. Edible. Up to 50cm in total length.

Distribution — Japan, Ryukyu, East China Sea, Indo-Pacific-Atlantic, South Africa

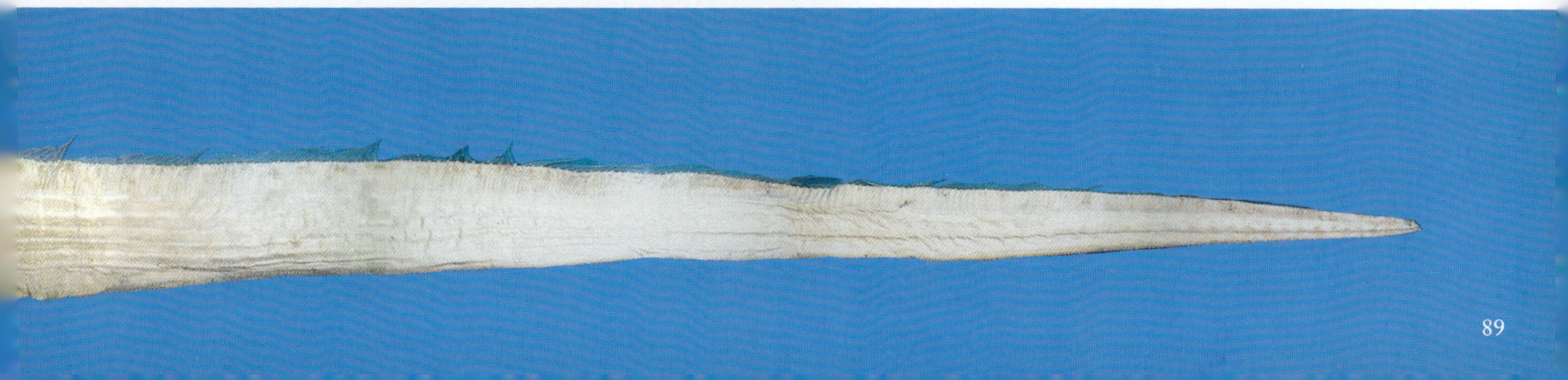

Zeidae

Zeus japonicus Valenciennes

John Dory, Target Dory, Target Fish, Target Perch

Diagnosis — Inhabits sandy-mud bottoms of rather deep waters. Body is compressed. Mouth is directed diagonally and extensible. A single black blotch below lateral line. Scales are remarkably small, sink under skin. Dorsal fin has 10 spines and 22–23 rays, anal fin has 4 spines and 21–23 rays. Spawns in spring. Feeds on small fish and crustacea. Edible. Up to 50cm in total length.

Distribution — Japan, Ryukyu, Indo-Pacific, South Africa

▲ Extended mouth of Target Fish.

Atherinidae

Hypoatherina woodwardi (Jordan et Starks)

[*Allanetta woodwardi* (Jordan et Starks)]

Narrowstripe Silverside

Diagnosis — A gregarious species inhabiting coastal waters. Body is semi-transparent, has a narrow dark band. Mouth is large. Scales are hard, margins are rounded, not easily removable. Anus is situated between pelvic fins. 4–6 spines on first dorsal fin, 1 spine and 8–10 rays on second. Anal fin has 1 spine and 10–13 rays. 40–41 lateral scales. Edible. Up to 10cm in total length.

Distribution — Southern Japan, Ryukyu

Mugilidae

Mugil cephalus Linnaeus

[*M. constantiae* Günther]

[*M. cephalotus* Günther]

[*M. oeur* Forskål]

Common Mullet, Jumping Mullet, Striped Mullet, Sea Mullet, Fatback, Flathead Mullet, Haarder, Mangrove Mullet, River Mullet, Springer

Diagnosis — Widely distributed species inhabiting warm waters, also in coastal waters, bays and rivers. Body is cylindrical. Snout is rounded. Lateral lines are absent. Adipose eyelids on eyes. Scales are hard, easily removable. 4 spines on first dorsal fin, 1 spine and 8 rays on second. Anal fin has 3 spines and 8 rays. 40–43 longitudinal scales. Spawns in deep waters between fall and winter. Edible. Up to 1m in total length.

Distribution — Japan, Ryukyu, Indo-Pacific-Atlantic, South Africa

Mugilidae

Crenimugil crenilabis (Forskål) ⇨

Warty-nosed Mullet, Fringelip Mullet, Wartylipped Mullet, Wartynose Mullet

Diagnosis — Resembles *Mugil cephalus*, but distinguished by eyes lacking adipose eyelids; yellow pectoral fins; thick upper lip; many fleshy knobs around mouth. 4 spines on first dorsal fin, 1 spine and 8 rays on second. Anal fin has 3 spines and 8−9 rays. 36−41 longitudinal scales. Edible. Up to 50cm in total length.

Distribution — Japan, Ryukyu, Indo-Pacific, Red Sea, South Africa

Mugilidae

⇦ *Crenimugil heterocheilos* (Bleeker)

River Mullet (New Common Name)

Diagnosis — Inhabits bays and fresh water areas of rivers. Body is somewhat blunt. Caudal fin is large. Eyes have no developed adipose eyelids. Upper lip is thick, lower is thin. Pectoral fins are yellowish, have a black spot on bases. First dorsal fin has 4 spines, second has 9−10 rays. Anal fin has 3 spines and 8−9 rays. 35−39 longitudinal scales. Edible. Up to 45cm in total length.

Distribution — Ryukyu, West Pacific

Mugilidae

Valamugil seheli (Forskål) ⇨

Blue-spot Mullet, Blue-tail Mullet, Sand Mullet, Long-finned Mullet

Diagnosis — Large species inhabiting shallow waters. Body is brilliantly silver. Pectoral fins are yellowish. Caudal fin is falcated. Eyes have no developed adipose eyelids. First dorsal fin has 4 spines, second has 8−9 rays. Anal fin has 3 spines and 8−9 rays. 38−40 longitudinal scales. Edible. Up to 50cm in total length.

Distribution — Japan, Ryukyu, Formosa, Indo-West Pacific

Mugilidae

⇦ *Valamugil cunnesius* (Valenciennes)

Longarm Mullet, Elsewhere-longfin Grey Mullet

Diagnosis — Small species inhabiting bays and brackish areas. Eyes have developed adipose eyelids. Operculum and pectoral fins are yellowish, have a black spot on pectoral bases. First dorsal fin has 4 spines, situated in the middle of dorsal margin. Second has 9 rays. Anal fin has 3 spines and 9 rays. 31−34 longitudinal scales. Edible. Up to 35cm in total length.

Distribution — Japan, Ryukyu, Formosa, Indo-West Pacific

Mugilidae

Liza macrolepis (Smith) ⇨

[*Mugil smithii* Günther]

[*M. borneensis* Bleeker]

[*M. troscheli* Bleeker]

Borneo Mullet, Largescale Mullet, Troschel's Mullet

Diagnosis — A littoral species. Resembles *Mugil cephalus*, but distinguished by deeper body. Adipose eyelids are absent. 4 spines on first dorsal fin, situated somewhat posteriorly, closer to base of caudal fin than to tip of snout. 1 spine and 8 rays on second. Anal fin has 3 spines and 9−10 rays. 30−35 longitudinal scales. Edible. Up to 30cm in total length.

Distribution — Southern Japan, Ryukyu, Indo-Pacific, Persian Gulf, South Africa

Polynemidae

Polydactylus sexfilis (Valenciennes) ⇨

Golden Six-thread Tassel Fish, Hawaiian Moi, Six-fingered Threadfin

Diagnosis — A littoral species. Body is deep. 6 black-tipped free rays on pectoral fins. 8 spines on first dorsal fin, 1 spine and 12–13 rays on second. Anal fin has 3 spines and 11–12 rays, margin is black. 60–65 lateral scales. Edible. Up to 90cm in total length.

Distribution — Ryukyu, Bonin Islands, Indo-Pacific, Hawaii, Mauritius

Mugilidae

⇩ ***Liza vaigiensis*** (Quoy et Gaimard)

[*Mugil waigiensis* (Quoy et Gaimard)]

Squaretail Mullet, Large-scaled Mullet, Diamond-scaled Mullet

Diagnosis — Inhabits shallow coral reefs and bays. Juveniles ascend rivers. Head is broad and compressed. Adipose eyelids are absent. 4 spines on first dorsal fin, 1 spine and 8 rays on second. Anal fin has 3 spines and 8–9 rays. Pectoral fins are black. Margin of caudal fin is truncated and black. 26–27 longitudinal scales. Scales are rough. Edible. Up to 50cm in total length.

Distribution — Ryukyu, Indo-Pacific, South Africa

Mugilidae

⇧ ***Valamugil buchanani*** (Bleeker)

Buchanan's Blue-tail Mullet, Bluetail Mullet

Diagnosis — Inhabits bays and brackish areas. Resembles *V. seheli*, but distinguished by fewer longitudinal scales; sickle-like, tapering second dorsal and anal fins. Eyes have no developed adipose eyelids. First dorsal fin has 4 spines, second has 7–9 rays. Anal fin has 3 spines and 9 rays. 33–37 longitudinal scales. Edible. Up to 50cm in total length.

Distribution — Ryukyu, Indo-West Pacific, South Africa

Polynemidae

Polydactylus plebejus (Broussonet) ⇩

[*Polynemus plebius* Lacépède]

[*P. lineatus* Lacépède]

[*P. taeniatus* Günther]

[*Polydactylus plebeius* (Bonnaterre)]

Bastard Mullet, Common Tassel Fish, Common Threadfin, King Salmon, Puttynose, Puttynose Perch, Striped Threadfin, Striped Salmon

Diagnosis — Inhabits shallow waters and river mouths. Snout is protruded. Mouth is opened toward ventral side. Pectoral fins are situated low, 5 separate inner rays serve as a sensory organ to find foods. Feeds on crustacea and polychaeta. Eyes are large, adipose eyelids are well-developed. 8 spines on first dorsal fin, 1 spine and 13–14 rays on second. Anal fin has 2–3 spines and 11–12 rays. Caudal fin is deeply falcated. 58–65 lateral scales. Edible. Up to 55cm in total length.

Distribution — Japan, Ryukyu, Formosa, Korea, Indo-Pacific, South Africa

Polynemidae

⇧ ***Eleutheronema tetradactylum*** (Shaw)

Four-thread Tassel Fish, Colonial Salmon, Cooktown Salmon, Blind Tassel-fish, Blue Salmon, Burnett Salmon, Giant Threadfin, Kingfish

Diagnosis — Inhabits muddy bottoms of rather deep waters, also found in brackish waters. Snout is protruded. Mouth is opened on ventral side. Pelvic and anal fins are yellowish. 4 free rays on pectoral fins. Scales are long and small. 8 spines on first dorsal fin, 1–2 spines and 13–15 rays on second. Anal fin has 2 spines and 15–17 rays. 78–80 lateral scales. Edible. Up to 2m in total length.

Distribution — Formosa, China, Indo-West Pacific, Persian Gulf

Polynemidae
Polydactylus sextarius
(Bloch et Schneider) ⇨

Blackspot Threadfin, Bastard Mullet

Diagnosis — Inhabits rather deep waters. Snout is protruded. Mouth is opened on ventral side. 8 spines on first dorsal fin, 1 spine and 12−13 rays on second. Anal fin has 2−3 spines and 12−13 rays. 6 free rays on pectoral fins. 44−50 lateral scales, a black blotch on origin of lateral line. Edible. Up to 30cm in total length.

Distribution — Formosa, South China Sea, Malay, Indian Ocean, South Africa

Sphyraenidae
⇦⇧ *Sphyraena barracuda* (Walbaum)
[*S. picuda* Bloch et Schneider]

Dingo-fish, Giant Barracuda, Giant Sea-pike, Banded Barracuda, Great Barracuda, Short Barracuda

Diagnosis — Inhabits upper layers of open sea in the world. Mouth is large; teeth are sharp. Caudal fin is double emarginated. Scales are large; 75−87 pored lateral scales. 5 spines on first dorsal fin, 1 spine and 9 rays on second. Anal fin has 2 spines and 7−8 rays. Juveniles inhabit bays. Sometimes attacks man. Often has ciguatera toxin. Up to 2m in total length.

Distribution — Southern Japan, Ryukyu, Indo-Pacific-Atlantic, Red Sea, Caribbean Sea

▼ Huge Barracuda is dangerous to man by attack and ciguatera poisoning. (Courtesy of Mr. J. Tapu)

Sphyraenidae
Sphyraena jello Cuvier ⇧⇩
[*S. putnamiae* Jordan et Seale] ⇩

Pickhandle Barracuda, Banded Barracuda, Dingo-fish, Pick Handle, Giant Sea Pike, Indian Barracuda, Slender Sea-pike Snoek

Diagnosis — Body is elongated; snout is pointed. Mouth is large; lower jaw is protruded, 2 rows of sharp teeth in upper jaw, 1 row in lower. 5 spines on first dorsal fin, 1 spine and 9 rays on second. Anal fin has 2 spines and 8 rays. Scales are small, 123−134 pored lateral scales. More than 10 cross stripes on body. Edible but sometimes causes ciguatera poisoning. Up to 1.8m in total length.

Distribution — Southern Japan, Ryukyu, Formosa, Indo-Pacific, Persian Gulf, Red Sea, South Africa

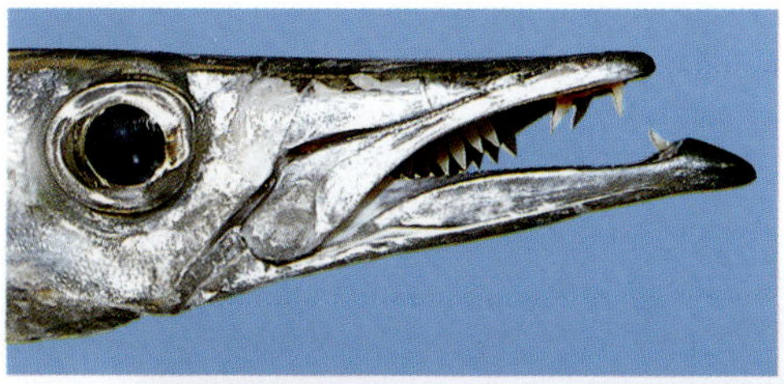

◀ Mouth has large canines.

Sphyraenidae

⇦ ***Sphyraena novaehollandiae*** Günther

Shortfinned Pike, Snook Barracuda, Shortfinned Barracuda

Diagnosis — Body has 2 yellowish longitudinal bands. Pelvic fins originate rather anterior to origin of first dorsal fin. Upper jaw is short. First dorsal fin has 5 spines, second has 1 spine and 9 rays. Anal fin has 2 spines and 8 rays. 140−150 lateral scales. Edible. Up to 50cm in total length.

Distribution — Ryukyu, Indo-Pacific, Hawaii, South Africa

Sphyraenidae

Sphyraena pinguis Günther ⇨

[*S. obtusata* Cuvier]

Brown Barracuda, Red Barracuda, Good Salmon

Diagnosis — A gregarious species inhabiting upper layers of temperate and tropical waters. Body is slightly deep. Back is reddish-brown. Eyes are large. Pelvic fins are situated anterior to first dorsal fin. 5 spines on first dorsal fin, 1 spine and 9 rays on second. Anal fin has 2 spines and 8 rays. 88−95 pored lateral scales. Spawns in early summer. Edible. Up to 60cm in total length.

Distribution — Japan, Ryukyu, Formosa, Korea, Indo-Pacific, Red Sea, East Africa

Sphyraenidae

Sphyraena japonica Cuvier ⇨

[*S. africana* Gilchrist et Thompson]

[*S. acutipinnis* Day]

[*S. natalensis* von Bonde]

Japanese Barracuda, Sea Pike, Snoek

Diagnosis — Inhabits upper layers of bays. First dorsal fin originates slightly anterior to pelvic fins. 1 spine on opercles. 5 spines on first dorsal fin, 1 spine and 9 rays on second. Anal fin has 2 spines and 8 rays. Caudal fin is yellowish, other fins are black. 115−136 pored lateral scales. Edible. Large ones often attack man. Up to 2m in total length.

Distribution — Japan, Formosa, Indo-Pacific, Persian Gulf, South Africa

▼ Canines of Sea-pike Barracuda.

Sphyraenidae

⇩ ***Sphyraena forsteri*** Cuvier ⇨ ⇘

[*S. jello* Cuvier et Valenciennes]

Sea-pike Barracuda, Forster's Barracuda

Diagnosis — Found in schools in coral reefs. Head length is 5 times shorter than standard length. Scales are large. Knobs on opercles. Pectoral fins are yellowish, other fins are not black. 5 spines on first dorsal fin, 1 spine and 9 rays on second. Anal fin has 2 spines and 8−9 rays. 110−133 pored lateral scales. Edible but sometimes causes ciguatera poisoning. Up to 60cm in total length.

Distribution — Japan, Ryukyu, Indo-Pacific, Hawaii

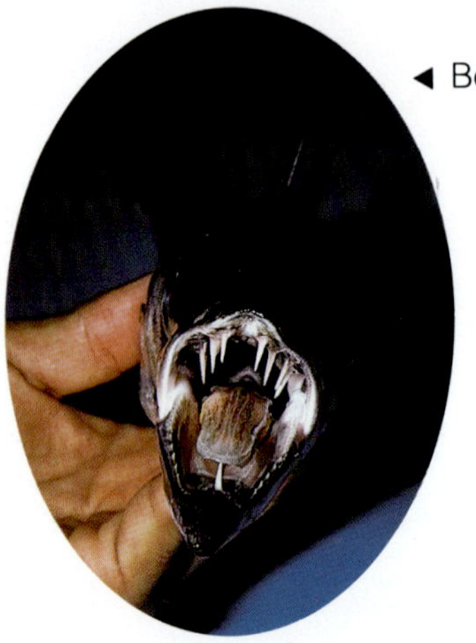

◀ Body is cylindrical.

Ammodytidae

⇧ ***Embolichthys mitsukurii*** (Jordan et Evermann)

Mitsukuri's Sand Lance

Diagnosis — Inhabits sandy-mud bottoms of shallow waters, moves into sand when alarmed. Body is elongated; snout is pointed. Lower jaw is protruded. Pelvic fins have 1 spine and 3 rays, dorsal fin has 40−43 rays, anal fin has 14−17 rays. Edible. Up to 16cm in total length.

Distribution — Japan, Formosa

Ammodytidae

⇧ ***Ammodytes personatus*** Girard

Japanese Sand Lance, Pacific Sand Lance, Sand Launce

Diagnosis — Inhabits sandy-graval bottoms of bays. Swarms at the surface during the day, moves into sand during the night to sleep. Body is elongated. Pelvic fins are absent. orsal fin has 51−59 rays, anal fin has 28−32 rays. Edible, also used for bait. Up to 25cm in total length.

Distribution — Japan

Apogonidae

Apogon aroubiensis (Hombron et Jacuinot) ⇨

[*Amia aroubiensis* (Hombron et Jacuinot)]

Broadband Cardinalfish

Diagnosis — Inhabits shallow rocky reefs. Nocturnal. Body has wide dark brown bands. First dorsal fin has 7 spines, second has 1 spine and 9 rays. Anal fin has 2 spines and 8 rays. 27 pored lateral scales. Up to 10cm in total length.

Distribution — Japan, Ryukyu, Formosa, Pacific

Apogonidae

Apogon cyanosoma Bleeker ⇨⇩

Goldstriped Cardinalfish, Pale-lined Cardinalfish

Diagnosis — A gregarious species inhabiting shallow coral reefs and rocky reefs. Nocturnal. Feeds on small fish and benthos. First dorsal fin has 7 spines, second has 1 spine and 9 rays. Anal fin has 2 spines and 8 rays. 26 pored lateral scales. Up to 9cm in total length.

Distribution — Japan, Ryukyu, Indo-Pacific, Red Sea, East Africa

Apogonidae

⇩ ***Cheilodipterus macrodon*** (Lacépède)

Big-toothed Cardinalfish, Big-toothed Percelle, Eight-lined Cardinalfish, Largetooth Cardinalfish

Diagnosis — Inhabits shallow coral reefs. Nocturnal. Jaws have pointed canines. Feeds on small fish and benthos. First dorsal fin has 6 spines, second has 1 spine and 9 rays. Anal fin has 2 spines and 8 rays. 23−26 pored lateral scales. Up to 24cm in total length.

Distribution — Japan, Ryukyu, Indo-Pacific, Red Sea, East Africa

Apogonidae
Apogon kallopterus Bleeker ⇨
[*A. snyderi* Jordan et Evermann]
[*A. frenatus* Valenciennes]
Iridescent Cardinalfish, Spinyhead Cardinal
Diagnosis — Inhabits shoreline reefs. Nocturnal. Body is not striped. Fins are transparent. First dorsal fin has 7 spines, second has 1 spine and 9 rays. Anal fin has 2 spines and 8 rays. 26 pored lateral scales. Up to 15cm in total length.
Distribution — Japan, Ryukyu, Bonin Islands, Indo-Pacific, Hawaii, Red Sea, South Africa

Apogonidae
⇧ *Apogon endekataenia* Bleeker
[*Apogon schlegeli* Bleeker]
Eleven-band Cardinalfish
Diagnosis — Inhabits rather deep rocky reefs. Body has 6–8 stripes. Caudal peduncles have a black spot. First dorsal fin has 7 spines, second has 1 spine and 9 rays. Anal fin has 2 spines and 8 rays. 27–28 pored lateral scales. Up to 11cm in total length.
Distribution — Japan, Ryukyu, Formosa, Indo-Pacific, Red Sea

Apogonidae
⇦ *Apogon amboinensis* Bleeker
[*Amia amboinensis* (Bleeker)]
Amboina Cardinalfish
Diagnosis — Inhabits brackish areas near river mouths. Body is semitransparent. Caudal peduncles have a black spot. First dorsal fin has 6 spines, second has 1 spine and 9 rays. Anal fin has 2 spines and 8 rays. 26–27 pored lateral scales. Up to 10cm in total length.
Distribution — Ryukyu, Formosa, West Pacific, East Africa

Apogonidae
⇦ *Apogon trimaculatus* Cuvier
Dark-banded Cardinalfish (New Common Name)
Diagnosis — Inhabits hollows in rocks. Body is deep, tail is extended. Eyes are large. Scales are also large. A black splotch is present on opercles. Tip of caudal fin is rounded. First dorsal fin has 6 spines, second has 1 spine and 9 rays. Anal fin has 2 spines and 8 rays. 25–26 pored lateral scales. Up to 13cm in total length.
Distribution — Ryukyu, West Pacific

Apogonidae
Cheilodipterus quinquelineatus Cuvier ⇨
[*Paramia quinquelineata* (Cuvier et Valenciennes)]
[*Jadamga quinquelineata* (Cuvier et Valenciennes)]
Five-lined Cardinalfish, Five-striped Percelle, Toothed Cardinal Fish
Diagnosis — Inhabits shoreline reefs and coral reefs. Active at night. Jaws have canines. Feeds on zooplankton, small fish and benthos. Body has 5 brown stripes. First dorsal fin has 6 spines, second has 1 spine and 9 rays. Anal fin has 2 spines and 8 rays. 23–26 pored lateral scales. Up to 12cm in total length.
Distribution — Japan, Ryukyu, Bonin Islands, Indo-Pacific, Hawaii, Red Sea, South Africa

Apogonidae

⇦ ***Apogon notatus*** (Houttuyn)

Blackspot Cardinalfish

Diagnosis — A gregarious species inhabiting shallow rocky reefs. Posterior head and caudal peduncles have a black spot. Spawns between June and September. Larvae are bred in male's mouth. First dorsal fin has 7 spines, second has 1 spine and 9 rays. Anal fin has 2 spines and 8 rays. 26 pored lateral scales. Up to 10cm in total length.

Distribution — Japan, Ryukyu, Formosa, Indo-Pacific

Apogonidae

Apogon lineatus (Temminck et Schlegel) ⇨

Cardinal Fish, Indian Perch, Indian Parch, Vertical-striped Cardinalfish

Diagnosis — Inhabits rather deep muddy-sand bottoms. Body has several cross bands. Spawns between July and September. Juvenile are bred in male's mouth. First dorsal fin has 7 spines, second has 1 spine and 9 rays. Anal fin has 2 spines and 8 rays. 25–26 pored lateral scales. Up to 10cm in total length.

Distribution — Japan, Formosa, South China Sea, West Pacific

Apogonidae

Apogon kiensis Jordan et Snyder ⇨

[*A. quadrifasciatus* Cuvier et Valenciennes]

Rifle Cardinal, Rifle Cardinalfish

Diagnosis — Inhabits muddy-sand bottoms of bays. Has brown stripes between eyes and caudal fin. Second dorsal and anal fins also have brown stripes. First dorsal fin has 7 spines, second has 1 spine and 9 rays. Anal fin has 2 spines and 8 rays. 25 pored lateral scales. Up to 8cm in total length.

Distribution — Japan, Formosa, China, Philippines, South Africa

Apogonidae

⇧ ***Apogon semilineatus*** Temminck et Schlegel

Indian Little Perch, Bottom Perch, Barface Cardinalfish, Indian Little Fish

Diagnosis — A gregarious species inhabiting bays and shoreline reefs bordering the ocean. Resembles *A. notatus*, but distinguished by lack of a black spot in the posterior head. Spawns between July and September. Hatched larvae are bred in male's mouth. First dorsal fin has 7 spines, second has 1 spine and 9 rays. Anal fin has 2 spines and 8 rays. 25 pored lateral scales. Up to 12cm in total length.

Distribution — Japan, Formosa, China, Philippines

Apogonidae

⇩ ***Apogon niger*** Döderlein

Black Cardinalfish

Diagnosis — Inhabits muddy-sand bottoms in bays. Body is deep. Eyes are large. First dorsal fin has 7 spines, second has 1 spine and 9 rays. Anal fin has 2 spines and 8 rays. 26 pored lateral scales. Up to 8cm in total length.

Distribution — Japan, Ryukyu, Formosa, South China Sea

Apogonidae
Apogon ishigakiensis
Ida et Moyer ⇨

New Guinea Cardinalfish

Diagnosis — Found in schools in sea grasses of shallow bays. Body is yellow. First dorsal fin has 7 spines, margin is white, second has 1 spine and 9 rays. Anal fin has 2 spines and 8 rays. 25–26 lateral scales. Aquarium fish. Up to 6cm in total length.

Distribution — Ryukyu, West Pacific

Apogonidae
⇩ *Siphamia elongata* Lachner

Slender Cardinalfish

(New Common Name)

Diagnosis — Inhabits sandy bottoms. Body is elongated; snout is pointed. First dorsal fin has 7 spines, second has 1 spine and 9 rays. Anal fin has 2 spines and 8 rays. 22–23 lateral scales. Up to 8cm in total length.

Distribution — Formosa

Apogonidae
Sphaeramia nematoptera (Bleeker) ⇩

Wide-band Orbiculate Cardinalfish

Diagnosis — A gregarious species inhabiting coral reefs not deeper than 20m. Body is deep, has a wide cross band. Posterior half of body is profusely dotted with brown. First dorsal fin has 7 spines, second has 1 spine and 9 rays. Anal fin has 2 spines and 9 rays. 26–27 pored lateral scales. Feeds on crab and small fish. Aquarium fish. Up to 12cm in total length.

Distribution — Ryukyu, Indo-Pacific

Apogonidae
Sphaeramia orbicularis
(Kuhl et Van Hasselt) ⇨

Orbiculate Cardinalfish

Diagnosis — Inhabits shallow coral reefs. Young are found also in mangrove swamps. Body is deep, has a black band in the middle. First dorsal fin has 6 spines, second has 1 spine and 9 rays. Anal fin has 2 spines and 8–9 rays. 23–24 pored lateral scales. Feeds on crab and small fish. Aquarium fish. Up to 12cm in total length.

Distribution — Ryukyu, Indo-West Pacific, Red Sea

Lobotidae
⇦ *Lobotes surinamensis* (Bloch)

[*L. auctorum* Günther]

[*L. pacificus* Jordan et Evermann]

Black Grunt, Black Perch, Blackish Tripletail, Brown Tripletail, Dormeur, Dusky Perch, Flasher, Jumping Cod, Jumpfish, Strawberry Bass, Triple-fin

Diagnosis — Distributed in warm waters. Inhabits rather deep waters. Body is deep and compressed. Eyes are small; snout is short. Lower jaw protruded, jaws have conical teeth. Caudal margin is rounded and white. Dorsal fin has 12 spines and 15–16 rays, anal fin has 3 spines and 11–12 rays. 45 pored lateral scales. Edible. Up to 1m in total length.

Distribution — Japan, Formosa, Indo-Pacific-Atlantic, South Africa

Pempheridae

⇧ ***Pempheris japonica*** Döderlein

Blackfin Sweeper, Oblique-mouth Sweeper

Diagnosis — Found in schools in shoreline waters. Body is flat and dark-brown. Scales are small. 70–80 lateral scales covering as far as caudal fin margin. Anterior parts of dorsal and anal fins are black. Dorsal fin has 6–7 spines and 10–12 rays, anal fin has 3 spines and 35–37 rays. Nocturnal. Spawns between June and July. Up to 18cm in total length.

Distribution — Japan, Bonin Islands, Philippines

Pempheridae

Pempheris xanthoptera Tominaga ⇧ ⇨

Black-margined Sweeper (New Common Name)

Diagnosis — Found in schools among shallow shoreline reefs and coral reefs. Scales are large, readily removable. Pectoral fin bases have no black spots. Dorsal, anal, caudal margins are black. Dorsal fin has 5–6 spines and 8–10 rays, anal fin has 3 spines and 34–41 rays. 45–54 lateral scales. Up to 18cm in total length.

Distribution — Japan, Ryukyu, Bonin Islands

Trichodontidae

⇧ ***Arctoscopus japonicus*** (Steindachner)

Japanese Sandfish, Sailfin Sandfish, Sandfish

Diagnosis — A cold water species inhabiting muddy-sand bottoms deeper than 100m. Swarms in coastal waters between November and December, spawns in shallow seagrass beds. Pectoral fins are large; anal fin base is long. First dorsal fin has 8–14 spines, second has 12–15 rays. Anal fin has 29–32 rays. Have no scales and lateral lines. Edible. Up to 22cm in total length.

Distribution — Northern Japan, Japan Sea, Sakhalin, Kamchatka, Alaska

Pempheridae

⇩ ***Pempheris oualensis*** Cuvier

[*P. otaitensis* Cuvier]

[*P. adustus* Bleeker]

Blacktip Sweeper, Caroline Sweeper, Oualan Sweeper, Sweeper

Diagnosis — A gregarious species inhabiting shallow coral reefs. Scales are large, readily removable. Pectoral fin bases have a black spot. Jaws have well-developed teeth protruding from mouth. Dorsal fin has 6 spines and 9 rays, anal fin has 3 spines and 38–42 rays. 62–68 lateral scales. Up to 20cm in total length.

Distribution — Ryukyu, Pacific, South Africa

Pempheridae

⇩ ***Parapriacanthus ransonneti*** Steindachner

[*P. peryciformes* Franz]

Slender Sweeper

Diagnosis — A gregarious species inhabiting shallow rocky reefs. Body is elongated. Jaws have a row of teeth. Emits light by a luminous organ situated between pectoral region and belly. Dorsal fin has 5 spines and 8–10 rays, anal fin has 3 spines and 18–20 rays. 63–80 lateral scales. For fertilizer and feed. Up to 10cm in total length.

Distribution — Japan, Ryukyu, Formosa, Korea, Micronesia

Cepolidae
Acanthocepola indica (Day) ⇩
Indian Bandfish

Diagnosis — Resembles *A. limbata*, but distinguished by thick body, standard length is 7−8 times longer than depth. Dorsal fin has 85 rays, a black spot in the anterior. Anal fin has 96 rays, margin is black. Edible. Up to 60cm in total length.

Distribution — Japan, Formosa, Indian Ocean

Cepolidae
⇧ *Acanthocepola limbata* (Valenciennes)
Bandfish, Blackspot Bandfish

Diagnosis — Inhabits holes dug up in muddy-sand bottoms about 100m deep. Body is long and narrow. Dorsal and anal fins continue into caudal fin. Dorsal fin has 102−104 rays, a black spot in the anterior. Anal fin has 105−107 rays. Feeds on benthos. Edible. Up to 50cm in total length.

Distribution — Japan, Ryukyu, Formosa, Indo-Pacific, South Africa

▼Large Adult

▲ Young

▼Adult

Oplegnathidae
Oplegnathus fasciatus (Temminck et Schlegel) ⇧⇗⇨
Japanese Parrot Fish, Rock Bream, Striped Beakperch

Diagnosis — Inhabits singly in shoreline reefs 15−45m deep. Spawns between April and July. Larvae move with floating seaweeds, feed on zooplankton. Young have black bands on the body, but disappear when the fish grows. In large adult male, margin of mouth turns black. Dorsal fin has 11−12 spines and 17−18 rays, anal fin has 3 spines and 12−13 rays. Edible. Up to 80cm in total length.

Distribution — Japan, Ryukyu, Formosa, Korea, China, East China Sea

▲ Large Adult

▲ Young

▼Adult

▲ Half-bred Parrotfish (*O. punctatus* × *O. fasciatus*).

Oplegnathidae

Oplegnathus punctatus

(Temminck et Schlegel)

[*Hoplognathus punctatus*
(Temminck et Schlegel)]

Rock Porgy, Spotted Parrot Fish, Stonewall Perch

Diagnosis — Resembles *O. fasciatus*, but distinguished by fast growth, reticulated-patterned young. Patterns disappear in adult. In large adult male, edges of mouth turn white. Teeth are strong. Feeds mostly on shellfish, crab, sea urchin, barnacle. Dorsal fin has 12 spines and 15 rays, anal fin has 3 spines and 13 rays. Edible. Up to 90cm in total length.

Distribution — Japan, Ryukyu, Formosa, Korea, China, South China Sea

▲ Edges of mouth turn white in large adult.

Mullidae

Parupeneus pleurostigma (Bennett)

[*Pseudupeneus pleurostigma* (Bennett)]

Spotted Goatfish, Blackspot Goatfish, Redmullet, Ryukyu Goatfish, Salmonet, Surmullet

Diagnosis — Inhabits coral reefs. Body has a large black and a light-colored blotches in the middle top. Lower part of second dorsal fin is black. First dorsal fin has 8 spines, second has 9 rays, anal fin has 7 rays. 28–30 lateral scales. Edible but may cause ciguatera poisoning. Up to 20cm in total length.

Distribution — Ryukyu, Indo-Pacific, Hawaii, South Africa

Mullidae

Parupeneus trifasciatus (Lacépède)

[*P. moana* Jordan et Seale]

[*P. multifasciatus* (Quoy et Gaimard)]

Five-barred Goatfish, Threebar Goatfish, Red and Black Banded Goatfish

Diagnosis — Inhabits shallow shoreline reefs and coral reefs. Body is shallow, has 2 black blotches in the posterior. Barbels are long. Eyes are crossed by a brown band. First dorsal fin has 8 spines, second has 9 rays. Anal fin has 7 rays. 27–30 lateral scales. Edible but may cause ciguatera poisoning. Up to 30cm in total length.

Distribution — Japan, Ryukyu, Indo-Pacific, Hawaii, Easter Island

Mullidae

Parupeneus cyclostomus (Lacépède)

[*P. chryserydros* (Lacépède)]

[*Upeneus luteus* Cuvier et Valenciennes]

[*Mullus oxycephalus* Bleeker]

Yellow-tailed Goatfish, Yellowsaddle Goatfish, Salmonet, Redmullet

Diagnosis — Abundant in coral reefs. Body is deep; snout is protruded; barbels are long enough to reach pelvic fins. Head has blue and yellow lines. Second dorsal fin has a pale blotch in the posterior. First dorsal fin has 8 spines, second has 9 rays, anal fin has 7 rays. 27–30 lateral scales. Edible but may cause ciguatera poisoning. Up to 35cm in total length.

Distribution — Japan, Ryukyu, Indo-Pacific, Hawaii, Red Sea, South Africa

Mullidae

Parupeneus bifasciatus (Lacépède)

[*Mullus bifasciatus* Lacépède]

[*Upeneus bifasciatus* Günther]

Two-saddled Goatfish, Two-barred Goatfish

Diagnosis — Inhabits sandy bottoms among shallow coral reefs. Body is deep, has 3 wide bands. First dorsal fin has 8 spines, second has 9 rays. Anal fin has 7 rays. 30 lateral scales. Feeds on fish, shellfish, crab, shrimp, polychaeta. Edible. Up to 40cm in total length.

Distribution — Ryukyu, Indo-Pacific

Mullidae

Parupeneus barberinoides (Bleeker)

Bicolor Goatfish, Swarthy-headed Goatfish

Diagnosis — Inhabits coral reefs. Anterior half of body is dark brown, posterior is light-yellow with a black splotch. Pelvic fins are also dark brown. First dorsal fin has 8 spines, second has 9 rays. Anal fin has 7 rays. 28–30 lateral scales. Edible. Up to 25cm in total length.

Distribution — Japan, Ryukyu, West Pacific

Mullidae

Parupeneus porphyreus (Jenkins) ⇨ ⇩

[*Pseudupeneus porphyreus* Jenkins]

Cardinal Goatfish, Unicolorhead Goatfish, White-striped Goatfish

Diagnosis — Inhabits shallow coral reefs. Body is deep. Scales have a white spots. Upper caudal peduncles have an indistinct black spot, second dorsal fin has a white spot in the posterior. Snout has 2−3 white lines extending backward. First dorsal fin has 8 spines, second has 9 rays. Anal fin has 7 rays. Edible. Up to 40cm in total length.

Distribution — Japan, Ryukyu, Pacific, Hawaii

Mullidae

⇦ ***Parupeneus spilurus*** (Bleeker) ⇧

[*Upeneus spilurus* Bleeker]

[*Pseudupeneus spilurus* (Bleeker)]

Japanese Goatfish

Diagnosis — Found in schools in shallow coral reefs. Body is deep, red with 2 dark-colored longitudinal bands. Upper caudal peduncles have a black splotch. First dorsal fin has 8 spines, second has 9 rays, anal fin has 7 rays. 27−30 lateral scales. Edible. Up to 35cm in total length.

Distribution — Japan, Ryukyu, Korea, Pacific

Mullidae

Parupeneus indicus (Shaw) ⇨

[*Pseudupeneus indicus* (Shaw)]

[*Upeneus russelli* Cuvier]

[*U. malabaricus* Valenciennes]

Indian Goatfish, Redmullet, Salmonet, Surmullet, Yellow-spot Goatfish

Diagnosis — Inhabits coral reefs. Body is deep, has an oblong yellow blotch in the upper middle. Caudal peduncles have a black splotch. Head has blue lines. First dorsal fin has 8 spines, second has 9 rays. Anal fin has 7 rays. 27−31 lateral scales. Edible. Up to 40cm in total length.

Distribution — Japan, Ryukyu, Indo-Pacific, South Africa

Mullidae

⇦ ***Parupeneus fraterculus*** (Valenciennes) ⇩

[*Pseudupeneus fraterculus* (Cuvier et Valenciennes)]

[*Parupeneus pleurotaenia* (Playfair)]

[*Upeneus natalensis* G. et T.]

Blacksaddle Goatfish, Redmullet, Twospot Goatfish, Salmonet, Surmullet

Diagnosis — Body is deep; barbels are long enough to reach gill slits. Second dorsal fin has a pale blotch in the posterior. Upper caudal peduncles have a black spot. First dorsal fin has 8 spines, second has 9 rays, anal fin has 7 rays. 28−30 lateral scales. Edible. Up to 30cm in total length.

Distribution — Japan, Ryukyu, Formosa, China, Indo-Pacific, South Africa

Mullidae

⇦ ***Parupeneus pleurospilos*** (Bleeker) ⇩

Barface Goatfish, Redmullet, Salmonet, Surmullet, Red-spot Goatfish

Diagnosis — Found singly or in schools in coastal coral rubble or muddy-sand bottoms. Body is deep. Jaws have a row of teeth. Barbels are long enough to reach pelvic fins. Lateral lines have a red spot. Has a blue line extending from snout to eyes. First dorsal fin has 8 spines, second has 9 rays. Anal fin has 7 rays. 25−30 lateral scales. Edible. Up to 30cm in total length.

Distribution — Southern Japan, Ryukyu, Indo-Pacific, Red Sea, South Africa

Mullidae

Mulloidichthys vanicolensis (Valenciennes) ⇧ ⇨

[*Mulloides vanicolensis* (Cuvier et Valenciennes)]

[*Mulloidichthys auriflamma* (Forskål)]

[*Mullus japonicus* Houttuyn]

Yellowfin Goatfish, Flame Goatfish, Goldenbanded Goatfish, Salmonet, Japanese Goatfish, Yellow Goatfish, Vanikolo Goatfish

Diagnosis — Found in schools in coral reefs. Body is elongated, reddish with a yellow longitudinal band in the middle. When the fish dies, this band fades away, turns red. First dorsal fin has 7−8 spines, second has 9 rays, anal fin has 7−8 rays. 35−42 lateral scales. Edible but may cause ciguatera poisoning. Up to 38cm in total length.

Distribution — Southern Japan, Ryukyu, China, Indo-Pacific, Hawaii, Red Sea, South Africa

Mullidae

⇧ ***Upeneus bensasi*** (Temminck et Schlegel) ⇨

Goatfish, Redmullet, Salmonet, Stripedfin Goatfish, Surmullet

Diagnosis — Inhabits coastal muddy-sand bottoms. Body is shallow. Barbels are yellow. Caudal margin is black, upper lobe has 3−5 stripes. First dorsal fin has 7 spines, second has 9 rays, anal fin has 7 rays. 28−33 lateral scales. Edible. Up to 20cm in total length.

Distribution — Japan, Ryukyu, Formosa, Korea, China, Indo-West Pacific, Persian Gulf, South Africa

Mullidae

Parupeneus sp. ⇧

Gold-lined Goatfish

(New Common Name)

Diagnosis — Body is deep with a wide yellow band on dorsal side. Body and fins have no black spots or bands. Both dorsal fins are situated close to each other. Suborbital space is wide, interorbital space is swollen. Edible. Up to 23cm in total length.

Distribution — Formosa

Mullidae

Upeneus vittatus (Forskål) ⇨

Bar-tailed Goatfish, Redmullet, Salmonet, Surmullet, Yellow Banded Goatfish, Yellow Striped Goatfish

Diagnosis — Inhabits shallow coral reefs. Body has 4–5 yellow bands. Caudal lobes have black cross stripes. First dorsal fin has 8 spines, second has 9 rays. Anal fin has 7 rays. 32–38 lateral scales. Edible. Up to 30cm in total length.

Distribution — Southern Japan, Ryukyu, Formosa, China, Indo-Pacific, Persian Gulf, Red Sea, South Africa

Mullidae

⇩ *Upeneus moluccensis* (Bleeker)

Golden Banded Goatfish, Goldband Goatfish, Mullet

Diagnosis — Inhabits muddy-sand bottoms of bays. Body has a yellow wide band. Caudal fin has 7 oblique stripes on the upper lobe. Barbels are hard. First dorsal fin has 8 spines, second has 9 rays conspicuously apart from each other. Anal fin has 7 rays. 33–38 lateral scales. Edible. Up to 17cm in total length.

Distribution — Southern Japan, Ryukyu, Formosa, Indo-West Pacific

Mullidae

⇧ *Parupeneus barberinus* (Lacépède)

[*Pseudupeneus barberinus* (Lacépède)]

Dash-dot Goatfish, Redmullet, Rosy Goatfish, Salmonet, Surmullet

Diagnosis — Inhabits coral reefs. Body is deep, has a black band on dorsal side. Snout is protruded. Caudal peduncles have a black splotch. First dorsal fin has 8 spines, second has 9 rays. Anal fin has 7 rays. 28–30 lateral scales. Edible. Up to 50cm in total length.

Distribution — Japan, Ryukyu, China, Indo-Pacific, Red Sea, South Africa

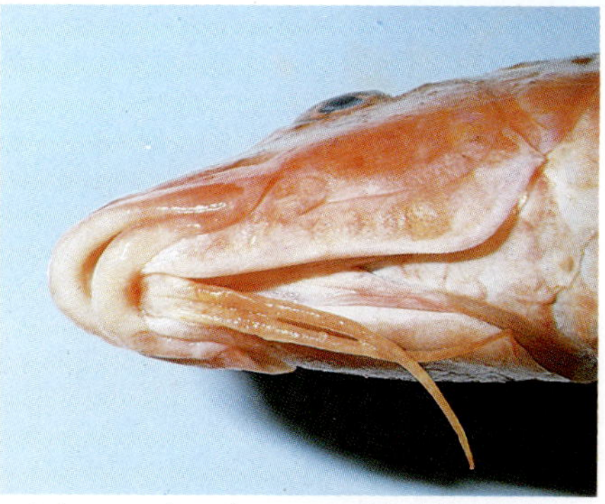

▲ Lower jaw has a pair of barbels.

Mullidae

Upeneus tragula Richardson ⇧

Bar-tailed Goatfish, Black Striped Goatfish, Golden Striped Goatfish, Mottled Goatfish, Red Mullet, Salmonet, Spotted Red-Mullet, Surmullet

Diagnosis — Body is elongated. Eyes are crossed by a black band. Head and body are profusely dotted with brown. Caudal lobes have cross stripes. First dorsal fin has 8 spines, second has 9 rays. Anal fin has 7 rays. 30–33 lateral scales. Edible. Up to 25cm in total length.

Distribution — Southern Japan, Ryukyu, Indo-Pacific, Persian Gulf, South Africa

Mullidae

Mulloidichthys flavolineatus (Lacépède) ⇧⇨

[*Mulloides samoensis* Günther]

Goat-fish, Samoan Goatfish, Yellowstripe Goatfish

Diagnosis — Found in schools in coral reefs. Body is elongated, has a wide yellow band on dorsal side. Distinguished from *M. vanicolensis* by a black spot in the middle of band. First dorsal fin has 7–8 spines, a black spot in the base, second has 9 rays, anal fin has 6–7 rays. 35–39 lateral scales. Edible but may cause ciguatera poisoning. Up to 30cm in total length.

Distribution — Japan, Ryukyu, Indo-Pacific, Red Sea, South Africa

Malacanthidae
Malacanthus brevirostris
Guichenot ⇧

[*M. hoedtii* Bleeker]

Banded Blanquillo, Blanquillo, Quakerfish

Diagnosis — Found in pairs in holes dug up in sandy bottoms of coral reefs. Body is elongated; snout is short. Caudal fin has 2 black stripes in the middle. Dorsal fin is single, has 1–4 spines and 52–60 rays, anal fin has 1 spine and 46–55 rays. 137–181 lateral scales. Aquarium fish. Up to 32cm in total length.

Distribution — Japan, Ryukyu, Indo-Pacific, Hawaii, Red Sea

Malacanthidae
⇧ *Hoplolatilus fourmanoiri* Smith

Spotted Tilefish

Diagnosis — Inhabits sandy and rubbly bottoms of coral reefs. Body is elongated; snout is short. Opercles have a black spot on the top. Caudal fin is black in the middle. Aquarium fish. Up to 20cm in total length.

Distribution — Philippines

Malacanthidae
⇦ *Hoplolatilus marcosi* Burgess

Marcos Tilefish (New Common Name)

Diagnosis — Inhabits sandy and rubbly bottoms of coral reefs. Found in pairs in holes dug up under rubbles. Body is elongated, has a wide red stripe; snout is remarkably short. Aquarium fish. Up to 20cm in total length.

Distribution — Philippines

Malacanthidae
⇦ *Hoplolatilus* sp.

Yellow Tilefish (New Common Name)

Diagnosis — Inhabits sandy and rubbly bottoms of coral reefs. Body is elongated, light-brown, has no distinct spots. Snout is remarkably short. Aquarium fish. Up to 20cm in total length.

Distribution — Philippines

Malacanthidae
⇧ *Malacanthus latovittatus* ⇧⇨
(Lacépède)

[*Oceanops latovittata* Jordan et Seale]

Blue Blanquillo, Blue Whiting, Eye of the Sea, Striped Blanquillo

Diagnosis — Inhabits singly in sand bottoms among coral reefs and rocky reefs. Body is elongated; snout is long. Dorsal fin is single, has 4 spines and 43–45 rays, anal fin has 1 spine and 39–40 rays. 125 lateral scales. Edible. Up to 45cm in total length.

Distribution — Japan, Ryukyu, Indo-Pacific, South Africa

Branchiostegidae

Branchiostegus auratus (Kishinouye) ⇨

[*B. japonicus auratus* (Kishinouye)]

[*Latilus argentatus* Cuvier et Valenciennes]

Yellow Horsehead

Diagnosis — Inhabits sandy bottoms 100–150m deep. Head is well-developed, has a silvery band extending from eyes to upper jaw. Upper lobe of caudal fin is striped with yellow, lower lobe is profusely dotted with yellow. Dorsal fin has 7 spines and 15 rays, anal fin has 2 spines and 12 rays. Edible. Up to 30cm in total length.

Distribution — Japan, Formosa, East China Sea

Branchiostegidae

⇦ ***Branchiostegus japonicus*** (Houttuyn)

[*Latilus auratus* Kishinouye]

Blanquillo, Horse-head, Japanese Tilefish, Red Horsehead, Sweet Perch, Tilefish

Diagnosis — Inhabits in holes dug up in muddy-sand bottoms 50–150m deep. Body is compressed; head is well-developed; eyes have a triangular silvery blotch in the behind. Spawns in coastal waters between September and December. Larvae feed on zooplankton, grow in 12 months. Dorsal fin has 7 spines and 15 rays, anal fin has 2 spines and 12 rays. Edible. Up to 45cm in total length.

Distribution — Japan, Formosa, Korea, Philippines, South China Sea

Branchiostegidae

Branchiostegus sp. ⇧

Yellow-lined Tilefish (New Common Name)

Diagnosis — Inhabits in holes dug up in sandy bottoms 100–200m deep. Moves less often. Body is whitish. Caudal fin has 2 yellow stripes in the middle. Feeds on fish and crustacea. Dorsal fin has 7 spines and 15 rays, anal fin has 2 spines and 12 rays. Edible. Up to 40cm in total length.

Distribution — Ryukyu

Bramidae

Taractichthys steindachneri (Döderlein) ⇧⇨⇩

[*Taractis longipinnis* (Lowe)]

[*Taractes princeps* Johnson]

Bigscale Pomfret

Diagnosis — Offshore species inhabiting warm waters. Body is compressed and deep. Scales are hard and large. Dorsal fin has 31–34 rays, anal fin has 23–26 rays. In young, fins extend long. Caudal peduncles have no keels. Interorbital space is slightly convex. Caudal margin is white. Edible. Muscle has much wax, may cause wax-poisoning. Up to 85cm in total length.

Distribution — Japan, Indo-North Pacific-North Atlantic, Hawaii, South Africa

▲ Small teeth of Pomfret.

Bramidae

⇧ *Brama japonica* Hilgendorf

[*Lepidotus brama* (Bonnaterre)]

Angelfish, Black Bream, Bulleye, Pacific Pomfret, Sea Bream, Smallscale Pomfret

Diagnosis — Inhabits bottoms 150−300m deep. Feeds on fish, squid and crustacea. Body depth is one half of standard length. Interorbital space is convex. Dorsal fin has 33−36 rays, anal fin has 27−30 rays and is scaled. Body is silvery-white, turns dark brown when the fish dies. Edible. Up to 50cm in total length.

Distribution — Japan, North Pacific

Bramidae

Eumegistus illustris Jordan et Jordan ⇩

Sweeper Pomfret

Diagnosis — Inhabits deep waters over 300m. Body is compressed and deep. Tail extends long. Scales are not spinous. Dorsal fin has 33−35 rays, anal fin has 24 rays, is white at end. Caudal lobes are also white. Edible. Up to 60cm in total length.

Distribution — Japan, Ryukyu, Hawaii, New Guinea

Emmelichthyidae

⇦ *Erythrocles schlegeli* (Richardson) ⇩

Bonnetmouth, Pacific Bonnetmouth

Diagnosis — Inhabits rocky reefs more than 100m deep. Body is elongated. Caudal fin is falcated. Upper lip is thick and extensible. Caudal peduncles have a keel. First dorsal fin has 10 spines, second has 1 spine and 11 rays. Anal fin has 3 spines and 10 rays. 65−73 lateral scales. Edible but may cause ciguatera poisoning. Up to 1m in total length.

Distribution — Japan, Ryukyu, Korea, Indo-Pacific, Hawaii, Red Sea, South Africa

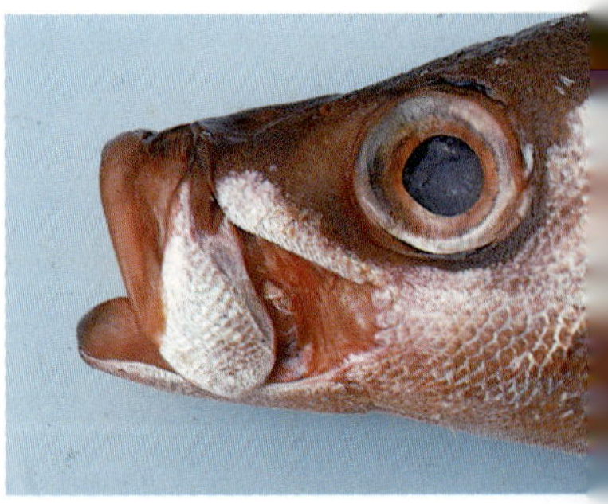

▲ Mouth is extensible.

Emmelichthyidae

⇧ *Emmelichthys nitidus* Richardson

Red Sea-haarder, Bonnetmouth

Diagnosis — Inhabits waters about 100m deep. Body is elongated. Caudal fin is falcated. In young, back has 8−9 stripes. Mouth is extensible. Feeds on zooplankton. First dorsal fin has 7−9 spines, second has 1 spine and 9−10 rays. Anal fin has 3 spines and 10 rays. 100 lateral scales. Edible. Up to 50cm in total length.

Distribution — Ryukyu, Formosa, Philippines, Australia, South Africa

Emmelichthyidae

Erythrocles scintillans (Jordan et Thompson) ⇩

Bonnetmouth, Boga

Diagnosis — Resembles *E. schlegeli*, but distinguished by somewhat smaller and scarlet. Caudal peduncles have no keels. First dorsal fin has 11 spines, second has 1 spine and 10−11 rays. Anal fin has 3 spines and 10−11 rays. 69−72 lateral scales. Edible. Up to 40cm in total length.

Distribution — Japan, Ryukyu, Hawaii

Priacanthidae

Pristigenys multifasciata Yoshino et Iwai ⇨

Red-striped Bigeye Porgy (New Common Name)

Diagnosis — Inhabits rocky reefs more than 100m deep. Resembles *P. niphonia*, but distinguished by rougher scales, many red stripes on body. Dorsal fin has 10 spines and 11–12 rays. Anal fin has 3 spines and 11–12 rays. 30–32 lateral scales. Edible. Up to 27cm in total length.

Distribution — Ryukyu, Izu Islands, South China Sea

Priacanthidae

⇦ ***Pristigenys niphonia*** (Cuvier)

[*P. niphonius* (Cuvier)]

Japanese Big-eye Porgy, Big-eye Porgy

Diagnosis — Inhabits rocky reefs rather deep coastal waters. Body is oval. Eyes are large. Mouth is directed upward. Scales are rough. Fin-spines are large and stout. Dorsal fin has 10 spines and 10–11 rays, anal fin has 3 spines and 9–10 rays. 30–36 lateral scales. Edible. Up to 30cm in total length.

Distribution — Japan, Ryukyu, Formosa, Indo-West Pacific, East Africa

Kuhliidae

⇧ ***Kuhlia sandvicensis*** (Steindachner)

[*Moronopsis sandvicensis* Steindachner]

Hawaiian Flag-tail Fish

Diagnosis — Found in schools in rivers, brackish waters and coral reefs. Resembles *K. mugil*, but has no black bands on tail. Caudal fin is falcated, margin is white. Eyes are large. Nocturnal. Edible. Up to 25cm in total length.

Distribution — Hawaii, Marcus Island, Polynesia

Kuhliidae

Kuhlia rupestris (Lacépède) ⇨ ⇘

[*Dules fuscus* Cuvier]

[*D. marginatus* Cuvier]

[*D. guamensis* Valenciennes]

[*D. vanicolensis* Valenciennes]

[*D. haswelli* Macleay]

[*Kuhlia malo* Lesson]

Buffalo Bream, Jungle Perch, Mountain Trout, Perch, Rock Flag Tail

Diagnosis — Inhabits rivers and brackish waters. Feeds on aquatic insect and crustacea. Mouth is large. Abdominal side is dotted with black. Dorsal fin has 10 spines and 10–12 rays, fin ray has a black blotch. Anal fin has 3 spines and 9–10 rays. Caudal lobes are rounded with black blotches. Edible. Up to 45cm in total length.

Distribution — Ryukyu, Indo-West Pacific, South Africa

Kuhliidae

⇦ ***Kuhlia marginata*** (Cuvier)

[*K. marginatus* (Cuvier)]

Spotted Flagtail, Sesele

Diagnosis — Inhabits rivers and brackish waters. Eyes are large, longer than snout length. Margins of dorsal, anal and caudal fins are black. Dorsal fin has 10 spines and 10–12 rays, anal fin has 3 spines and 11–12 rays. 48–53 lateral scales. Feeds on aquatic insect and crustacea. Edible but may cause ciguatera poisoning. Up to 25cm in total length.

Distribution — Japan, Ryukyu, Formosa, Indo-Pacific, Hawaii, South Africa

Kuhliidae

Kuhlia mugil Bloch et Schneider ⇨

[*K. taeniura* (Cuvier et Valenciennes)]
[*Perca argentea* Bennett]
[*Dulus taeniurus* Cuvier]
[*Dules bennetti* Bleeker]
[*Moronopsis taeniurus* Bleeker]

Barred Flagtail, Flagfish, Flag Tail

Diagnosis — Littoral fish inhabiting shallow shoreline reefs and coral reefs. Body is silvery-white. Caudal fin has black bands. Dorsal fin has 10 spines and 9–11 rays, anal fin has 3 spines and 10–11 rays. 48–56 lateral scales. Edible. Up to 30cm in total length.

Distribution — Japan, Ryukyu, Indo-Pacific, Hawaii, Mexico, South Africa

Scombropidae

⇦ *Scombrops boops* (Houttuyn) ⇨ ⇩

Japanese Bluefish

Diagnosis — During young stage inhabits bay shores and tide pools, when grown, moves out to deep, offshore rocky bottoms. Mouth is large, has sharp canines. Eyes are also large. Adult mouth is black inside. First dorsal fin has 8–9 spines, second has 1 spine and 12–14 rays. Anal fin has 3 spines and 11–13 rays. 50–57 lateral scales. Edible. Flesh contains much fat, may cause histamine poisoning. Up to 1.5m in total length.

Distribution — Japan, Ryukyu, Korea, East China Sea

▲ Bluefish has sharp canines.

Pentacerotidae

⇩ *Pentaceros japonicus* Döderlein

Boarfish, Japanese Boarfish, Spearfin Boarhead

Diagnosis — Inhabits waters more than 100m deep. Body is deep; snout is pointed. Head is covered with exposed bones. Pelvic fin membranes are black. Dorsal fin has 11 spines and 13–15 rays, anal fin has 4–5 spines and 8–10 rays. 46–56 lateral scales. Young inhabits upper layer of water. Up to 25cm in total length.

Distribution — Japan, Ryukyu, Bonin Islands

Pentacerotidae

Evistias acutirostris (Temminck et Schlegel) ⇧

Banded Boarhead

Diagnosis — Inhabits rather deep rocky reefs. Body is deep; snout is protruded. Dorsal fin is large with 4 spines and 26–29 rays. Anal fin has 3 spines and 13 rays. Lower jaw bristles with small barbels. In young, body has black round blotches. 62–65 lateral scales. Up to 55cm in total length.

Distribution — Japan, Bonin Islands

▼ Young Adult

▲ Adult

Percichthyidae

Niphon spinosus Cuvier ⇧⇗

Sawedged Perch

Diagnosis — Inhabits rocky reefs approximately 200m deep. Body is fusiform, resembles *Lateolabrax japonicus*. Opercles have a large, stout spine in corner. Scales are small. In young, body has a brown band. Spawns in shallow waters in summer. Dorsal fin has 13 spines and 10−11 rays, anal fin has 3 spines and 6−8 rays. 84−93 pored lateral scales. Edible. Liver may cause poisoning. Up to 1m in total length.

Distribution — Japan, Ryukyu, Korea, China, East China Sea, Philippines

Percichthyidae

⇩ ***Lateolabrax japonicus*** (Cuvier)

Bass, Common Sea Bass, Japanese Sea Bass, Japanese Sea Perch, Sea Perch

Diagnosis — Inhabits shoreline reefs. Body is elongated; snout is pointed. Mouth is large; lower jaw is long. In some individuals, back has black spots. Dorsal fin has 12−15 spines and 12−14 rays, anal fin has 3 spines and 7−9 rays. 71−86 pored lateral scales. Spawns in rocky reefs of open sea. Young ascends rivers. Carnivorous. Edible. Up to 1.5m in total length.

Distribution — Japan, Formosa, Korea, China, Hong Kong, South China Sea

Percichthyidae

Acropoma japonicum Günther ⇧

[*A. cynodon* Regan]

Firefly-fish, Lanternbelly

Diagnosis — Luminous fish inhabiting deep waters. Body is fusiform, covered with rough scales. Anus is situated closer to pelvic fins than to anal fin. First dorsal fin has 8 spines, second has 1 spine and 10 rays. Anal fin has 3 spines and 7 rays. 43−45 pored lateral scales. Up to 18cm in total length.

Distribution — Japan, Ryukyu, Formosa, Indo-West Pacific, South Africa

Percichthyidae

⇦ ***Lateolabrax latus*** Katayama

Japanese Plain Sea Bass (New Common Name)

Diagnosis — Resembles *L. japonicus*, but distinguished by deeper body, 1 row of scales on underside of lower jaw. Back has no black spots. Young does not move into fresh water areas. Dorsal fin has 12 spines and 15−16 rays, anal fin has 3 spines and 9−10 rays. 71−76 pored lateral scales. Edible. Up to 90cm in total length.

Distribution — Japan

Percichthyidae

Stereolepis gigas Ayres ⇨

Smallmouth Stone-bass (New Common Name)

Diagnosis — Inhabits waters more than 400m deep. Body is laterally compressed and somewhat deep. Scales are small and ctenoid. Maxillary is broad and short. Dorsal fin has 11 spines and 10 rays, anal fin has 3 spines and 8 rays. 80 pored lateral scales. Edible. Noted for causing Vitamin A poisoning. Liver yields liver oil. Up to 2.2m in total length.

Distribution — Northern Japan, Japan Sea, California

Percichthyidae

⇦ ***Neoscombrops pacificus*** Mochizuki

Disguised Bluefish (New Common Name)

Diagnosis — Inhabits waters more than 60m deep. Body is fusiform. Resembles *Scombrops boops*, but distinguished by somewhat deeper body, straight abdominal side. Pectoral fins are large. First dorsal fin has 9 spines, second has 1 spine and 10 rays. Anal fin has 3 spines and 7 rays. 47–51 pored lateral scales. Edible. Up to 35cm in total length.

Distribution — Izu Islands, Ryukyu, Samoa

Percichthyidae

Doederleinia berycoides (Hilgendorf) ⇨

Blackthroat Seaperch

Diagnosis — Inhabits waters more than 100m deep. Resembles *Scombrops boops*, but distinguished by reddish body, large pectoral fins. Caudal fin is not compressed. Eyes are large. Mouth is black inside. Scales are large, easily removable. Dorsal fin has 9 spines and 10 rays, anal fin has 3 spines and 6–8 rays. 41–46 pored lateral scales. Edible. Up to 40cm in total length.

Distribution — Japan, Korea

Serranidae

⇦ ***Variola louti*** (Forskål) ⇩

⇙ [*Serranus punctulatus* Quoy et Gaimard]

Lunartail Rockcod, Painted Coral-trout, Coral Cod, Moontail Rock Cod, Striped Orange Rockcod, Lunartail Grouper, Coronation Trout

Diagnosis — Inhabits rather deep shoreline reefs and coral reefs. Body is fusiform. Caudal lobes extend long. Body is reddish and purplish-brown, whole body is profusely dotted with small spots. Dorsal fin has 9 spines and 13–14 rays, anal fin has 3 spines and 8 rays. 64–73 pored lateral scales. Carnivorous. Edible but often causes ciguatera poisoning. Up to 85cm in total length.

Distribution — Japan, Ryukyu, Indo-Pacific, Red Sea, South Africa

▲ Adult

Serranidae

⇦ ***Plectropomus truncatus*** Fowler

Squaretail Grouper

Diagnosis — Resembles *P. leopardus*, but distinguished by truncated and white margined caudal fin. Body is covered with large blue spots. Dorsal fin has 8 spines and 11 rays, anal fin has 3 spines and 7–8 rays. 85–95 pored lateral scales. Edible but may cause ciguatera poisoning. Up to 1m in total length.

Distribution — Ryukyu, Indo-Pacific, Red Sea, East Africa

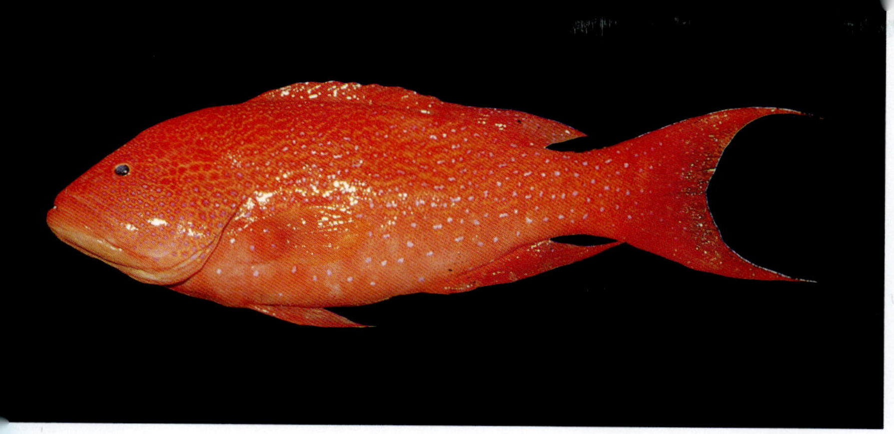

Serranidae
Variola albimarginata Baissac

White-margined Lunartail Rockcod (New Common Name)

Diagnosis — Inhabits outer slopes of rather deep coral reefs. Resembles *V. louti*, but caudal margin is white, anal rays are long enough to reach caudal base. Pelvic fins are somewhat short. Edible. Up to 60cm in total length.

Distribution — Southern Japan, Ryukyu, Indo-Pacific

Serranidae
Plectropomus leopardus (Lacépède)

[*Holocentrus leopardus* Lacépède]

Leopard Coraltrout

Diagnosis — Inhabits shallow rocky reefs and coral reefs. Body is fusiform, profusely dotted with blue. Caudal fin is large. Body varies widely in color. Lower jaw has canines. Dorsal fin has 8 spines and 11 rays, anal fin has 3 spines and 8 rays. 81–91 pored lateral scales. Edible but may cause ciguatera poisoning. Up to 1m in total length.

Distribution — Japan, Ryukyu, Indo-Pacific, Hawaii, Red Sea, South Africa

▲ Young Adult ▶

Serranidae
Plectropomus melanoleucus (Lacépède)

Coral Trout

Diagnosis — Body is fusiform. Caudal margin is emarginated. Pectoral fins are yellow. Body is reddish-brown or greenish-brown, dotted with blue. Dorsal fin has 8 spines and 11 rays, anal fin has 3 spines and 8 rays. 81–90 pored lateral scales. Edible but may cause ciguatera poisoning. Up to 1m in total length.

Distribution — Japan, Ryukyu, Indo-Pacific

Serranidae
Plectropomus oligacanthus (Bleeker)

[*Paracanthius oligacanthus* (Bleeker)]

Vermiculate Coral Trout

Diagnosis — Inhabits shallow coral reefs and rocky reefs. Head has blue lines, body has blue spots. Caudal fin is emarginated. Dorsal fin has 8 spines and 11–12 rays, anal fin has 3 spines and 8 rays. 83–90 pored lateral scales. Edible but may cause ciguatera poisoning. Up to 60cm in total length.

Distribution — Indo-West Pacific

Serranidae
Cephalopholis aurantius (Valenciennes) ⇧⇗
Garrupa, Orange Rock Cod, Spotted Red Rockcod

Diagnosis — Inhabits rather deep waters. Body is deep, reddish or dark brown; snout is pointed. Interorbital space is depressed. Caudal margin is rounded. Dorsal fin has 9 spines and 14–15 rays, anal fin has 3 spines and 9 rays. 50–58 pored lateral scales. Edible. Up to 60cm in total length.

Distribution — Southern Japan, Ryukyu, Indo-Pacific, South Africa

Serranidae
⇧ *Cephalopholis igarashiensis* Katayama
Sevenband Grouper

Diagnosis — Inhabits rather deep waters. Body is deep, has 7 yellow cross bands. In young, dorsal fin ray has a black blotches. Dorsal fin has 9 spines and 13–14 rays, anal fin has 3 spines and 9 rays. 58–65 pored lateral scales. Edible. Up to 40cm in total length.

Distribution — Japan, Ryukyu, Formosa, South China Sea, Philippines

Serranidae
Cephalopholis pachycentron (Valenciennes) ⇧
Brown-banded Rock-cod, Brownbarred Rockcod, Brown Coral-cod, Garrupa, Spotface Rockcod, Spotted-faced Rock Cod

Diagnosis — Inhabits rather deep shoreline reefs and coral reefs. Body is somewhat deep, has 8 cross bands. Caudal margin is rounded and white. Fins are dark. Dorsal fin has 9 spines and 15–16 rays, anal fin has 3 spines and 8 rays. 46–50 pored lateral scales. Edible. Up to 30cm in total length.

Distribution — Southern Japan, Ryukyu, Formosa, Indo-Pacific

Serranidae
Cephalopholis obtusaurus Evermann et Seale ⇨
Long-finned Orange Reefcod (New Common Name)

Diagnosis — Body is somewhat deep and yellowish-red. Interorbital space is depressed. Pectoral fins are long enough to reach origin of anal fin. Caudal margin is rounded. Dorsal fin has 9 spines and 15 rays, anal fin has 3 spines and 9 rays. 48–51 pored lateral scales. Edible. Up to 40cm in total length.

Distribution — Ryukyu, Pacific

Serranidae
⇦ *Cephalopholis urodelus* (Schneider) ⇨
Banded-tail Coral-cod, Flag-tailed Rock Cod, Flagtail Rockcod

Diagnosis — Body is red; tail is dark. Upper and lower parts of caudal fin has white oblique bands. Dorsal fin has 9 spines and 14–16 rays, anal fin has 3 spines and 9 rays. 55–60 pored lateral scales. Edible. Up to 25cm in total length.

Distribution — Southern Japan, Ryukyu, Indo-Pacific

Serranidae

Cephalopholis miniatus (Forskål)

Rock Cod, Blue Spot Rock-cod, Coral Cod, Coral Trout, Vermilion Coral-cod, Coral Grouper

Diagnosis — Abundant about coral reefs. Body is red, profusely dotted with distinct blue. Caudal margin is rounded. Dorsal fin has 9 spines and 15–16 rays, anal fin has 3 spines and 9 rays. 47–51 pored lateral scales. Edible but may cause ciguatera poisoning. Up to 45cm in total length.

Distribution — Japan, Ryukyu, Bonin Islands, Indo-Pacific, Persian Gulf, Red Sea, South Africa

Serranidae

Cephalopholis sexmaculatus (Rüppell)

[*C. coatesi* Whitley]

Sixgirdle Grouper, Sixspot Grouper

Diagnosis — Body is yellowish-red, profusely dotted with blue; back has 6 black blotches. Pectoral fins are somewhat long. Dorsal fin has 9 spines and 15 rays, anal fin has 3 spines and 9 rays. 47–48 pored lateral scales. Edible. Up to 50cm in total length.

Distribution — Southern Japan, Ryukyu, Indo-Pacific

Serranidae

Cephalopholis argus Schneider

[*Epinephelus myriaster* Cuvier]

[*E. kendalli* Evermann et Seale]

Peacock Coralcod, Blue-spotted Argus, Blue-spotted Grouper, Black Rock-cod, Garrupa, Peacock Rock-cod

Diagnosis — Inhabits shallow coral reefs. Body is somewhat deep, has 10 black bands, is profusely dotted with blue. Fins are dark blue; pectoral fins are short. Dorsal fin has 9 spines and 15–16 rays, anal fin has 3 spines and 9 rays. 45–51 pored lateral scales. Edible but noted for many cases of ciguatera poisoning. Up to 50cm in total length.

Distribution — Southern Japan, Ryukyu, China, Indo-Pacific, Hawaii, Persian Gulf, Red Sea, South Africa

Serranidae

Saloptia powelli Smith

Orange Bodied Grouper

Diagnosis — Inhabits rather deep waters. Body has no spots. Abdomen is pale. Dorsal, anal and caudal fins are yellow. Caudal fin is emarginated. Dorsal fin has 8 spines and 11–12 rays, anal fin has 3 spines and 8 rays. 71–74 pored lateral scales. Edible. Up to 40cm in total length.

Distribution — Ryukyu, South China Sea, Polynesia

Serranidae

Cephalopholis boenak (Bloch)

Blue-lined Coral Cod, Blue-lined Rock-cod

Diagnosis — Body is somewhat deep, blackish-brown, has thin blue lines; snout is pointed. Caudal margin is rounded. Dorsal fin has 9 spines and 15–17 rays, anal fin has 3 spines and 8–9 rays. 44–53 pored lateral scales. Edible. Up to 35cm in total length.

Distribution — Japan, Ryukyu, Indo-Pacific

Serranidae

⇦ ***Aethaloperca rogaa*** (Forskål)

[*Cephalopholis rogaa* (Forskål)]

Dusky Rock-cod, Red-flushed Rock-cod, Redmouth Grouper

Diagnosis — Body is compressed and deep. Caudal margin is truncated and white. Pectoral fins are long. Abdominal side has a white band. Dorsal fin has 9 spines and 17–18 rays, anal fin has 3 spines and 9 rays. 50–55 pored lateral scales. Edible. Up to 60cm in total length.

Distribution — Japan, Ryukyu, Indo-Pacific, Red Sea, South Africa

Serranidae

Anyperodon leucogrammicus (Valenciennes) ⇨ ⇩

[*Serranus urophthalmus* Bleeker]

White-lined Rockcod, Whitelined Grouper, Slender Grouper

Diagnosis — Head is elongated; snout is pointed. Lower jaw has no canines. Body has white bands, profusely dotted with blackish-red spots. Caudal margin is rounded. Dorsal fin has 11 spines and 15 rays, anal fin has 3 spines and 9 rays. 65–66 pored lateral scales. Edible but may cause ciguatera poisoning. Up to 50cm in total length.

Distribution — Ryukyu, Indo-Pacific, Red Sea, East Africa

Serranidae

⇩ ***Epinephelus truncatus*** Katayama

Darkred Grouper, Red-tipped Rock-cod

Diagnosis — Inhabits rather deep waters. Resembles *E. fasciatus*, but distinguished by red margin of dorsal fin membrane, truncated caudal fin, more pored scales on lateral line (70–72). Dorsal fin rays and upper part of caudal fin are dark-colored. Dorsal fin has 11 spines and 16–17 rays, anal fin has 3 spines and 8 rays. Edible. Up to 40cm in total length.

Distribution — Southern Japan, Ryukyu, Bonin Islands, Indo-Pacific

Serranidae

Epinephelus fasciatus (Forskål) ⇧ ⇨

[*E. marginalis* Bloch]

[*E. emoryi* Schultz]

Blacktipped Grouper, Banded Reefcod, Redbarred Rockcod, Blacktip Rockcod, Striped Grouper, Reef Cod

Diagnosis — Inhabits shoreline reefs, and coral reefs. Body is reddish, has 5–6 cross bands. Dorsal fin has 11 spines and 15–17 rays, margin of fin membrane is black. Caudal margin is rounded. Anal fin has 3 spines and 8 rays. 52–57 pored lateral scales. Edible but noted for cases of ciguatera poisoning. Up to 40cm in total length.

Distribution — Japan, Ryukyu, Indo-Pacific, Red Sea, South Africa

Serranidae

Epinephelus hoedtii (Bleeker) ⇨⇩

[*E. kohleri* Schultz]

Purple Rockcod

Diagnosis — Inhabits rather deep rocky reefs. Body is deep and thick, is densely covered with black speckles. Caudal fin is truncated, margin is black. Mouth has a pair of canines. Dorsal fin has 11 spines and 16–17 rays, anal fin has 3 spines and 8 rays. 65–70 pored lateral scales. Edible but noted for some cases of ciguatera poisoning. Up to 1.5m in total length.

Distribution — Japan, Ryukyu, Bonin Islands, Indo-Pacific

Serranidae

Epinephelus flavocaeruleus (Lacépède) ⇩

[*Holocentrus flavo-caeruleus* Lacépède]

[*Serranus flavocaeruleus* (Lacépède)]

[*S. hoedtii* Günther]

Purple Rock-cod, Blue-and-yellow Reef-cod

Diagnosis — Inhabits rocky reefs approximately 100m deep. Body is deep. Resembles Purple Rockcod, *E. hoedtii*, but distinguished by yellow fins. Jaws have 1 or 2 pairs of canines. Dorsal fin has 11 spines and 16–17 rays, anal fin has 3 spines and 8 rays. Edible. Up to 80cm in total length.

Distribution — Bonin Islands, China, Indo-Pacific, South Africa

Serranidae

Gracila albomarginata (Fowler et Bean) ⇧

Blue-striped Grouper (New Common Name)

Diagnosis — Inhabits rather deep waters. Body is somewhat deep, has many blue bands; head is small. Caudal fin is emarginated. Interorbital space is convex. In young, body is reddish. Dorsal fin has 9 spines and 15 rays, anal fin has 3 spines and 9 rays. 68–73 pored lateral scales. Edible. Up to 50cm in total length.

Distribution — Ryukyu, Indo-Pacific

Serranidae

Epinephelus salmonoides (Lacépède) ⇨⇩⇙

[*E. samoides* (Lacépède)]

Malabar Reef-cod

Diagnosis — Resembles *E. malabaricus*, but distinguished by black blotches on body, also on lower jaw. Body has 5 dark cross bands. Dorsal fin has 11 spines and 15 rays, anal fin has 3 spines and 8 rays. 57–58 pored lateral scales. Edible. Up to 90cm in total length.

Distribution — Japan, Ryukyu, Indo-Pacific, Red Sea, South Africa

◀ Adult

▼Young Adult

Serranidae

Epinephelus fasciatomaculatus (Peters)

Six-banded Reefcod (New Common Name)

Diagnosis — Resembles *E. awoara* or *E. diacanthus*, but distinguished by fin margins are not yellow, preoperculum lacking 2 spines, large-slightly convex eyes. Pectoral fins are dark. Dorsal fin has 11 spines and 16–17 rays, anal fin has 3 spines and 8 rays. 47–53 pored lateral scales. Edible. Up to 30cm in total length.

Distribution — Japan, Ryukyu, Formosa, South China Sea, Philippines

Serranidae

Epinephelus rhyncholepis (Bleeker)

White-dotted Grouper

Diagnosis — Inhabits shallow shoreline reefs and coral reefs. Resembles *E. fasciatus*, but body is dark with brown cross bands. Pectoral fins are dark; caudal margin is rounded. Dorsal fin has 11 spines and 17 rays, anal fin has 3 spines and 8 rays. 47–52 pored lateral scales. Edible. Up to 40cm in total length.

Distribution — Southern Japan, Ryukyu, Indo-Pacific

Serranidae

Epinephelus septemfasciatus (Thunberg)

Sevenband Grouper, True Bass

Diagnosis — Inhabits rather deep rocky reefs. Body has 7 dark cross bands. Caudal margin is rounded and white. Pectoral fins are short, does not extend beyond dark fourth cross band. Dorsal fin has 11 spines and 14–16 rays, anal fin has 3 spines and 9–10 rays. 64–71 pored lateral scales. Edible. Up to 1m in total length.

Distribution — Japan, Ryukyu, East China Sea, Indo-Pacific-Atlantic

Serranidae

Epinephelus awoara (Temminck et Schlegel)

Banded Grouper, Green Hata, Yellow Grouper

Diagnosis — Inhabits rather deep rocky reefs. Body has 6 dark bands. Margins of fins are yellow. Dorsal fin has 11 spines and 15–16 rays, anal fin has 3 spines and 8 rays. 50–53 pored lateral scales. Edible. Up to 45cm in total length.

Distribution — Japan, Ryukyu, Formosa, South China Sea

Serranidae

Epinephelus sexfasciatus (Valenciennes) ⇨

Six-banded Rock-cod

Diagnosis — Body is shallow, has 6 dark bands. Dorsal, anal and caudal fins are dotted with black. Caudal margin is rounded. Dorsal fin has 11 spines and 15 rays, anal fin has 3 spines and 8 rays. Edible. Up to 30cm in total length.

Distribution — South China Sea, Indo-West Pacific

Serranidae

⇦ *Epinephelus malabaricus* (Schneider)

Malabar Rock-cod

Diagnosis — Body is fusiform and thick. Resembles *E. salmonoides*, but has brown spots on body, has no spots on lower jaw. Interorbital space is slightly projected. Dorsal fin has 11 spines and 15 rays, anal fin has 3 spines and 8 rays. 64–66 pored lateral scales. Edible. Up to 90cm in total length.

Distribution — Ryukyu, Indo-Pacific, Red Sea

Serranidae

Epinephelus dictyophorus (Bleeker) ⇨

[*E. fuscus* Fourmanoir]

Honeycomb Fin Rock Cod

Diagnosis — Body has reticulate patterns, 3 broad cross bands; fins have large black splotches. Caudal margin is rounded. Jaws have 2 pairs of canines. Dorsal fin has 11 spines and 16–17 rays, anal fin has 3 spines and 8 rays. 46 pored lateral scales. Edible. Up to 45cm in total length.

Distribution — Ryukyu, South China Sea, Indo-West Pacific

Serranidae

Epinephelus heniochus Fowler ⇨

[*E. hata* Katayama]

Barcheek Grouper

Diagnosis — Inhabits rather deep waters. Body is light-brown, profusely dotted with black. Cheeks have 3 brown stripes. Caudal margin is rounded. Dorsal fin has 11 spines and 14–15 rays, anal fin has 3 spines and 8 rays. 58–62 pored lateral scales. Edible. Up to 40cm in total length.

Distribution — Southern Japan, Formosa, South China Sea

Serranidae

⇦ *Epinephelus magniscuttis* Postel

Black-spotted Grouper

(New Common Name)

Diagnosis — Inhabits coral reefs. Resembles *E. epistictus*, but distinguished by small black spots not forming rows but dotting back. Dorsal fin has 11 spines and 14–15 rays, anal fin has 3 spines and 8 rays. Edible. Up to 1.5m in total length.

Distribution — New Caledonia, South West Pacific

Serranidae
Epinephelus tauvina (Forskål) ⇩

[*E. elongatus* Schultz]
[*E. salonotus* Smith et Smith]
[*E. chewa* Morgans]

Giant Grouper, Spotted Sea Bass, Black Sea Bass, Bullnose Bass, Estuary Grouper, Great Reef-cod, Spotted River Cod, Variegated Rock Cod

Diagnosis — Body is somewhat shallow; snout is pointed; mouth is large. Back has dark blotches, spots on body are rough, not reticulate-patterned. Dorsal fin has 11 spines and 15–16 rays, anal fin has 3 spines and 8 rays. 62–66 pored lateral scales. Edible but noted for cases of ciguatera poisoning. Up to 2.2m in total length.

Distribution — Southern Japan, Ryukyu, Indo-Pacific, Hawaii, Persian Gulf, Red Sea, South Africa

Serranidae
Epinephelus corallicola (Valenciennes) ⇧

[*E. spilotus* Schultz]

Coral Rock Cod, Duskyfin Grouper

Diagnosis — Head is rounded; mouth is large; maxillary groove is black. Dorsal fin has a black blotch in the middle of base. Pectoral fins are black. Dorsal fin has 11 spines and 16–17 rays, anal fin has 3 spines and 8 rays. 49–53 pored lateral scales. Edible. Up to 45cm in total length.

Distribution — Ryukyu, Indo-Pacific

Serranidae
⇦ *Epinephelus megachir* (Richardson)

Garrupa, Longfin Rockcod, Long-finned Grouper, Spotted Grouper, Rock Cod, Longpectoral Honeycomb Grouper

Diagnosis — Body is somewhat elongated; snout is pointed. Interorbital space is slightly convex. Pectoral fins are large; Caudal margin is rounded. Whole body is profusely blotched with black. Dorsal fin has 11 spines and 17 rays, anal fin has 3 spines and 8 rays. 47–51 pored lateral scales. Edible. Up to 45cm in total length.

Distribution — Japan, Ryukyu, Indo-Pacific, South Africa

Serranidae
Epinephelus fario (Thunberg) ⇨

[*E. maculatus* Bloch]

Trout Rockcod, Garrupa, Black-saddled Grouper, Trout Reef-cod

Diagnosis — Inhabits rocky reefs of coastal waters. Mouth is large; snout is pointed. Back has 3 black blotches. Caudal margin is rounded and white. Dorsal fin has 11 spines and 15–17 rays, anal fin has 3 spines and 8 rays. 47–49 pored lateral scales. Edible but may cause ciguatera poisoning. Up to 50cm in total length.

Distribution — Japan, Ryukyu, Formosa, China, Indo-Pacific, South Africa

Serranidae
Epinephelus areolatus (Forskål) ⇨

[*Serranus angularis* Cuvier et Valenciennes]

Areolated Grouper, Areolated Reef Cod, Green-spotted Rock-cod, Rock-bass, Rock Perch, Squaretail Rockcod, Yellowspotted Rock Cod

Diagnosis — Resembles *E. chlorostigma*, but distinguished by large and somewhat indistinct reticulate pattern. Caudal margin is emarginated. Dorsal fin has 11 spines and 15–17 rays, anal fin has 3 spines and 8 rays. 51–53 pored lateral scales. Edible. Up to 45cm in total length.

Distribution — Southern Japan, Ryukyu, Indo-Pacific, Persian Gulf, Red Sea, South Africa

Serranidae
Epinephelus microdon
(Bleeker)

[*E. fuscoguttatus* (Forskål)]

[*Serranus microdon* Bleeker]

Black Rock Cod, Brown-marbled Reef-cod, Mottled Grouper, Smalltooth Grouper

Diagnosis — Body is fusiform. Caudal margin is rounded. Back has 5 dark blotches. Mouth is large. Upper margin of opercles are curved sharply. Dorsal fin has 11 spines and 14–15 rays, anal fin has 3 spines and 8 rays. 49–53 pored lateral scales. Edible but may cause ciguatera poisoning. Up to 1m in total length.

Distribution — Southern Japan, Ryukyu, Bonin Islands, Indo-Pacific, Red Sea, South Africa

Serranidae
Epinephelus fuscoguttatus (Forskål)

[*E. microdon* (Bleeker) subsp.]

[*Serranus horridus* Cuvier et Valensiennes]

Black Sea Bass, Black Rock Cod, Marbled Sea Bass, Brown-marbled Reef-cod, Brownmarbled Grouper

Diagnosis — Resembles *E. microdon*, but distinguished by deeper body. Body is yellowish-brown. Caudal margin is rounded. Dorsal fin has 11 spines and 14–15 rays, anal fin has 3 spines and 8 rays. 60 pored lateral scales. Carnivorous. Edible but often causes ciguatera poisoning. Up to 1m in total length.

Distribution — Ryukyu, Indo-Pacific, Red Sea, South Africa

Serranidae
Epinephelus tukula Morgans

Potato Bass

Diagnosis — Body is large, has brown blotches. Fins are profusely dotted with small spots. Head is straight; interorbital space is convex. Dorsal fin has 11 spines and 13–15 rays, anal fin has 3 spines and 7–9 rays. About 70 pored lateral scales. Edible. Up to 2m in total length.

Distribution — Japan, Ryukyu, Formosa, South China Sea, West Indian Ocean

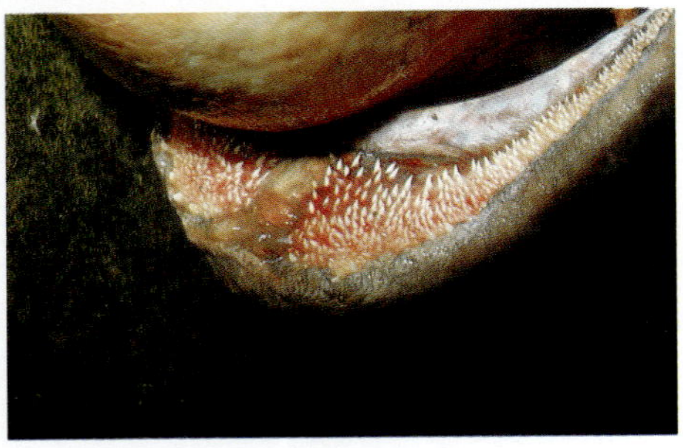

▲Teeth are villiform.

Serranidae

Epinephelus moara (Temminck et Schlegel)

Kelp Bass, Kelp Grouper, Saladfish

Diagnosis — Inhabits rather deep rocky reefs. Body is fusiform, has 6 dark bands. Mouth is large. Caudal margin is rounded. Dorsal fin has 11 spines and 14−15 rays, anal fin has 3 spines and 8 rays. 64−67 pored lateral scales. Edible. Up to 1.8m in total length, 63kg in weight.

Distribution — Japan, Ryukyu, Formosa, China, South China Sea

Serranidae

Epinephelus caeruleopunctatus (Bloch)

Garrupa, Ocellated Rock Cod, White-spotted Reef-cod

Diagnosis — Inhabits coral reefs. In adult, snout is pointed. Body is profusely dotted with large and small white. Margins of fins are white. Dorsal fin has 11 spines and 16−17 rays, anal fin has 3 spines and 8 rays. 52−54 pored lateral scales. Edible. Up to 75cm in total length.

Distribution — Southern Japan, Ryukyu, Indo-Pacific, South Africa

Serranidae

Epinephelus maculatus (Bloch)

Whitemottled Grouper

Diagnosis — Upper part of body is covered with reticulate blotches, abdominal side has black spots. In young, body is blotched with white. In adult, white parts are retained. Caudal margin is somewhat rounded. Dorsal fin has 11 spines and 15−16 rays, anal fin has 3 spines and 8 rays. 49−52 pored lateral scales. Edible. Up to 60cm in total length.

Distribution — Southern Japan, Ryukyu, Indo-Pacific

Serranidae

Epinephelus chlorostigma (Valenciennes)

[*E. geoffroyi* Klunzinger]

Brown-spotted Grouper, Reef Cod, Brown Spotted Reef Cod, Garrupa,

Diagnosis — Reticulate patterns of body are small and distinct. Caudal fin is truncated. Dorsal fin has 11 spines and 16−18 rays, anal fin has 3 spines and 8 rays. 49−52 pored lateral scales. Edible. Up to 75cm in total length.

Distribution — Japan, Ryukyu, Indo-Pacific, Persian Gulf, Red Sea, South Africa

Serranidae
Epinephelus merra (Bloch)

Bloch's Grouper, Honeycomb Cod, Honeycomb Rock Cod, Mahon, Matty, Mottled Grouper, Tassellated Rock Cod, Wire-netting Cod

Diagnosis — Abundant about shoreline reefs and coral reefs. Body is small, covered with reticulate patterns. Resembles *E. hexagonatus*, but has black spots on pectoral fins, fewer pored scales (48–50) on lateral line. Dorsal fin has 11 spines and 16–17 rays, anal fin has 3 spines and 8 rays. Edible but noted for cases of ciguatera poisoning. Up to 45cm in total length.

Distribution — Japan, Ryukyu, Indo-Pacific, South Africa

Serranidae
Epinephelus hexagonatus (Schneider)

Hexagonalspot Grouper

Diagnosis — Inhabits shoreline reefs and coral reefs. Body has regular reticulate patterns; back has 5 black blotches. Resembles to *E. merra*, but distinguished by pectoral fins lacking black spots, more pored scales on lateral line (60–62). Dorsal fin has 11 spines and 15–17 rays, anal fin has 3 spines and 8 rays. Edible but may cause ciguatera poisoning. Up to 60cm in total length.

Distribution — Ryukyu, Izu Islands, Bonin Islands, Indo-Pacific, South Africa

Serranidae
Epinephelus summana (Forskål)

Garrupa, Rock-cod, Speckled-finned Rock-cod, Summana Rock Cod, Whitespotted Rockcod, Summana Grouper

Diagnosis — Body is densely covered with white spots and wavy white lines. Caudal margin is rounded and white. Upper part of opercles are conspicuously distended. Pectoral fins are long. Dorsal fin has 11 spines and 14–15 rays, anal fin has 3 spines and 8 rays. 48–50 pored lateral scales. Edible. Up to 55cm in total length.

Distribution — Ryukyu, Bonin Islands, Indo-Pacific, Red Sea, South Africa

Serranidae
Epinephelus akaara (Temminck et Schlegel)

Garrupa, Red Grouper, Red Spotted Grouper

Diagnosis — Inhabits rocky reefs in coastal waters. Body is dotted with orange spots; back has dark blotches. Snout is pointed. Dorsal fin has 11 spines and 15–17 rays, anal fin has 3 spines and 8 rays. 50–53 pored lateral scales. Edible. Up to 40cm in total length.

Distribution — Japan, Korea, China, Philippines, India

Serranidae

Epinephelus cometae Tanaka

Broom Grouper

Diagnosis — Resembles *E. morrhua*, but distinguished by thinner brown bands running parallel with abdominal side. Caudal margin is rounded. Dorsal fin has 11 spines and 14–15 rays, anal fin has 3 spines and 8 rays. 58–64 pored lateral scales. Edible. Up to 80cm in total length.

Distribution — Southern Japan, Ryukyu, Formosa, Indo-Pacific

Serranidae

Epinephelus lanceolatus (Bloch)

[*Promicrops lanceolatus* (Bloch)]

[*Oligorus goliath* de Vis]

Brindle Bass, Garrupa

Diagnosis — Large species. Body is covered with dark brown blotches; fins are yellowish. Dorsal fin has 11 spines and 14–15 rays, anal fin has 3 spines and 8 rays. 60–70 pored lateral scales. Edible. Enormous ones are dangerous to man. Up to 3.6m in total length.

Distribution — Indo-Pacific, Red Sea, South Africa

Serranidae

Epinephelus morrhua (Valenciennes)

Banded-cheek Reef-cod, Contour Rockcod, Curve-banded Grouper, Curveband Grouper

Diagnosis — Inhabits somewhat deep waters. Body has dark brown oblique bands running parallel on anterior surface, crossing body on posterior surface. Light-colored body is profusely dotted with black. Fins are also light-colored. Dorsal fin has 11 spines and 14–15 rays, anal fin has 3 spines and 8 rays. 61–67 pored lateral scales. Edible. Up to 60cm in total length.

Distribution — Japan, Ryukyu, Formosa, Indo-Pacific, South Africa

Serranidae

Epinephelus poecilonotus (Temminck et Schlegel)

Garrupa, Linearspot Grouper, Rock Cod

Diagnosis — Body is yellowish-brown, has 4 rows of irregular arch-like blotches. In young, body has black bands. Dorsal fin has 11 spines and 14–15 rays, anal fin has 3 spines and 8 rays. 60–65 pored lateral scales. Edible. Up to 60cm in total length.

Distribution — Southern Japan, Ryukyu, Formosa, Indian Ocean, Red Sea, South Africa

Serranidae

Epinephelus latifasciatus (Temminck et Schlegel)

Laterally-banded Grouper

Diagnosis — Inhabits rather deep waters. In young, body has 2 white bands which form rows of 3 black spots in adult. Caudal fin is profusely dotted with black. Interorbital space is convex. Dorsal fin has 11 spines and 12–14 rays, anal fin has 3 spines and 8 rays. 60–66 pored lateral scales. Edible. Up to 40cm in total length.

Distribution — Japan, Ryukyu, Formosa, Indo-Pacific, Persian Gulf

Serranidae

Cromileptes altivelis (Valenciennes) ⇩⇨

Barramundi Cod, Highfin Grouper, Hump-backed Rock Cod

Diagnosis — Body is deep; snout is pointed; head is small, top is conspicuously depressed. Jaws have no canines. Whole body is profusely blotched with black. Caudal margin is rounded. Dorsal fin has 10 spines and 17−19 rays, anal fin has 3 spines and 10 rays. Edible but may cause ciguatera poisoning. Up to 65cm in total length.

Distribution — Japan, Ryukyu, Formosa, China, Indo-Pacific

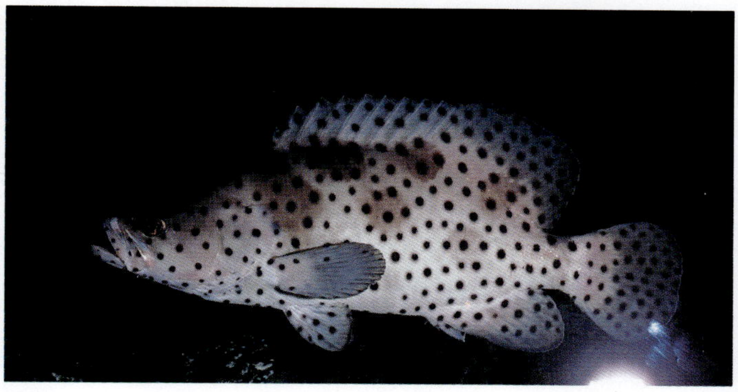

Serranidae

Giganthias immaculatus Katayama ⇨

Rose Porgy (New Common Name)

Diagnosis — Inhabits rather deep rocky reefs. Body is deep; snout is short; eyes are large. Caudal fin is large, lobes are pointed. Dorsal fin has 9 spines and 13 rays, anal fin has 3 spines and 8 rays. 42−44 pored lateral scales. Edible. Up to 40cm in total length.

Distribution — Izu Islands, Ryukyu, Formosa

Serranidae

⇦ *Plectranthias kamii* Randall

[*P. anthioides* (Günther)]

Yellowsnout Sea Bass

Diagnosis — Inhabits rather deep rocky reefs. Body is somewhat deep, has irregular yellow blotches. Caudal fin is emarginated, upper lobe slightly extends. Dorsal fin has 10 spines and 18 rays, third spine is remarkably long. Anal fin has 3 spines and 7 rays. 33−36 pored lateral scales. Edible. Up to 30cm in total length.

Distribution — Izu Islands, Ryukyu

Serranidae

Pseudanthias thompsoni (Fowler) ⇨

[*Anthias thompsoni* (Fowler)]

[*Caesioperca thompsoni* Fowler]

Hawaiian Anthias, Thompson's Grouper

Diagnosis — Found in schools about coastal coral reefs. Dorsal fin ray and caudal lobes extend like thread. Aquarium fish. Up to 20cm in total length.

Distribution — Hawaii

Serranidae

⇦ *Plectranthias yamakawai* Yoshino

Yamakawa's Sea Bass (New Common Name)

Diagnosis — Inhabits rather deep rocky reefs. Body is deep; snout is pointed. Caudal fin is emarginated. Back is profusely dotted with blue, has a red spot in the middle. Dorsal fin has 10 spines and 16−18 rays, anal fin has 3 spines and 7 rays. 31−33 pored lateral scales. Edible. Up to 25cm in total length.

Distribution — Ryukyu

Serranidae
⇦ ***Holanthias katayamai***
Randall, Maugé et Plessis
[*H. chrysostictus* (Günther)]

Rosy Sea Bass

Diagnosis — Inhabits rather deep rocky reefs. Body is compressd and deep. Dorsal fin has 10 spines and 16–17 rays, fin rays extend long. Caudal fin is falcated. Anal fin has 3 spines and 7–8 rays. Edible. Up to 25cm in total length.

Distribution — Southern Japan, Ryukyu, Guam, Indonesia

Serranidae
Sacura margaritacea
(Hilgendorf) ⇨⇨

Cherry Bass, Cherry Porgy

Diagnosis — Found in schools about rather deep rocky reefs. Body is deep. Caudal lobes extend long. In male, body is blotched with white. In female, dorsal fin has a black blotch. Transsexual. Dorsal fin has 10 spines and 16–18 rays, anal fin has 3 spines and 7 rays. Aquarium fish. Up to 20cm in total length.

Distribution — Japan

▲ Male

▲ Female

Serranidae
⇦ ***Caprodon schlegelii*** (Günther)

Schlegel's Red Bass

Diagnosis — Inhabits rather deep waters. Body is compressed and deep. Caudal fin is truncated, white at both lobes. In male, yellow stripes extend radially from eyes, dorsal fin has a black blotch. In female, back has 3–4 dark-colored blotches. Dorsal fin has 10 spines and 19–21 rays, anal fin has 3 spines and 7–9 rays. Edible. Up to 40cm in total length.

Distribution — Southern Japan, Ryukyu, Formosa, Pacific, Hawaii, Chile

Serranidae
Chelidoperca pleurospilus
(Günther) ⇨

Spotted Princess Perch
(New Common Name)

Diagnosis — Inhabits muddy-sand bottoms about 100m deep. Mouth is large; snout is pointed. Caudal fin is truncated. No scales are present between eyes. Dorsal fin has 10 spines and 10 rays, anal fin has 3 spines and 6 rays. 42–44 pored lateral scales. Up to 18cm in total length.

Distribution — Southern Japan, Formosa, Philippines

Serranidae

Liopropoma lunulatum (Guichenot) ⇨

[*Chorististium lunulatum* (Guichenot)]

[*Pikea lunulata* (Guichenot)]

Yellow-lined Yellow Bass (New Common Name)

Diagnosis — Inhabits rather deep rocky reefs. Body is yellowish-red or reddish. One yellow band runs between snout and tail. Marginal part of fins are yellow. Caudal fin is lunated. In some individuals, body has black spots. Dorsal fin is compressed, has 8 spines and 12 rays. Anal fin has 3 spines and 9 rays. 50–51 pored lateral scales. Up to 20cm in total length.

Distribution — Izu Islands, Ryukyu, West Indian Ocean

Serranidae

⇦ ***Liopropoma*** sp.

Double-colored Yellow Bass (New Common Name)

Diagnosis — Resembles *L. lunulatum*, but distinguished by truncated caudal fin. Back is yellow; abdomen is red. Dorsal fin is remarkably compressed, has 8 spines and 12 rays. Anal fin has 3 spines and 9 rays. Up to 20cm in total length.

Distribution — Formosa

Serranidae

Liopropoma maculatum (Döderlein) ⇨

[*Chorististium maculatum* (Döderlein)]

Brownspotted Yellow Bass

Diagnosis — Inhabits rather deep rocky reefs. Body is shallow, snout is pointed, head is small and compressed. Eyes are protruded. Body and fins are profusely dotted with red. Dorsal, anal, caudal fin margins are yellow. Dorsal fin has 8 spines and 12 rays, anal fin has 3 spines and 8 rays. 65–70 pored lateral scales. Edible. Up to 25cm in total length.

Distribution — Southern Japan, Ryukyu, Hawaii

Male

Female

Serranidae

⇦ ***Franzia squamipinnis*** (Peters) ⇩

[*F. ardens* Jordan et Thompson]

[*F. ruber* Tanaka]

[*F. affinis* Tanaka]

[*Anthias lepidolepis* Bleeker]

[*A. nobilis* Franz]

[*A. squamipinnis* (Peters)]

Orange Sea-perch, Sea Godie, Threadfin Red Bass

Diagnosis — Found in schools about shallow shoreline reefs and coral reefs. In male, upper part of pectoral fins have red blotches, third dorsal spine extends like thread. In female, body is yellowish-red, dorsal fin does not extend long. Transsexual. Dorsal fin has 10 spines and 17 rays, anal fin has 3 spines and 7 rays. Aquarium fish. Up to 17cm in total length.

Distribution — Japan, Ryukyu, Indo-Pacific, Red Sea, South Africa

Serranidae

⇦ ***Mirolabrichthys pascalus*** (Jordan et Tanaka) ⇧

[*M. tuka pascalus* (Jordan et Tanaka)]

[*Entomanthias pascalus* Jordan et Tanaka]

Forktail Sea Bass

Diagnosis — Found in schools about shoreline reefs. Body is fusiform; snout is pointed. Caudal fin is falcated with extending end. Male is larger than female, dorsal rays extend long. Dorsal fin has 10 spines and 15−17 rays, anal fin has 3 spines and 7−8 rays. Transsexual. Aquarium fish. Up to 17cm in total length.

Distribution — Japan, Ryukyu, West Pacific

◀Female

Serranidae

Mirolabrichtys dispar Herre ⇨

[*Anthias disper* (Herre)]

Madder Sea-perch (New Common Name)

Diagnosis — Found in schools among shallow coral reefs. Resembles *M. pascalus*, but distinguished by longer second dorsal spine and pelvic fins extended like thread. In male, dorsal fin is scarlet. Dorsal fin has 10 spines and 16−18 rays, anal fin has 3 spines and 7−8 rays. 55−63 pored lateral scales. Aquarium fish. Up to 10cm in total length.

Distribution — Ryukyu, West Pacific

Grammistidae

Diploprion bifasciatus Cuvier ⇧ ⇨

[*D. bifasciatum* Kuhl et Hasselt]

Yellow Emperor, Two-banded Sea Perch, Double-banded Perch, Sea Perch

Diagnosis — Inhabits shallow coral reefs. Body is deep. Mouth is directed upward. Dorsal fin is remarkably compressed, has 8 spines and 13−16 rays, anal fin has 2 spines and 12−13 rays. 80−88 pored lateral scales. Skin secretes Ichthyocrinotoxin destructive to other fishes. Aquarium fish. Up to 25cm in total length.

Distribution — Japan, Ryukyu, Formosa, China, Indo-Pacific

Grammistidae
Aulacocephalus temmincki
Bleeker ⇨ ⇩

Goldribbon Rockcod, Emerald Bass, Gold Ribbon Grouper

Diagnosis — Inhabits rather deep rocky reefs. Body is deep; snout is pointed with a yellow band between snout and caudal peduncles. Pelvic fins are short. Dorsal fin has 9 spines and 12 rays, anal fin has 3 spines and 9 rays. 76–82 pored lateral scales. Skin secretes Grammistin destructive to other fishes. Aquarium fish. Up to 35cm in total length.

Distribution — Japan, Ryukyu, Korea, China, Indo-Pacific, South Africa

Grammistidae
Pogonoperca punctata (Valenciennes) ⇨

Speckled Soap-fish

Diagnosis — Body is brown, profusely dotted with white. Caudal margin is rounded. Lower jaw has a flap. Scales are cycloid, sink under skin. Skin secretes Ichthyocrinotoxin producing strong bitterness. Dorsal fin has 8 spines and 12–13 rays, anal fin has 3 spines and 8 rays. 59–70 pored lateral scales. Up to 35cm in total length.

Distribution — Japan, Ryukyu, Indo-Pacific

⇩ Plesiopidae
Calloplesiops altivelis (Steindachner)

Comet

Diagnosis — Inhabits coral reefs. Body is compressed, profusely dotted with white. Fins are well-developed. Dorsal fin has 11 spines and 9–10 rays, posterior end has a white-margined black blotch. Anal fin has 3 spines and 9–10 rays. Lateral lines are not completed. Aquarium fish. Up to 16cm in total length.

Distribution — Ryukyu, Indo-West Pacific, Red Sea

Grammistidae
Grammistes sexlineatus (Thunberg) ⇧

[*G. sexlineatus sexlineatus* (Thunberg)]

[*Perca sexlineata* Thunberg]

Golden-striped Bass, Six-lined Perch, Goldstriped Rockcod, Golden-striped Grouper, Chain-gang Sea Bass, Redtop Grouper, Grammistid Perch, Goldstriped Soapfish

Diagnosis — Inhabits shallow coral reefs. Body is dark brown with white lines and spots. Lower jaw has a small flap. Dorsal fin has 7 spines and 12–14 rays, anal fin has 3 spines and 8–9 rays. 60–72 pored lateral scales. Skin secretes Ichthyocrinotoxin destructive to other fishes. Should not be eaten, death reported in some cases. Up to 30cm in total length.

Distribution — Japan, Ryukyu, Indo-Pacific, Red Sea, South Africa

Pseudochromidae
⇦ *Dampieria cyclophthalmus*
(Müller et Troschel)

[*D. spiloptera* Bleeker]
[*Labracinus cyclophthalmus*
(Müller et Troschel)]

Spottysail Dottyback, Unicolorsail Dottyback

Diagnosis — Inhabits shallow coral reefs. Jaws have 2−3 pairs of canines. Caudal margin is rounded. Dorsal fin has 2 spines and 25 rays, anal fin has 3 spines and 14 rays. Lateral lines are not completed, 53−56 and 19−24 pored scales. Carnivorous. Up to 20cm in total length.

Distribution — Southern Japan, Ryukyu, West Pacific

Pseudochromidae
Pseudochromis diadema
Lubbock et Randall ⇨

Royal Basslet

Diagnosis — Inhabits shallow coral reefs. Body is elongated and yellow. Head to back margin is reddish-purple. Feeds on zooplankton. Dorsal fin has 3 spines and 22 rays, anal fin has 3 spines and 12 rays. 23−28 pored lateral scales. Aquarium fish. Up to 6cm in total length.

Distribution — Philippines

Pseudochromidae
⇦ *Pseudochromis paccagnellae*
Axelrod

Double-coloured Basslet
(New Common Name)

Diagnosis — Inhabits shallow coral reefs. Anterior half of body is reddish-purple, posterior half is yellowish. Caudal margin is truncated. Dorsal fin has 3 spines and 22 rays, anal fin has 3 spines and 12 rays. Aquarium fish. Up to 8cm in total length.

Distribution — Solomon, New Hebrides, New Caledonia

Pseudochromidae
Pseudochromis porphyreus
Lubbock et Goldman ⇨

Purple Basslet (New Common Name)

Diagnosis — Inhabits shallow coral reefs. Body is elongated, beautiful reddish-purple in color. Caudal fin is truncated, transparent along margin. Feeds on zooplankton. Dorsal fin has 3 spines and 22 rays, anal fin has 3 spines and 12 rays. 21−23 pored lateral scales. Aquarium fish. Up to 5cm in total length.

Distribution — Ryukyu, West Pacific

Centropomidae

Psammoperca waigiensis

(Cuvier) ⇨

Dwarf Palmer, Glass-eyed Perch, Jewel-eye, Reef Barramundi, Sand Bass, Sand Perch, Waigeu Perch, Waigiu Sand Perch

Diagnosis — Inhabits shallow coral reefs. Body is shallow; snout is pointed. Margin of preoperculum is spineless. Lateral lines are long enough to reach caudal margin. First dorsal fin has 7 spines, second has 1 spine and 13–14 rays. Anal fin has 3 spines and 9 rays. 48–49 pored lateral scales. Edible. Up to 40cm in total length.

Distribution — Ryukyu, Indo-Pacific

Glaucosomatidae

⇩ *Glaucosoma hebraicum*

Richardson

[*G. bürgeri* Richardson]

Westralian Jewfish

Diagnosis — Inhabits rather deep waters. Body is compressed and deep. Caudal fin is truncated. Dorsal fin rays are high, dorsal fin has 8 spines and 11 rays, anal fin has 3 spines and 9–10 rays. 49–52 pored lateral scales. Edible. Up to 50cm in total length.

Distribution — Southern Japan, Ryukyu, Formosa, South China Sea, Thailand, Australia

Centropomidae

⇧ *Lates calcarifer* (Bloch)

[*Holocentrus calcarifer* Bloch]

[*Lates darwiniensis* Macleay]

Cock-up, Giant Perch

Diagnosis — Tropical, brackish fish also found in freshwater areas. Body is deep; snout is pointed; head is small. Eyes are situated at end of head. Lateral line is long enough to reach caudal margin. First dorsal fin has 7 spines, second has 1 spine and 11 rays. Anal fin has 3 spines and 8 rays. Resembles *L. japonicus*, but distinguished by shallower body, 17–18 pectoral fin rays, fewer pored lateral scales (54–57). Edible. Up to 1.5m in total length.

Distribution — Formosa, China, Indo-West Pacific, Persian Gulf

Sciaenidae

⇦ *Nibea japonica*

(Temminck et Schlegel)

Japanese Croaker

Diagnosis — Inhabits rather deep muddy-sand bottoms. Body is long and compressed. Mouth is large, lower jaw is slightly projected, has a row of teeth. Dorsal fin has 11 spines and 27 rays, anal fin has 2 spines and 6 rays. 50 lateral scales. Edible. Up to 1.8m in total length.

Distribution — Southern Japan, Formosa, China, South China Sea

Sciaenidae

⇦ ***Miichthys miiuy*** (Basilewsky)

[*M. imbricatus* (Matsubara)]

Brown Croaker, Nibe Croaker, Roncadore, Croaker

Diagnosis — Inhabits rather deep muddy-sand bottoms. Body is slightly elongated. Lower jaw is short, jaws have canines. Dorsal fin has 10 spines and 27–30 rays, anal fin has 2 spines and 7 rays. 53–56 lateral scales. Spawns between June and September. Edible. Up to 90cm in total length.

Distribution — Formosa, Korea, Yellow Sea, East China Sea

Sciaenidae

⇦ ***Nibea mitsukurii*** (Jordan et Snyder)

Blue Drum, Nibe Croaker

Diagnosis — Inhabits rather deep muddy-sand bottoms. Body is slightly elongated; snout is roundish. Jaws are equal in length, have 2 rows of teeth. Back has black oblique bands. Produces noise by vibration of bladder. Dorsal fin has 9 spines and 27–31 rays, anal fin has 2 spines and 7 rays. 60 lateral scales. Edible. Up to 80cm in total length.

Distribution — Japan, East China Sea

Sciaenidae

⇦ ***Argyrosomus argentatus*** (Houttuyn)

[*Johnius argentatus* (Houttuyn)]

Croaker, Roncadore, Silver Jewfish, White Chinese Croaker, White Croaker

Diagnosis — Inhabits rather deep muddy-sand bottoms. Body is slightly deep. Lower jaw is rather short; jaws have 2 rows of teeth. Opercle has a black blotch on top. Dorsal fin has 11 spines and 25–28 rays, anal fin has 2 spines and 7–8 rays. 43–52 lateral scales. Spawns between May and August. Produces noise by vibration of bladder. Edible. Up to 70cm in total length.

Distribution — Japan, Formosa, Korea, China, Indo-West Pacific

Sciaenidae

⇦ ***Nibea albiflora*** (Richardson)

Croaker, Roncadore, Spotted Maigre, Yellow Drum

Diagnosis — Inhabits rather deep muddy-sand bottoms. Body has dark oblique bands. Lower jaw is short, jaws have 2 rows of teeth, underside of lower jaw has 5 pores. Dorsal fin has 11–12 spines and 29–30 rays, anal fin has 2 spines and 7–8 rays. 47–53 lateral scales. Spawns between April and July. Edible. Up to 45cm in total length.

Distribution — Japan, Korea, China, East China Sea

Sciaenidae

Atrobucca nibe (Jordan et Thompson) ⇨

[*Argyrosomus nibe* (Jordan et Thompson)]

Black Chinese Croaker, Black Croaker, Black Mouth Croaker

Diagnosis — Inhabits rather deep muddy-sand bottoms. Jaws are equal in length, lower jaw has 2 rows of teeth. Mouth is black inside. Dorsal fin has 11 spines and 27–32 rays, anal fin has 2 spines and 7 rays. 49–55 lateral scales. Spawns between April and June. Edible. Up to 50cm in total length.

Distribution — Japan, Formosa, Yellow Sea, East China Sea

Sciaenidae

⇦ *Pennahia macrophthalmus* (Bleeker)

Big-eyed Croaker

Diagnosis — Inhabits rather deep muddy-sand bottoms. Body is slightly deep and silvery-white. Mouth is large, bent obliquely upward. Dorsal fin has 9–10 spines and 23–25 rays, anal fin has 2 spines and 7–8 rays. Edible. Up to 20cm in total length.

Distribution — Philippines

Sillaginidae

Sillago parvisquamis Gill ⇨

Bay Whiting, Blue Whiting

Diagnosis — Inhabits muddy-sand bottoms of bays. Larger than *S. japonica*, body is bluish. First dorsal fin has 12–13 spines, second has 1 spine and 21–23 rays with rows of black spots on fin rays. Anal fin has 2 spines and 22–23 rays. 78–82 pored lateral scales. Edible. Up to 50cm in total length.

Distribution — Japan, Formosa, Philippines

Sillaginidae

Sillago maculata Quoy et Gaimard ⇨

[*S. maculata burrus* Richardson]

Spotted Whiting, Blotchy Sillago, Diver Whiting, Smelt, Sand Borer, Spotted Asohos, Trumpeter Whiting, Winter Whiting

Diagnosis — Inhabits muddy-sand bottoms of coastal waters. Body has 2 rows of dark oblique blotches. Pectoral fin bases are black. Cheek scales are cycloid. First dorsal fin has 11–12 spines, second has 1 spine and 19–20 rays. Anal fin has 2 spines and 18–19 rays. 69–72 pored lateral scales. Edible. Up to 36cm in total length.

Distribution — Ryukyu, Formosa, Indo-West Pacific, South Africa

Sillaginidae

Sillago japonica Temminck et Schlegel ⇧ ⇨

[*S. sihama* (Forskål)]

Japanese Whiting, Silver Whiting

Diagnosis — Inhabits muddy-sand bottoms of coastal waters and bays. Scales on interorbital space and cheeks are ctenoid. First dorsal fin has 10–12 spines, second has 1 spine and 21–22 rays. Anal fin has 2 spines and 21–24 rays. 70–73 pored lateral scales. Feeds on crustacea and polychaeta. Spawns between June and September. Edible. Up to 35cm in total length.

Distribution — Japan, Formosa, Korea, China, Indo-Pacific, Persian Gulf, Red Sea, South Africa

Girellidae

Girella punctata Gray

Girella, Largescale Blackfish, Rudder Fish

Diagnosis — Inhabits shoreline reefs of coastal waters. Youngs are found in tide-pools. Scales have dark spots. Dorsal fin has 15 spines and 13 rays, anal fin has 3 spines and 12 rays. 50−56 lateral scales. Spawns between February and June. Omnivorous. Edible. Up to 55cm in total length.

Distribution — Japan, Formosa, China, East China Sea, Philippines

Girellidae

Girella melanichthys (Richardson)

Smallscale Blackfish

Diagnosis — Inhabits rocky reefs bordering open seas. Resembles *G. punctata*, but distinguished by black opercle margins, black pectoral fin bases. Caudal lobes are pointed. Scales are small. Dorsal fin has 14 spines and 14 rays, anal fin has 3 spines and 13 rays. 62−66 lateral scales. Omnivorous. Spawns between November and December. Edible. Up to 70cm in total length.

Distribution — Japan, Ryukyu, Formosa, East China Sea

Girellidae

Girella mezina Jordan et Starks

Yellowstriped Blackfish

Diagnosis — Inhabits shoreline reefs of coastal waters. Youngs are found in tide-pools, has a white band. Body is compressed. Opercles are covered with scales. Upper lip is broad. Dorsal fin has 14 spines and 14 rays, anal fin has 3 spines and 11 rays. 46−56 lateral scales. Edible. Up to 45cm in total length.

Distribution — Japan, Ryukyu, Formosa, East China Sea

Kyphosidae

⇧ ***Kyphosus bigibbus*** Lacépède

Bluefish, Chub, Grey Chub, Rudderfish, Striped Drummer, Worldwide-grey Chub

Diagnosis — Resembles *K. cinerascens*, but distinguished by shallow dorsal and anal rays. In some individuals, body is yellowish. Dorsal fin has 10−11 spines and 10−12 rays, anal fin has 3−4 spines and 10−12 rays. 51−54 pored lateral scales. Edible. Up to 75cm in total length.

Distribution — Japan, Bonin Islands, Indo-Pacific, Red Sea, South Africa

Gerreidae

Gerres abbreviatus Bleeker ⇨

Blue-backed Silver-biddy, Deep-bodied Silver-biddy, Silver-biddy, Short Silver Belly

Diagnosis — Body is deep. Pectoral fins are long. Pelvic and anal fins are yellowish. Dorsal fin has 9 spines and 10 rays, anal fin has 3 spines and 7 rays. 37−39 pored lateral scales. Edible. Up to 30cm in total length.

Distribution — Ryukyu, Indo-West Pacific

Kyphosidae

⇩ ***Kyphosus cinerascens*** (Forskål)

Rudderfish, Topsail Drummer, Ash Colored Rudderfish, Chub, Ashen Drummer, Blue Chub, Bluefish

Diagnosis — Body is oval, has many black stripes. Pectoral fins are short. Dorsal and anal fin rays are expanded. Omnivorous. Dorsal fin has 11 spines and 12 rays, anal fin has 3 spines and 11 rays. 50−56 lateral scales. Edible but has records of ciguatera poisoning. Up to 50cm in total length.

Distribution — Japan, Ryukyu, Indo-West Pacific, Hawaii, Red Sea, South Africa

Kyphosidae

⇦ ***Kyphosus lembus*** (Cuvier)

[*K. vaigiensis* (Quoy et Gaimard)]

Bluefish, Chub, Large-tailed Drummer, Brassy Chub, Lembus Rudderfish, Rudderfish

Diagnosis — Inhabits shoreline reefs. Body is oblong; snout is not pointed. Dorsal fin has 10−11 spines and 14 rays, anal fin has 3 spines and 12−13 rays. 53−55 lateral scales. Omnivorous. Spawns between May and June. Edible. Up to 70cm in total length.

Distribution — Japan, Ryukyu, China, Indo-Pacific, Hawaii, Red Sea, South Africa

Gerreidae

⇦ ***Gerres filamentosus*** Cuvier

[*G. punctatus* Cuvier]

[*Pertica filamentosa* (Cuvier)]

Long Rayed Silver Biddy, Silver Biddy, Spotted Mojarras, Spotted Silver-biddy, Threadfin Pursemouth, Thread Fin Silver Belly

Diagnosis — Body is deep, has dark cross bands. Second spine of dorsal fin extends remarkably long. Dorsal fin has 11 spines and 10 rays, anal fin has 3 spines and 7 rays. 42−47 pored lateral scales. Edible. Up to 30cm in total length.

Distribution — Southern Japan, Ryukyu, Indo-West Pacific, Persian Gulf, Red Sea, East Africa

Gerreidae

Gerres oyena (Forskål) ⇨

Black-tipped Silver-biddy, Lined Silverbiddy, Slenderspine Pursemouth, Slenderspine Mojarra

Diagnosis — Found in schools in muddy-sand bottoms of bays. Body is fusiform; tail is extended. Scales are easily removable. Pectoral fins are short. Dorsal fin has 9 spines and 10 rays, anal fin has 3 spines and 7 rays. 37–43 pored lateral scales. Edible. Up to 32cm in total length.

Distribution — Japan, Ryukyu, Indo-Pacific, Red Sea, South Africa

Gerreidae

⇦ *Gerres oblongus* (Cuvier)

Long-bodied Silver-biddy
(New Common Name)

Diagnosis — Body is oblong. Eyes are large. Pectoral fins are long. Dorsal fin has 9 spines and 10 rays, second spine is extended. Anal fin has 3 spines and 7 rays. 47–49 pored lateral scales. Edible. Up to 35cm in total length.

Distribution — Ryukyu, Indo-West Pacific

Gerreidae

⇦ *Gerres macrosoma* Bleeker

Bigmouth Mojarra, Large-bodied Silver-biddy

Diagnosis — Resembles *G. oyena*, but mouth is large, has many pored lateral scales (40–44). Pectoral fins are short. Dorsal fin has 9 spines and 10 rays, anal fin has 3 spines and 7 rays. Edible. Up to 25cm in total length.

Distribution — Ryukyu, Formosa, Indo-Pacific

Gerreidae

Gerres japonicus Bleeker ⇨

Japanese Mojarra

Diagnosis — Body is oblong, somewhat deep. Dorsal fin has 10 spines and 9 rays, anal fin has 3 spines and 7 rays. 38–42 pored lateral scales. Edible. Up to 25cm in total length.

Distribution — Japan, Ryukyu, Formosa, China, Philippines

Gerreidae

Gerres sp. ⇨

Striped Silver Biddy

Diagnosis — Resembles *G. japonicus*, but distinguished by indistinct dark bands on body. Dorsal fin has 9 spines and 9 rays, second dorsal spine is extended. Anal fin has 3 spines and 7 rays. 39 lateral scales. Edible. Up to 25cm in total length.

Distribution — Ryukyu, Formosa, South China Sea

Gerreidae

⇦ *Pentaprion longimanus* (Cantor)

Taiwan Silver-biddy

(New Common Name)

Diagnosis — Body is oblong and rosy-silvery. Snout is pointed with high ridged top. Dorsal fin has 9–10 spines and 14–15 rays, spines are extended. Anal fin has 5–6 spines and 12–13 rays. 44–48 lateral scales. Edible. Up to 18cm in total length.

Distribution — Ryukyu, Formosa, Indo-West Pacific

Nemipteridae

Nemipterus marginatus (Valenciennes) ⇨ ⇩

Palefinned Threadfin Bream

Diagnosis — Resembles *N. peronii*, but distinguished by rounded snout, 2 yellow lines on body. Upper jaw has canines. Upper lobe of caudal fin is yellow and pointed. Dorsal fin has 10 spines and 9 rays, dorsal margin is yellow. Anal fin has 3 spines and 7 rays. 46–48 lateral scales. Edible. Up to 30cm in total length.

Distribution — Ryukyu, South China Sea, Indo-Pacific

Nemipteridae

Nemipterus virgatus (Houttuyn) ⇨

[*Euthyopteroma virgatum* (Houttuyn)]

[*Sparus sinensis* Lacépède]

Threadfin Bream, Besugo, Golden Thread, Gold Thread, Red Coat

Diagnosis — Inhabits rather deep muddy bottoms. Body is fusiform, somewhat deep, has 6–8 yellow bands. Origin of lateral line has a red blotch. Upper lobe of caudal fin is yellow, extends like thread. Dorsal fin has 10 spines and 9 rays, margin is yellow. Anal fin has 3 spines and 8 rays. 47–48 lateral scales. Carnivorous. Spawns between May and August. Edible. Up to 40cm in total length.

Distribution — Japan, Formosa, China, Korea, East China Sea, Hong Kong

Nemipteridae

Nemipterus tolu (Valenciennes) ⇨

[*N. mulloides* (Bleeker)]

[*Odontoglyphis tolu* (Valenciennes)]

Notched Butterfly-bream, Notched Threadfin-bream

Diagnosis — Inhabits muddy-sand bottoms of shoreline reefs. Body is fusiform. Resembles *N. peronii*, but distinguished by deeply-notched membrane of dorsal spines. Caudal lobes are not extended. Dorsal fin has 10 spines and 9 rays, anal fin has 3 spines and 7 rays. 47–50 lateral scales. Edible. Up to 30cm in total length.

Distribution — Ryukyu, Formosa, China, Indo-West Pacific, Persian Gulf, Red Sea

Nemipteridae

⇦ *Nemipterus metopias* (Bleeker)

Slender Threadfin Bream (New Common Name)

Diagnosis — Body is elongated and light-red, has indistinct stripes. Head is rounded. Upper lobe of caudal fin is extended. A yellow band below eyes. Dorsal fin has 10 spines, 9 rays and a yellow stripe. Anal fin has 3 spines and 7 rays, is lined with yellow spots. 47–48 lateral scales. Edible. Up to 25cm in total length.

Distribution — Ryukyu, South East Asia

Nemipteridae

⇦ *Nemipterus bathybius* (Snyder)

[*Euthyopteroma bathybium* (Snyder)]

Bottom Threadfin-bream

Diagnosis — Body is oblong, red, has 3 yellow bands. Tail is long. Upper lobe of caudal fin is yellow and extended. Dorsal fin has 10 spines and 9 rays, anal fin has 3 spines and 7 rays. 46–47 lateral scales. Edible. Up to 40cm in total length.

Distribution — Japan, Formosa, Korea, China, South China Sea, India

Nemipteridae

⇦ *Nemipterus peronii* (Valenciennes) ⇧

[*N. personii* (Cuvier et Valenciennes)]

Person's Butterfly Bream, Redspot Threadfin-bream

Diagnosis — Inhabits muddy-sand bottoms of coastal waters. Body is fusiform; light-red, has no stripes; tail extends long. Origin of lateral line has a red blotch. Caudal fin is falcated, lobes do not extend. Dorsal fin has 10 spines and 9 rays, anal fin has 3 spines and 7 rays. 49–51 lateral scales. Edible. Up to 35cm in total length.

Distribution — Ryukyu, Formosa, Indo-West Pacific

Nemipteridae

Nemipterus japonicus (Bloch) ⇨

Japanese Threadfin Bream

Diagnosis — Inhabits rather deep waters. Body is compressed, somewhat deep, has 10 yellow lines. Caudal lobes extend long. Anterior part of lateral lines have a red blotch. Dorsal fin has 10 spines and 8–9 rays, anal fin has 3 spines and 7 rays. 44–45 lateral scales. Edible. Up to 25cm in total length.

Distribution — Formosa, South China Sea, Indian Ocean

Nemipteridae

⇦ *Scolopsis vosmeri* (Bloch)

Monocle Bream, Silverflash Spinecheek, Whitecheek,

Diagnosis — Body is oval, deep, purplish-brown. Opercles are crossed by broad white bands. Dorsal fin has 10 spines and 8–9 rays, is yellow. Anal fin has 3 spines and 7 rays. 40–42 lateral scales. Edible. Up to 25cm in total length.

Distribution — Ryukyu, Formosa, China, Indo-Pacific, Red Sea, South Africa

Nemipteridae

Scolopsis personatus (Cuvier) ⇨

Blue-nosed Monocle-bream, Masked Spine-cheek

Diagnosis — Inhabits muddy-sand bottoms of bays. Body is oblong; snout is pointed. Eyes are large, a yellowish-white line runs from snout to eyes. Dorsal fin has 10 spines and 9 rays, anal fin has 3 spines and 7 rays. 41–44 lateral scales. Edible. Up to 20cm in total length.

Distribution — Ryukyu, Indo-Pacific

Nemipteridae

⇦ *Scolopsis eriomma* (Jordan et Richardson)

[*Parascolopsis eriomma* (Jordan et Richardson)]

Rosy Monocle Bream, Shimmering Spinecheek

Diagnosis — Inhabits shoreline reefs. Body is oblong; tail extends somewhat long; back is red except yellow middle part. Dorsal and caudal fins are red. Dorsal fin has 10 spines and 9 rays, anal fin has 3 spines and 7 rays. 36 lateral scales. Edible. Up to 30cm in total length.

Distribution — Southern Japan, Ryukyu, Formosa, Indo-Pacific, South Africa

Nemipteridae

⇦⇩ *Scolopsis inermis* (Temminck et Schlegel)

Redbelt Monoclebream

Diagnosis — Inhabits shoreline reefs. Body is oblong, has 4 reddish-brown bands. Snout is pointed. Pectoral fins are yellow. Dorsal fin has 10 spines and 9 rays, anal fin has 3 spines and 7 rays. 34–36 lateral scales. Edible. Up to 35cm in total length.

Distribution — Japan, Ryukyu, Indo-West Pacific

Nemipteridae

⇩ *Pentapodus macrurus* (Bleeker)

Yellowstriped Whiptail

Diagnosis — Inhabits outer slope of coral reefs. Body is fusiform, has 2 yellow bands; snout is pointed. Caudal lobes extend long. Jaws have canines. Dorsal fin has 10 spines and 9 rays, anal fin has 3 spines and 7 rays. 45–47 lateral scales. Edible. Up to 35cm in total length.

Distribution — Ryukyu, South East Asia, Indian Ocean

Nemipteridae

Scolopsis ciliatus (Lacépède) ⇩

Saw-jawed Monocle-bream

Diagnosis — Inhabits shallow coral reefs. Body is oval; back has a white band. Upper jaw is indented. Dorsal fin has 10 spines and 8–9 rays, anal fin has 3 spines and 7–8 rays. 40 lateral scales. Edible. Up to 20cm in total length.

Distribution — Ryukyu, Indo-Pacific

Nemipteridae

⇧ *Pentapodus microdon* (Bleeker)

Small-toothed Threadfin-bream, Smalltooth Whiptail

Diagnosis — Resembles *P. macrurus*, but distinguished by a yellow band running from eyes. Caudal lobes do not extend, lower lobe is reddish. Upper jaw has canines. Edible. Up to 20cm in total length.

Distribution — Ryukyu, Indonesia, Melanesia

Nemipteridae

Scolopsis cancellatus (Cuvier) ⇩

Green-lined Spine-cheek, Latticed Monocle-bream, Monocle-bream, Whitestriped Monoclebream

Diagnosis — Inhabits shallow coral reefs. Body is oblong. Eyes are large, upper surfaces are swollen. Back has 3–4 dark bands. Dorsal and caudal fins are dark-red. Dorsal fin has 10 spines and 9 rays, anal fin has 3 spines and 7 rays. 41–44 lateral scales. Edible. Up to 25cm in total length.

Distribution — Ryukyu, Indo-Pacific

Nemipteridae

Scolopsis dubiosus Weber ⇨ ⇩

Yellow-streaked Monocle-bream

Diagnosis — Inhabits muddy-sand bottoms of bays. Body is oblong, has dark band-like blotches. Snout is pointed. Around eyes are blue and yellow-striped. Caudal fin is yellow, lobes are pointed. Dorsal fin has 10 spines and 9 rays, anal fin has 3 spines and 7 rays. 42−45 lateral scales. Edible. Up to 40cm in total length.

Distribution — Ryukyu, Formosa, Indo-Pacific

Nemipteridae

Pentapodus nagasakiensis (Tanaka) ⇦ ⇧

Nagasaki Whiptail

Diagnosis — Body is elongated, has a yellow band; snout is pointed. Upper jaw has a pair of canines. Dorsal fin has 10 spines and 9 rays, anal fin has 3 spines and 7 rays. 44−45 lateral scales. Edible. Up to 25cm in total length.

Distribution — Japan, Ryukyu

Nemipteridae

Scolopsis bilineatus (Bloch) ⇨ ⇩

Bridled Spine-cheek, Monocle Bream, Two-line Monocle-bream, Two-striped Scolopsis, Yellow-finned Spine-cheek, Yellowfin Spinecheek

Diagnosis — Inhabits shoreline reefs and coral reefs. Body is oval; snout is pointed. A black-margined white band runs from snout to back. Anterior part of dorsal fin is yellow, posterior part is black. Anterior part of anal fin is also black. Dorsal fin has 10 spines and 9 rays, anal fin has 3 spines and 7 rays. 44−46 lateral scales. Edible. Up to 25cm in total length.

Distribution — Southern Japan, Ryukyu, Indo-Pacific

Sparidae

Evynnis cardinalis (Lacépède) ⇧

Crimson Sea-bream, Godlen-skinned Porgy, Red-fin Pargo, Threadfin Porgy

Diagnosis — Body is compressed and deep. Third and fourth dorsal spines extend like thread. Pectoral fin bases and margin of opercles are red. Dorsal fin has 12 spines and 10 rays, anal fin has 3 spines and 9 rays. 58–64 lateral scales. Edible. Up to 40cm in total length.

Distribution — Southern Japan, Ryukyu, Formosa, East China Sea

Sparidae

⇦ *Argyrops bleekeri* Oshima ⇧

Bowen Snapper, Fryingpan Snapper, Longspined Red Bream, Long-spined Snapper, Redfin, Steamboat Bream

Diagnosis — Body is compressed, remarkably deep. Dorsal fin spines extend like thread. In young, body has 6 brown bands. Dorsal fin has 11 spines and 10 rays, anal fin has 3 spines and 8 rays. 50–53 lateral scales. Edible. Up to 50cm in total length.

Distribution — Southern Japan, Ryukyu, Formosa, South China Sea

Sparidae

Acanthopagrus latus (Houttuyn) ⇨

Black Sea-bream, Yellowfin Bream, Japanese Silver Bream

Diagnosis — Inhabits bays and river mouths. Resembles *A. schlegeli*, but distinguished by grayish-brown body, yellow anal fin, pelvic fin and lower caudal lobe. Scales above lateral line are in 4 rows. Jaws have molar teeth. Dorsal fin has 11 spines and 11 rays, anal fin has 3 spines and 8 rays. 43–48 lateral scales. Spawns between September and November. Transsexual. Edible. Up to 50cm in total length.

Distribution — Southern Japan, Formosa, Indo-West Pacific, Persian Gulf, Red Sea, East Africa

Sparidae

Evynnis japonica Tanaka ⇧ ⇨

Crimson Sea Bream

Diagnosis — Inhabits waters more than 100m deep. Resembles *Pagrus major*, but distinguished by long third and fourth dorsal spines, red opercular membranes and dorsal spine membranes. Body is profusely dotted with blue. Caudal margin is not black. Jaws have molar teeth. Dorsal fin has 12 spines and 10 rays, anal fin has 3 spines and 9 rays. 54–61 lateral scales. Spawns between fall and winter. Edible. Up to 45cm in total length.

Distribution — Japan, Korea, Formosa, East China Sea

▼ Mouth has molar teeth.

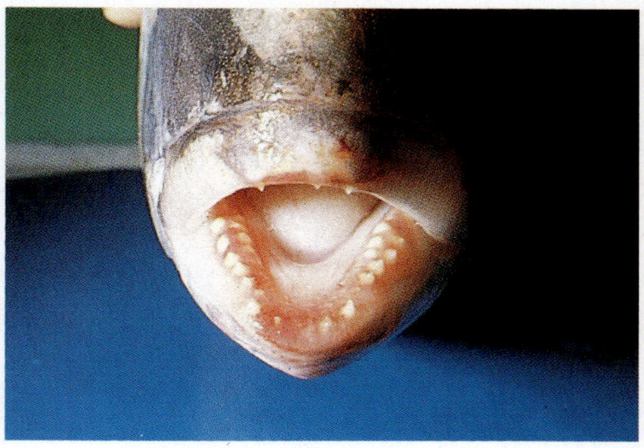

▼ Cycloid scale of Sea Bream.

Sparidae

Pagrus major (Temminck et Schlegel)

Genuin Porgy, Red Pargo, Red Porgy, Red Sea Bream, Red Tai, Red Sea Bream Snapper, Sea Bream, Snapper Sea Bream

Diagnosis — Inhabits coastal waters, also found in rather deep rocky reefs and sand gravel bottoms. Body is compressed and deep, profusely dotted with small blue. Tail extends somewhat long. Caudal margin is black. Jaws have molar teeth. Dorsal fin has 12 spines and 10 rays, anal fin has 3 spines and 8 rays. 53–59 lateral scales. Spawns between March and June. Grows fully in 3 years, lives as long as 50 years. Carnivorous. Up to 1.1m in total length.

Distribution — Japan, Ryukyu, Formosa, Korea, China, Indo-Pacific

Sparidae

Dentex sp.

Yellowfin Red Sea-bream

Diagnosis — Inhabits bottoms about 100m deep. Resembles *D. tumifrons*, but distinguished by remarkably yellowish body, yellow dorsal and anal fins. Body has small blue spots. Interorbital space is covered with scales. Dorsal fin has 12 spines and 10 rays, anal fin has 3 spines and 8 rays. 48–50 lateral scales. Edible. Up to 40cm in total length.

Distribution — Ryukyu

Sparidae

Dentex tumifrons (Temminck et Schlegel)

[*Taius tumifrons* (Temminck et Schlegel)]

Golden Tai, Red Sea-bream, Yellow Porgy, Yellow Sea Bream

Diagnosis — Inhabits muddy-sand bottoms more than 100m deep. Resembles *Pagrus major*, but distinguished by deep remarkably reddish body. Snout and back have yellow blotches. Pectoral fins are red. Jaws have canines and conical teeth. Dorsal fin has 12 spines and 10 rays, anal fin has 3 spines and 8 rays. 46–50 lateral scales. Spawns between summer and fall. Transsexual. Edible. Up to 40cm in total length.

Distribution — Japan, Formosa, Korea, China, East China Sea

Sparidae

Acanthopagrus schlegeli (Bleeker)

Black Sea Bream, Gilthead, Black Porgy

Diagnosis — Inhabits muddy-sand bottoms of bays. Young are also found in freshwater areas. Body is oval; snout is pointed. Scales above lateral lines are in 6−7 rows. Jaws have canines and molars. Omnivorous. Transsexual. Spawns between April and June. Dorsal fin has 11−12 spines and 11 rays, anal fin has 3 spines and 8 rays. 48−56 lateral scales. Edible but has records of Ichthyootoxic poisoning. Up to 55cm in total length.

Distribution — Japan, Formosa, Korea, China

Sparidae

Acanthopagrus australis (Günther)

Australian Black Porgy (New Common Name)

Diagnosis — Resembles *A. schlegeli*, but distinguished by deep body, straight margin of head. Scales above lateral lines are in 5 rows. Pectoral, pelvic and anal fins are yellowish. Anal spines are long. Dorsal fin is black with 11 spines and 11 rays, anal fin has 3 spines and 8 rays. 44−47 lateral scales. Edible. Up to 55cm in total length.

Distribution — Ryukyu, West Pacific

Sparidae

Acanthopagrus berda (Forskål)

Black Bream, Mud Bream, Picnic Bream, Pikey Bream, River Bream, River Perch, Sly Bream, Yellowfin Sea Bream

Diagnosis — Inhabits coastal waters and bays. Resembles *A. schlegeli*, but distinguished by deep body, 4 rows of scales above lateral lines. Pelvic and anal fins are dark. Jaws have molar teeth. Dorsal fin has 11 spines and 11 rays, anal fin has 3 spines and 8 rays. 43−52 lateral scales. Edible. Up to 50cm in total length.

Distribution — Ryukyu, Formosa, Indo-West Pacific, Red Sea, East Africa

Sparidae

Acanthopagrus sivicolus Akazaki

River Bream, Mud Bream, Sly Bream, Yellowfin Sea Bream, Goldsilk Sea Bream, Pickey Bream, Picnic Bream, Black Bream

Diagnosis — Inhabits bays and river mouths. Resembles *A. schlegeli*, but body is somewhat deep, eyes are large. Opercular margins are black. Scales above lateral lines are in 5 rows. Pelvic and anal fins are dark. Jaws have molars. Dorsal fin has 11 spines and 11 rays, anal fin has 3 spines and 8 rays. 46−52 lateral scales. Good for breeding. Spawns between May and June. Edible but has records of Ichthyootoxic poisoning. Up to 50cm in total length.

Distribution — Southern Japan, Ryukyu

Sparidae

Sparus sarba Forskål

[*Rhabdosargus sarba* (Forskål)]

Silver Bream, Flatbream, River Stumpnose, Yellowfin Bream, Natal Stumpnose

Diagnosis — Inhabits about shoreline reefs, also found in bays. Body is compressed, silvery-white and deep. Snout is yellow and rounded. Jaws have molar teeth and canines. Pectoral, pelvic, anal fins are yellowish. Dorsal fin has 11 spines and 13 rays, anal fin has 3 spines and 11 rays. 53–63 lateral scales. Edible. Eggs are poisonous. Up to 46cm in total length.

Distribution — Japan, Ryukyu, Formosa, China, Korea, Indo-West Pacific, Red Sea, South Africa

Banjosidae

Banjos banjos (Richardson)

Banjofish

Diagnosis — Inhabits rather deep muddy-sand bottoms. Body is deep, light-brown, has indistinct dark bands. Dorsal spines are strong and long. Dorsal fin ray has a black blotch. Caudal margin is white. Dorsal fin has 10 spines and 11 rays, anal fin has 3 spines and 7 rays. 48 lateral scales. Edible. Up to 30cm in total length.

Distribution — Japan, Ryukyu, Formosa, China, East China Sea

Lethrinidae

Lethrinus lentjan (Lacépède)

[*L. lentjanus* (Lacépède)]

Spangled Emperor, Redspot Emperor

Diagnosis — Resembles *L. nebulosus*, but distinguished by somewhat smaller body, white spots on scales. Opercular margins and pectoral fin bases are red. In young, upper part of pectoral fins have a black blotch. Dorsal fin has 10 spines and 9 rays, anal fin has 3 spines and 8 rays. 47–48 lateral scales. Edible but has records of ciguatera poisoning. Up to 50cm in total length.

Distribution — Southern Japan, Ryukyu, Formosa, China, Indo-Pacific, Hawaii, Red Sea, East Africa

Lethrinidae

Lethrinus kallopterus Bleeker

Yellow-spotted Emperor

Diagnosis — Body is oblong; snout is rounded. Fins are red; caudal margin is rounded. Dorsal fin has 10 spines and 9 rays, anal fin has 3 spines and 8 rays. 47–48 lateral scales. Edible. Up to 80cm in total length.

Distribution — Southern Japan, Ryukyu, Indo-Pacific, Persian Gulf, East Africa

Lethrinidae

Lethrinus harak (Forskål)

[*L. rhodopterus* Bleeker]

Blackspot Emperor, Rose-finned Pigface Bream

Diagnosis — Body is somewhat deep, yellowish-brown with a large dark blotch on middle surface. Caudal fin is reddish and pointed. Dorsal fin has 10 spines and 9 rays, anal fin has 3 spines and 8 rays. 46–48 lateral scales. Edible but has records of ciguatera poisoning. Up to 60cm in total length.

Distribution — Southern Japan, Ryukyu, Indo-Pacific, Hawaii, Red Sea, South Africa

Lethrinidae

Lethrinus nebulosus (Forskål)

[*L. choerorhynchus* (Schneider)]

Blue Emperor, Green Snapper, Morwong, Parrot-fish, Pearl-spotted Porgy, Pearly Lentjan, Pig-face Bream, Sand Bream, Scavenger, Snuppy, Spangled Emperor

Diagnosis — Body is compressed, somewhat deep. Snout is pointed. Pectoral fins have scales inside. Scales have whitish-blue spots. Teeth are molar, mouth is red inside. Dorsal fin has 10 spines and 9 rays, anal fin has 3 spines and 8 rays. 45–48 lateral scales. Good for breeding. Edible. Up to 1m in total length.

Distribution — Japan, Ryukyu, Formosa, Korea, Indo-Pacific, Persian Gulf, Red Sea, South Africa

Lethrinidae

Lethrinus ornatus Valenciennes

Ornate Emperor, Ornate Pigface Bream, Pig-face Bream, Yellow-striped Emperor

Diagnosis — Body is compressed and deep with several yellow bands. Surface around eyes and opercular margins are red. Caudal fin is reddish and pointed. Pectoral and anal fins are yellow. Jaws have molars. Dorsal fin has 10 spines and 9 rays, anal fin has 3 spines and 8 rays. 46–47 lateral scales. Edible. Up to 45cm in total length.

Distribution — Southern Japan, Ryukyu, Formosa, Indo-Pacific

Lethrinidae

Lethrinus haematopterus Temminck et Schlegel

Adjutant, Common Lentjan, Japanese Scavenger

Diagnosis — Body is compressed and deep; snout is projected; head has no scales. In some individuals, opercular margins are red. Dorsal fin has 10 spines and 9 rays, anal fin has 3 spines and 8 rays. 47–48 lateral scales. Edible. Up to 1m in total length.

Distribution — Japan, Ryukyu, Pacific

Lethrinidae

Lethrinus mahsena (Forskål)

[*L. mahsenoides* Valenciennes]

White Spotted Pigface Bream, Yellow-tailed Emperor, Striped-snout Porgy, Tricky Snapper, Yellow Morwong

Diagnosis — Body is compressed and deep. Jaws have molars; snout is pointed; mouth is red inside. Scales have dark spots. Fins are reddish. Dorsal fin has 10 spines and 9 rays, anal fin has 3 spines and 8 rays. 46–48 lateral scales. Edible but has records of ciguatera poisoning. Up to 50cm in total length.

Distribution — Japan, Ryukyu, Indo-Pacific, Red Sea, East Africa

Lethrinidae

⇧ ***Lethrinus*** sp.

Diagonal Striped Snapper (New Common Name)
Diagnosis — Inhabits rather deep rocky reefs. Body is compressed and deep. Back has some 8 dark oblique bands. Edible. Up to 60cm in total length.
Distribution — Gulf of California

Lethrinidae

⇩ ***Lethrinus semicinctus*** Valenciennes

Reticulated Emperor (New Common Name)
Diagnosis — Body is somewhat shallow, has irregular dark blotches. Snout is pointed. Upper parts of eyes are protruded. Caudal fin is reddish, lobes are pointed. Dorsal fin has 10 spines and 9 rays, anal fin has 3 spines and 8 rays. 46–49 lateral scales. Edible. Up to 35cm in total length.
Distribution — Ryukyu, Formosa, West Pacific

Lethrinidae

⇦ ***Lethrinus microdon*** Valenciennes

Small-toothed Emperor, Small-toothed Pigface Bream
Diagnosis — Body is somewhat deep, greenish-brown with no red spots. Lips are white. 2–3 stripes run from eyes. Dorsal fin has 10 spines and 9 rays, anal fin has 3 spines and 8 rays. 47–49 lateral scales. Edible. Up to 60cm in total length.
Distribution — Southern Japan, Ryukyu, Formosa, Indo-West Pacific, Red Sea

Lethrinidae

Lethrinus amamianus Akazaki ⇨

Amami Emperor (New Common Name)
Diagnosis — Resembles *L. nebulosus*, but distinguished by brown body, black-spotted scales. Lips and pectoral fin bases are red. Caudal margin is rounded. Dorsal fin has 10 spines and 9 rays, anal fin has 3 spines and 8 rays. 45–49 lateral scales. Edible. Up to 70cm in total length.
Distribution — Southern Japan, Ryukyu, West Pacific

Lethrinidae

⇦ ***Wattsia mosambica*** (Smith)

[*Gnathodentex mossambicus* Smith]

Mozambique Large-eye Bream
Diagnosis — Body is oval. Fins are yellow. Caudal margin is rounded. Dorsal fin has 10 spines and 10 rays, anal fin has 3 spines and 10 rays, spines are stiff. 42–45 lateral scales. Edible. Up to 55cm in total length.
Distribution — Southern Japan, Ryukyu, South East Asia, South Africa

Lethrinidae

Lethrinus miniatus (Schneider)

[*L. chrysostomus* Richardson]

[*Lethrinella miniata* (Schneider)]

Long Snouted Bream, Long-snouted Porgy, Longnose Emperor, Long-nosed Pigface Bream, Emperor Bream, Scavenger

Diagnosis — Body is somewhat deep; snout extends long. Lips are red and thick; cheeks have no scales. Dorsal fin has 10 spines and 9 rays, anal fin has 3 spines and 8 rays. 47–48 lateral scales. Carnivorous. Edible but has records of ciguatera poisoning. Up to 1m in total length.

Distribution — Southern Japan, Ryukyu, China, Indo-Pacific, Red Sea, South Africa

Lethrinidae

Lethrinus rubrioperculatus Sato

[*L. variegatus* Cuvier et Valenciennes]

Variegated Emperor

Diagnosis — Inhabits shoreline reefs and coral reefs. Body is compressed, somewhat shallow. Snout is projected, lips are red. Opercular margins have a red blotch. Pectoral fins are long enough to reach anus. Dorsal fin has 10 spines and 9 rays, anal fin has 3 spines and 8 rays. 46–48 lateral scales. Carnivorous. Edible but has records of ciguatera poisoning. Up to 60cm in total length.

Distribution — Japan, Ryukyu, Indo-Pacific, East Africa

Lethrinidae

Lethrinus xanthochilus Klunzinger

[*L. variegatus* Valenciennes]

Snapper, Yellowlip Emperor

Diagnosis — Body is somewhat deep. Snout is pointed. Pectoral fin bases and posterior margins of opercles are red. Surface between upper lip and snout is red. Dorsal fin has 10 spines and 9 rays, anal fin has 3 spines and 8 rays. 47–48 lateral scales. Edible but has records of ciguatera poisoning. Up to 80cm in total length.

Distribution — Ryukyu, Indo-Pacific, Red Sea, East Africa

Lethrinidae

⇦ ***Lethrinus nematacanthus*** Bleeker ⇩

Lancer, Thread-fin Emperor

Diagnosis — Body is shallow with irregualr dark bands and orange stripes. Upper part of pectoral fins have a black blotch. Dorsal spines extend long. Dorsal fin has 10 spines and 9 rays, anal fin has 3 spines and 8 rays. 46−47 lateral scales. Edible. Up to 25cm in total length.

Distribution — Japan, Ryukyu, Indo-Pacific

Lethrinidae

⇦ ***Lethrinus amboinensis*** Bleeker

Amboina Porgy

Diagnosis — Inhabits coral reefs. Remebles *L. rubrioperculatus*, but distinguished by absence of a red spot on opercular margins and pectoral fin bases, no radial stripes run from eyes. Dorsal fin has 10 spines and 9 rays, anal fin has 3 spines and 8 rays. 46−49 lateral scales. Edible. Up to 70cm in total length.

Distribution — Ryukyu, Pacific

Lethrinidae

Gnathodentex aurolineatus (Lacépède) ⇩⇘

Goldlined Large-eye Bream, Glowfish, Gold-lined Large-eyed Bream, Gold-lined Sea-bream

Diagnosis — Found in schools about coral reefs. Body is oblong, has brown stripes. Snout is red and pointed. Eyes are large. Dorsal, anal, caudal fins are reddish. Dorsal fin has a white spot on lower surface of posterior end. Upper jaw has a row of serrated teeth. Dorsal fin has 10 spines and 10 rays, anal fin has 3 spines and 9 rays. 60−70 lateral scales. Edible but has records of ciguatera poisoning. Up to 45cm in total length.

Distribution — Japan, Ryukyu, Indo-Pacific, South Africa

Lethrinidae

Gymnocranius japonicus Akazaki ⇨ ⇩

Drab Large-eye Bream

Diagnosis — Inhabits rather deep rocky reefs. Body is oblong, snout is pointed. Scales have a dark spot. Caudal fin is slightly emarginated with red margin. Jaws have molars. Dorsal fin has 10 spines and 10 rays, anal fin has 3 spines and 10 rays. 49–50 lateral scales. Edible. Up to 50cm in total length.

Distribution — Southern Japan, Ryukyu, South China Sea

Lethrinidae

Gymnocranius microdon (Bleeker) ⇨

Long-bodied Sea-bream (New Common Name)

Diagnosis — Body is oblong, slightly elongated. Caudal lobes are pointed. Head is closely covered with small blue spots. Dorsal, anal, caudal fin margins are reddish. Dorsal fin has 10 spines and 10 rays, anal fin has 3 spines and 10 rays. 48–49 lateral scales. Edible. Up to 50cm in total length.

Distribution — Southern Japan, Ryukyu, Formosa, West Pacific

Lethrinidae

⇦ *Gymnocranius elongatus* Senta

Long-tailed Naked-head Seabream

Diagnosis — Body is oblong. Caudal fin is deeply falcated, margin is red. Dorsal fin has 10 spines and 10 rays, anal fin has 3 spines and 10 rays. 47–48 lateral scales. Edible. Up to 30cm in total length.

Distribution — Southern Japan, Ryukyu, Indo-Pacific

Lethrinidae

Gymnocranius lethrinoides (Bleeker) ⇨

Silver Large-eye Bream

Diagnosis — Body is oval and roundish. Caudal margin is rounded. Dorsal, caudal, anal fins are red. Dorsal fin has 10 spines and 10 rays, anal fin has 3 spines and 10 rays. 48 lateral scales. Edible. Up to 50cm in total length.

Distribution — Southern Japan, Ryukyu, West Pacific

Lethrinidae

⇦ ***Gymnocranius robinsoni*** (Gilchrist et Thompson) ⇩

Bambu Sole, Hookmouth, Rippled Large-eye Bream, Scavenger

Diagnosis — Body is oblong and compressed. Blue wavy lines run from snout to cheeks. Caudal lobes are pointed. Dorsal fin has 10 spines and 10 rays, anal fin has 3 spines and 10 rays. 45–49 lateral scales. Edible. Up to 80cm in total length.

Distribution — Southern Japan, Ryukyu, Indo-Pacific, Red Sea, East Africa

◄ Adult

▲ Young Adult

Young ►

Lethrinidae

⇧ ***Monotaxis grandoculis*** (Forskål) ⇨

Roundtooth Large-eye Bream, Big-eye, Bigeye Barenose, Porgy, Large-eyed Seabream, Roundtoothed Large-eye Bream

Diagnosis — Inhabits coral reefs. Body is oval; head bulges out. Eyes are large, has 4 rows of scales on lower surface. Lips are thick; jaws have molars. Dorsal fin has 10 spines and 10 rays, spines and rays can be folded back into groove. Anal fin has 3 spines and 9 rays. 44–47 lateral scales. In young, body has broad cross bands. Edible but has records of ciguatera poisoning. Up to 75cm in total length.

Distribution — Southern Japan, Ryukyu, Indo-Pacific, Persian Gulf, Red Sea, South Africa

Lethrinidae

Gymnocranius griseus (Temminck et Schlegel) ⇩⇨

Wise Porgy, Nake-headed Sea Bream, Scavenger, Ginkgo Fish, Grey Barenose, Large-eyed Bream, Meichi Porgy, Naked-headed Large-eyed Bream

Diagnosis — Inhabits rather deep rocky reefs. Body is oval, has irregular brown blotches. Eyes are crossed by brown band. Caudal lobes are pointed and reddish. Dorsal fin has 10 spines and 10 rays, anal fin has 3 spines and 10 rays. 46–48 lateral scales. Edible but has records of ciguatera poisoning. Up to 50cm in total length.

Distribution — Japan, Ryukyu, Formosa, China, Indo-Pacific, South Africa

▲ Young

Lutjanidae

⇦ *Lutjanus carponotatus* (Richardson)

Striped Sea Perch

Diagnosis — Found in schools about outer slopes of coral reefs. Body is somewhat deep with 6–8 yellow stripes. Fins are yellow. Dorsal fin has 10 spines and 15 rays, anal fin has 3 spines and 10 rays. 68 lateral scales. Edible. Up to 40cm in total length.

Distribution — Fiji, Australia

Lutjanidae

⇦ *Lutjanus vitta* (Quoy et Gaimard)

Blackstriped Snapper, Brown-striped Snapper, Olive-striped Snapper, One-band Sea-perch, Striped Sea Perch

Diagnosis — Body is fusiform and compressed. A longitudinal black band through eyes has a black blotch on middle surface. Interorbital space has scales. Dorsal fin has 10 spines and 13 rays, anal fin has 3 spines and 8 rays. 49–52 lateral scales. Edible but has records of ciguatera poisoning. Up to 40cm in total length.

Distribution — Japan, Ryukyu, Indo-Pacific, Red Sea, East Africa

Lutjanidae

Lutjanus lutjanus Bloch ⇨

Madras Snapper, Rosy Sea-perch, Rosy Snapper

Diagnosis — Inhabits coral reefs. Resembles *L. vittata*, but distinguished by slightly deeper body, reddish head. A longitudinal band crossing eyes is narrow and pale. Dorsal scale rows are oblique, lower scale rows are longitudinal. Dorsal fin has 10 spines and 13–14 rays, anal fin has 3 spines and 8 rays. 47–49 lateral scales. Edible. Up to 30cm in total length.

Distribution — Japan, Ryukyu, Indo-West Pacific

Lutjanidae

⇦ *Lutjanus monostigma* (Cuvier)

[*Mesoprion monostigma* Cuvier et Valenciennes]

Onespot Snapper, Red Snapper, One-spot Sea-perch

Diagnosis — Inhabits coral reefs. Resembles *L. russelli*, but distinguished by blackish, reddish-brown body, a smaller black blotch on lateral lines. Teeth are absent on tongue. Dorsal fin has 10 spines and 12–13 rays, anal fin has 3 spines and 8 rays. 50–55 lateral scales. Edible but has records of ciguatera poisoning. Up to 60cm in total length.

Distribution — Ryukyu, Formosa, Indo-West Pacific, Red Sea

Lutjanidae

Lutjanus rufolineatus (Valenciennes) ⇨

Rufous Sea-perch, Sevenstripe Snapper

Diagnosis — Resembles *L. caeruleovittatus*, but distinguished by orange-red body with 7 yellow stripes. Dorsal fin has 11 spines and 13 rays, anal fin has 3 spines and 8 rays. 50 lateral scales. Edible. Up to 30cm in total length.

Distribution — Southern Japan, Ryukyu, China, Indo-Pacific

Lutjanidae

Lutjanus kasmira (Forskål)

Blue-lined Snapper Fish, Blue and Yellow Snapper, Blue-banded Hussar, Blue-banded Sea Perch, Blue Banded Snapper, Moon Lighter, Bluestripe Snapper

Diagnosis — Inhabits shoreline reefs and coral reefs. Body is somewhat deep, has 4 blue stripes. Snout is projected. Dorsal and caudal margins are black. Dorsal fin has 10–11 spines and 14–15 rays, anal fin has 3 spines and 8 rays. 47–51 lateral scales. Edible but has records of ciguatera poisoning. Up to 40cm in total length.

Distribution — Japan, Ryukyu, Indo-Pacific, Hawaii, Persian Gulf, Red Sea, South Africa

Lutjanidae

Lutjanus spilurus (Bennett)

Blue-striped Snapper

Diagnosis — Found in schools about shoreline reefs and coral reefs. Resembles *L. kasmirà*, but distinguished by 5 blue stripes on body, 6 blue stripes on head. Dorsal and caudal margins are not black. Lateral lines have an indistinct dark blotch. Dorsal fin has 10 spines and 13–14 rays, anal fin has 3 spines and 8 rays. 47–49 lateral scales. Edible. Up to 30cm in total length.

Distribution — Ryukyu, Indo-West Pacific, Red Sea, South Africa

Lutjanidae

Lutjanus lineolatus (Rüppell)

Golden Striped Snapper, Yellow Snapper, Gold-striped Sea-perch, Red Sea Lined Snapper

Diagnosis — Body is fusiform. Eyes are crossed by a yellow band. Dorsal scale rows are oblique, lower scale rows are longitudinal. Dorsal fin has 11 spines and 12 rays, anal fin has 3 spines and 8 rays. 46–48 lateral scales. Edible. Up to 38cm in total length.

Distribution — Ryukyu, Formosa, Indo-West Pacific, Persian Gulf, Red Sea, South Africa

Lutjanidae
Lutjanus caeruleovittatus (Valenciennes)

Ninestripe Snapper, Golden-lined Snapper

Diagnosis — Inhabits shoreline reefs and coral reefs. Body is somewhat deep, has 8–9 yellow stripes; snout is pointed. Head is reddish-brown; eyes are large. Fins are yellow. Dorsal fin has 11 spines and 13 rays, anal fin has 3 spines and 8 rays. 46–47 lateral scales. Edible. Up to 36cm in total length.

Distribution — Southern Japan, Ryukyu, Indo-West Pacific

Lutjanidae
Lutjanus russelli (Bleeker)

[*Mesoprion russelli* Bleeker]

Russell's Snapper, Russell's One-spot Snapper, Fingermark Bream, Moses Perch, Red Bream

Diagnosis — Inhabits shallow coral reefs and rocky reefs. Body is somewhat deep, dark-brown with a black blotch above lateral lines. Snout is pointed. In young, body has 4 black bands. Dorsal and caudal fins are dark; pectoral, pelvic, anal fins are yellow. Dorsal fin has 10 spines and 14–15 rays, anal fin has 3 spines and 8 rays. 47–50 lateral scales. Edible but has records of ciguatera poisoning. Up to 60cm in total length.

Distribution — Japan, Ryukyu, Formosa, Korea, Indo-Pacific, Red Sea, South Africa

Lutjanidae
Lutjanus fulviflamma (Forskål)

Blackspot Sea Perch, Dory Snapper, Five-spot Snapper, Golden Snapper, Johnny, Finger-mark Bream, Goldflame Snapper, Moses Perch, Spot Snapper

Diagnosis — Inhabits shallow coral reefs and bays. Resembles *L. ruselli*, but distinguished by reddish body with 5–6 yellow bands. A black blotch slightly below lateral lines. Caudal fin is yellow. Dorsal fin has 10 spines and 12–13 rays, anal fin has 3 spines and 8 rays. 50–52 lateral scales. Edible but has records of ciguatera poisoning. Up to 35cm in total length.

Distribution — Ryukyu, Indo-Pacific, Persian Gulf, Red Sea, South Africa

Lutjanidae

Lutjanus stellatus Akazaki

Spotted Snapper (New Common Name)

Diagnosis — Inhabits deeper shoreline reefs. Body is somewhat deep, yellowish-brown. Surface above lateral lines have a white spot. In young, blue lines run from eyes to mouth. Fins are yellow. Dorsal fin has 10 spines and 13–16 rays, anal fin has 3 spines and 8 rays. Posterior end of anal fin is rounded. 47–50 lateral scales. Edible but has records of ciguatera poisoning. Up to 45cm in total length.

Distribution — Japan, Ryukyu

Lutjanidae

Lutjanus erythropterus (Bloch)

[*L. sanguineus* (Cuvier)]

Firespot Snapper, Red Striped Snapper, Snapper

Diagnosis — Inhabits rather deep rocky reefs. Body is deep and reddish. Snout is projected. Upper part of caudal peduncles have a black blotch. Dorsal fin has 11 spines and 14 rays, anal fin has 3 spines and 9 rays. 46–48 lateral scales. Edible. Up to 60cm in total length.

Distribution — Southern Japan, Ryukyu, South China Sea, Indo-Pacific

Lutjanidae

Lutjanus fulvus (Schneider)

[*L. vaigiensis* (Quoy et Gaimard)]

[*Diacope fulva* Cuvier]

Waigiu Snapper, Yellow-margined Sea Perch, Red and Green Snapper Fish, Flame-colored Snapper

Diagnosis — Inhabits shallow coral reefs. Body is deep; snout is pointed; eyes are large. Dorsal and caudal fins are reddish-brown with white margin. Pectoral, pelvic, anal fins are yellow. Dorsal fin has 10 spines and 13–15 rays, anal fin has 3 spines and 8 rays. 47–51 lateral scales. Edible but has records of ciguatera poisoning. Up to 65cm in total length.

Distribution — Southern Japan, Ryukyu, Indo-Pacific, South Africa

Lutjanidae
Lutjanus gibbus (Forskål)

Hunched Snapper, Red Snapper, Humpback Snapper, Paddle-trail, Humped Red Snapper

Diagnosis — Inhabits shoreline reefs and coral reefs. Body is deep and red. In young, body is blueish. Snout is pointed. Upper lobe of caudal fin is rounded, lower is pointed. Pectoral fins are long. Dorsal fin has 10 spines and 13–15 rays, anal fin has 3 spines and 8–9 rays. 46–53 lateral scales. Carnivorous. Edible but has records of ciguatera poisoning. Up to 80cm in total length.

Distribution — Southern Japan, Ryukyu, Indo-Pacific, Hawaii, Red Sea, South Africa

▲ Young

Lutjanidae
Macolor niger (Forskål)

Black-and-white Sea-perch, Black Beauty, Mottled Snapper

Diagnosis — Body is oval and blackish. Head is roundish. In young, body has white blotches. Jaws have canines. Dorsal fin has 10 spines and 13 rays, anal fin has 3 spines and 10 rays. 52–53 lateral scales. Edible. Up to 75cm in total length.

Distribution — Ryukyu, Indo-Pacific, Red Sea, East Africa

▲ Adult

▼ Upper jaw has sharp canines.

Lutjanidae
Lutjanus decussatus (Cuvier)

Checkered Snapper, Cross-banded Snapper

Diagnosis — Inhabits coral reefs. Body is somewhat shallow, has 5 reddish-brown bands. Caudal peduncles have a large black blotch. Dorsal fin has 10 spines and 13 rays, anal fin has 3 spines and 8 rays. 45–50 lateral scales. Edible. Up to 30cm in total length.

Distribution — Ryukyu, Indo-Pacific

Lutjanidae

⇦ *Lutjanus bohar* (Forskål) ⇩

White-spotted Red Snapper, Two-spot Snapper, Two-spot Sea-perch, Large Red Mumea, Red Mumea, Red Snapper, Twinspot Snapper

Diagnosis — Inhabits rather deep rocky reefs and coral reefs. Body is somewhat deep and roundish. Body color varies from red to dark. Caudal margin is black. Mouth is large with sharp-pointed canines. Dorsal fin has 10 spines and 13−14 rays, anal fin has 3 spines and 8 rays. 48−55 lateral scales. Carnivorous. Feeds on fish and crab. Edible but has records of frequent ciguatera poisoning. Up to 1.1m in total length.

Distribution — Southern Japan, Ryukyu, Indo-Pacific, Red Sea, South Africa

▼Young Adult

Lutjanidae

⇩ *Lutjanus argentimaculatus* (Forskål) ⇨

Silver Snapper, Gray Snapper, Creek Red Bream, Dog Bream, Red Bass, Mangrove Jack, Purple Sea Perch, Red Bream, Red Salmon, River Roman, Rock Barramundi

Diagnosis — Inhabits shoreline reefs. Young are found in brackish areas. Body is somewhat deep and reddish-brown. Scales have a small black spot. In young, body has dark cross bands. Lips are thick; teeth are canine-like. Dorsal fin has 10 spines and 13−14 rays, anal fin has 3 spines and 8 rays. 48−50 lateral scales. Edible but has records of ciguatera poisoning. Up to 90cm in total length.

Distribution — Southern Japan, Ryukyu, Formosa, China, Indo-Pacific, Hawaii, Red Sea, South Africa

◀Adult

Lutjanidae

Lutjanus rivulatus (Cuvier)

[*Diacope rivulatus* Cuvier et Valenciennes]

Maori Snapper, Blue-lined Snapper, Blue-spotted Seaperch, Flute Porgy, Maitland Fish, Maori Bream, Maori Sea-perch, Speckled Snapper

Diagnosis — Inhabits rather deep rocky reefs. Body is deep; snout is pointed. Head has blue wavy lines. Lateral lines have an indistinct white spot. Dorsal fin has 10 spines and 15−16 rays, anal fin has 3 spines and 8−9 rays. 46−49 lateral scales. Edible but has records of ciguatera poisoning. Up to 80cm in total length.

Distribution — Ryukyu, Indo-West Pacific, South Africa

▼Young Adult

▲Adult

▼Young Adult

Lutjanidae

Symphorus nematophorus (Bleeker)

[*Glabrilutjanus nematophorus* (Bleeker)]

Chinaman Fish, Threadfin Snapper, Thread-finned Sea Perch, Threadfin Sea-perch

Diagnosis — Inhabits rather deep rocky reefs and coral reefs. Body is oval; head is roundish, profusely dotted with black spots. In young, body has some 10 blue stripes, dorsal rays are extended long. Dorsal fin has 10 spines and 15−16 rays, anal fin has 3 spines and 9 rays. 52−55 lateral scales. Edible but has records of ciguatera poisoning. Up to 1m in total length.

Distribution — Ryukyu, Formosa, West Pacific

▲ Adult

Young Adult ▶

Lutjanidae

⇩ ***Lutjanus sebae*** (Cuvier) ⇨

[*Diacope sebae* Cuvier et Valenciennes]

Emperor Snapper, Emperor Bream, Government Bream, King Snapper, Queenfish, Red Emperor, Red Kelp, Seba's Snapper, Bourgeois Jenoar

Diagnosis — Body is deep, has 3 reddish-brown bands which are indistinct in adult. Snout is pointed. Upper jaw has a pair of canines. Dorsal and anal margins are blackish. Caudal lobes are also blackish. Dorsal fin has 11 spines and 15–17 rays, anal fin has 3 spines and 10–11 rays. 46–53 lateral scales. Edible but has records of ciguatera poisoning. Up to 1m in total length.

Distribution — Southern Japan, Ryukyu, Formosa, Indo-Pacific, Hawaii, Red Sea, South Africa

▲ Young Adult

▲ Young Adult

Lutjanidae

Lipocheilus carnolabrum (Chan) ⇩

Tang's Snapper

Diagnosis — Inhabits waters more than 100m deep. Body is somewhat deep and yellowish-brown. In young, body has 5 cross bands. Dorsal and anal rays have no scales. Lips are thick. Dorsal fin has 10 spines and 10 rays, anal fin has 3 spines and 8 rays. 49–54 pored lateral scales. Edible. Up to 60cm in total length.

Distribution — Ryukyu, Indo-West Pacific

Lutjanidae

⇩ ***Lutjanus johnii*** (Bloch)

Blackspot Snapper, John's Sea-pearch, John's Snapper, Moses Pearch, Red Bream, Spotted-scale Sea Pearch

Diagnosis — Resembles *L. rusellii*, but distinguished by larger black splotches on lateral lines, 5–6 yellow stripes on body. Caudal margin is slightly lunated. Dorsal fin has 10 spines and 14 rays, anal fin has 3 spines and 8–9 rays. 48–50 lateral scales. Edible. Up to 70cm in total length.

Distribution — Ryukyu, South China Sea, Indo-West Pacific

Young Adult ▶

Lutjanidae

⇩ ***Symphonichthys spilurus*** (Günther)

Blue-and-gold Striped Snapper

Diagnosis — Inhabits sandy bottoms among coral reefs. Body is deep, forehead is vertical. Upper part of caudal peduncles have a black blotch. In young, a black band runs from eyes to tail. Dorsal fin has 10 spines and 17–18 rays, anal fin has 3 spines and 11 rays. 53–59 pored lateral scales. Edible. Up to 60cm in total length.

Distribution — Ryukyu, West Pacific

Lutjanidae
⇦ *Etelis marshi* Jenkins
Yellowstriped Ruby Snapper

Diagnosis — Inhabits waters more than 200m deep. Body is fusiform. Caudal lobes are short. Mouth is large. Pectoral fins are long. Dorsal fin has 10 spines and 11 rays, anal fin has 3 spines and 8 rays. 47−52 pored lateral scales. Edible. Up to 1m in total length.

Distribution — Izu Islands, Ryukyu, Indo-Pacific, Hawaii

Lutjanidae
Etelis radiosus Anderson ⇨
Bigmouth Ruby Snapper (New Common Name)

Diagnosis — Resembles *E. carbunculus*, but distinguished by more reddish body, rather short caudal lobes. Mouth is large. Posterior end of upper jaw reaches surface under posterior margin of eyes. Flesh is reddish. Dorsal fin has 10 spines and 11 rays, anal fin has 3 spines and 8 rays. 50−53 pored lateral scales. Edible. Up to 1m in total length.

Distribution — Ryukyu, Indo-West Pacific

Lutjanidae
⇦ *Etelis carbunculus* Cuvier
[*E. coruscans* Valenciennes]

Ruby Snapper

Diagnosis — Inhabits rocky reefs more than 200m deep. Body is somewhat deep; tail extends long. Caudal fin is deeply falcated with extended lobes. Mouth is large. Dorsal fin has 10 spines and 11 rays, is compressed between spine and ray. Anal fin has 3 spines and 8 rays. 47−53 pored lateral scales. Carnivorous. Edible. Up to 1m in total length.

Distribution — Japan, Ryukyu, Indo-Pacific-Atlantic, Hawaii, South Africa

Lutjanidae
Aphareus furcatus (Lacépède) ⇨
[*A. flavivultus* (Jenkins)]

Olive Small-toothed Jobfish, Fork-tailed Snapper Fish

Diagnosis — Inhabits rather deep rocky reefs. Body is fusiform. Fins are yellow. Mouth is large, teeth are indistinct. Dorsal fin has 10 spines and 11 rays, anal fin has 3 spines and 8 rays. 72−75 pored lateral scales. Edible but has records of ciguatera poisoning. Up to 60cm in total length.

Distribution — Japan, Ryukyu, Indo-Pacific, Hawaii, East Africa

Lutjanidae
Randallichthys filamentosus (Fourmanoir) ⇨
[*Etelis nudimaxillaris* Yoshino et Araga]

Short-finned Ruby Snapper (New Common Name)

Diagnosis — Inhabits rather deep waters. Body is fusiform. Caudal lobes are extended. Pectoral fins are short. Pelvic fins are black. Dorsal fin has 10 spines and 11 rays, marginal part of spine is black. Anal fin has 3 spines and 9 rays. 48−49 lateral scales. Edible. Up to 50cm in total length.

Distribution — Ryukyu, Guam, New Caledonia, Hawaii

Lutjanidae

⇦ ***Aphareus rutilans*** Cuvier

Flame Snapper, Rusty Small-toothed Jobfish, Smalltoothed Jobfish

Diagnosis — Inhabits waters more than 100m deep. Body is fusiform. Caudal fin is deeply falcated, lobes extend long. Mouth is large. Resembles *A. furcatus*, but distinguished by monocolor body. In young, mouth has teeth which disappear in adult. Dorsal fin has 10 spines and 11 rays, anal fin has 3 spines and 8 rays. 70−75 pored lateral scales. Edible. Up to 1m in total length.

Distribution — Southern Japan, Bonin Islands, Ryukyu, Indo-Pacific, Hawaii, Red Sea

Lutjanidae

Pristipomoides multidens (Day) ⇨

[*P. argyrogammicus* (Valenciennes)]

Sharptooth Snapper

Diagnosis — Inhabits waters more than 100m deep. Body is fusiform and dark-purple. Head has blue and yellow wavy blotches. Posterior ends of dorsal and anal fins extend long. Dorsal fin has 10 spines and 11 rays, anal fin has 3 spines and 8 rays. 48−52 pored lateral scales. Edible. Up to 1m in total length.

Distribution — Southern Japan, Ryukyu, Indo-West Pacific, East Africa

Lutjanidae

⇦ ***Pristipomoides auricilla*** (Jordan, Evermann et Tanaka)

Yellow-speckled Snapper (New Common Name)

Diagnosis — Inhabits waters more than 100m deep. Body is fusiform, covered with yellow speckles. Caudal fin is yellow. Dorsal fin has 10 spines and 11 rays, anal fin has 3 spines and 8 rays. 70−74 pored lateral scales. Edible. Up to 50cm in total length.

Distribution — Southern Japan, Ryukyu, South China Sea, Guam, Hawaii

Lutjanidae

Pristipomoides sieboldii (Bleeker) ⇨

[*Chaetopterus sieboldii* Bleeker]

Crimson Snapper

Diagnosis — Inhabits waters more than 100m deep. Resembles *P. filamentosus*, but distinguished by slightly smaller, bright, blotcheless body. Pectoral fins are rosy and long. Tongue is covered with tooth patches. Dorsal fin has 10 spines and 11 rays, anal fin has 3 spines and 8 rays. 70−74 pored lateral scales. Edible. Up to 60cm in total length.

Distribution — Japan, Ryukyu, Indo-Pacific, Hawaii, Red Sea, East Africa

Lutjanidae

⇦ ***Pristipomoides flavipinnis*** Shinohara

Golden-eye Snapper (New Common Name)

Diagnosis — Inhabits waters more than 100m deep. Resembles *P. filamentosus*, but marginal parts of eyes and dorsal fin are yellow. Head has wavy blotches on back surface. Dorsal fin has 10 spines and 11 rays, anal fin has 3 spines and 8 rays. 59–62 pored lateral scales. Edible. Up to 80cm in total length.

Distribution — Ryukyu, West Pacific

Lutjanidae

Pristipomoides filamentosus (Valenciennes) ⇨

[*P. filamentosus roseus* (Castelnau)]

Crimson Snapper

Diagnosis — Inhabits waters more than 100m deep. Body is fusiform and dark-purple. Head is dotted with small black. Caudal margin is orange. Dorsal fin has 10 spines and 11 rays, anal fin has 3 spines and 8 rays. 60–65 pored lateral scales. Edible. Up to 90cm in total length.

Distribution — Southern Japan, Ryukyu, Indo-West Pacific

Lutjanidae

⇦ ***Tropidinius zonatus*** (Valenciennes)

Obliquebanded Snapper, Skewband Jobfish

Diagnosis — Body is oblong with 4 yellow bands. Interorbital space is elevated. Dorsal fin has 10 spines and 11 rays, anal fin has 3 spines and 8 rays. 63–67 pored lateral scales. Edible. Up to 50cm in total length.

Distribution — Japan, Izu Islands, Ryukyu, Indo-Pacific, South Africa

Lutjanidae

⇧⇩ ***Aprion virescens*** Valenciennes

Green Jobfish, Blue Snapper, King-snapper, Job-fish, Streaker, Slender Snapper, Blue-gray Snapper

Diagnosis — Inhabits shallow coral reefs. Body is elongated fusiform. Caudal fin is well-developed, pectoral fins are short. Surface in front of eyes are depressed like groove. Snout is rounded. Dorsal fin has 10 spines and 11 rays, anal fin has 3 spines and 8 rays. 48–50 pored lateral scales. Edible but has records of ciguatera poisoning. Up to 1m in total length.

Distribution — Southern Japan, Ryukyu, Indo-Pacific, South Africa

Lutjanidae

⇧ ***Tropidinius amoenus*** (Snyder)

[*Pristipomoides amoenus* (Snyder)]

Flower Flute Porgy, Flower Jobfish

Diagnosis — Inhabits waters more than 100m deep. Body is oblong, compressed, profusely dotted with blue. Dorsal fin origin has a large blue blotch. Caudal fin is orange. Dorsal fin has 10 spines and 11 rays, anal fin has 3 spines and 8 rays. 58–66 pored lateral scales. Edible. Up to 50cm in total length.

Distribution — Japan, Izu Islands, Ryukyu, Pacific, Hawaii

Lutjanidae
Paracaesio kusakarii Abe ⇧
Banded Fusilier

Diagnosis — Resembles *P. caeruleus*, but distinguished by indistinct cross bands on body. Caudal fin is yellowish, upper lobe is white. Dorsal fin has 10 spines and 10−11 rays, anal fin has 3 spines and 8 rays. 47−50 pored lateral scales. Edible. Up to 65cm in total length.

Distribution — Japan, Ryukyu, Bonin Islands, West Pacific

Lutjanidae
Paracaesio tumidus (Tanaka) ⇧ ⇧
[*P. xanthurus* (Bleeker)]

[*Scolopsis xanthurus* (Bleeker)]

[*Vegetichthys tumidus* Tanaka]

Yellowback Fusilier, Yellowtail Fusilier

Diagnosis — Inhabits rather deep coral reefs and rocky reefs. Body is oblong; snout is short. Back and caudal fin are yellow. Dorsal fin has 10 spines and 9−10 rays, spines are not stiff. Anal fin has 3 spines and 8 rays. 70−72 pored lateral scales. Edible. Up to 45cm in total length.

Distribution — Japan, Ryukyu, Indo-Pacific, South Africa

Lutjanidae
Paracaesio caeruleus (Katayama) ⇩
Blue Fusilier

Diagnosis — Inhabits waters more than 100m deep. Body is oblong and compressed. Back is dark-brown. Pectoral fins are long. Interorbital space is elevated. Dorsal fin has 10 spines and 10 rays, is not compressed. Anal fin has 3 spines and 8 rays. 47−50 pored lateral scales. Edible. Up to 55cm in total length.

Distribution — Japan, Ryukyu

Lutjanidae
Caesio erythrogaster Cuvier ⇩ ⇩
Fusilier, Red-bellied Fusilier, Yellow-tail, Yellow-tailed Caesio, Redbelly Fusilier

Diagnosis — Inhabits rather deep waters. Resembles *C. xanthonotus*, but distinguished by deep body, dark-brown dorsal and caudal fins. Dorsal fin has 10 spines and 15 rays, anal fin has 3 spines and 11 rays. 45−51 lateral scales. Edible. Up to 40cm in total length.

Distribution — Ryukyu, Indo-West Pacific

Lutjanidae
⇦ *Caesio xanthonotus* Bleeker ⇨
Redfin Fusilier

Diagnosis — Inhabits outer slopes of coral reefs. Body is blue; back and caudal fin are yellow. Pectoral fin bases are black. Caudal fin has no black blotches. Dorsal fin has 11 spines and 15 rays, anal fin has 3 spines and 12 rays. 55−60 lateral scales. Edible. Up to 40cm in total length.

Distribution — Izu Islands, Ryukyu, Indo-West Pacific, East Africa

Lutjanidae

↖ ***Caesio chrysozonus*** Cuvier ⇧

[*Pterocaesio chrysozona* (Cuvier et Valenciennes)]

[*Caesio chrysozona* Kuhl et Van Hasselt]

Bananafish, Black-tip Fusilier, Gold-banded Fusilier, Golden Caesio, Golden-banded Caesio, Golden Banded Fusilier

Diagnosis — Found in schools about outer slopes of coral reefs. Body is fusiform. Back has 2 yellow stripes, a lower yellow longitudinal stripe under lateral lines extending to the upper end of eyes. Dorsal fin has 10 spines and 15–16 rays, anal fin has 3 spines and 12 rays. 70–78 lateral scales. Edible. Up to 35cm in total length.

Distribution — Japan, Ryukyu, Indo-Pacific, Red Sea, East Africa

Lutjanidae

⇧ ***Caesio diagramma*** (Bleeker)

[*Caesio digramma* Bleeker]

Yellow Lateral-lined Caesio (New Common Name)

Diagnosis — Resemble *C. chrysozonus*, but distinguished by a lower yellow longitudinal stripe runs just bellow lateral lines, does not extend to the upper end of eyes. Dorsal fin has 10 spines and 14–15 rays, anal fin has 3 spines and 12 rays. Edible. Up to 25cm in total length.

Distribution — Southern Japan, Ryukyu, Philippines, Indonesia

Lutjanidae

⇦ ***Caesio pisang*** Bleeker ⇧

[*Pterocaesio pisang* (Bleeker)]

Slender Fusilier, Yellowstriped Fusilier

Diagnosis — Resembles *Pterocaesio chrysozona*, but distinguished by smaller, slender body, a yellow stripe on back. Caudal lobes are black. Dorsal fin has 10 spines and 15–16 rays, anal fin has 3 spines and 12 rays. 67–68 lateral scales. Edible. Up to 25cm in total length.

Distribution — Ryukyu, Indo-Pacific, East Africa

Lutjanidae

⇦ ***Caesio tile*** Cuvier ⇧

Bartail Fusilier

Diagnosis — Found in schools about coral reefs. Resembles *C. chrysozona*, but back is highly bluish with a longitudinal stripe. Caudal lobes have black stripes. Pectoral fin bases are black. Dorsal fin has 10–11 spines and 19–22 rays, anal fin has 3 spines and 12–13 rays. 65–70 lateral scales. Feeds on zooplankton. Edible. Up to 30cm in total length.

Distribution — Izu Islands, Ryukyu, Indo-Pacific, East Africa

Lutjanidae

Caesio caerulaureus Lacépède ⇨

Blue and Gold Fusilier, Caesio, Goldbanded Fusilier

Diagnosis — Found in schools about coral reefs. Body is somewhat deep; back has a broad yellow band. Caudal fin is dark with pale margin. Dorsal fin has 10 spines and 14–15 rays, anal fin has 3 spines and 11–12 rays. 48–56 lateral scales. Edible. Up to 40cm in total length.

Distribution — Southern Japan, Ryukyu, Indo-Pacific, Hawaii, South Africa

▲Young Adult

Pomadasyidae

Parapristipoma trilineatum (Thunberg)

Chicken Grunt, Threeline Grunt

Diagnosis — Found in schools rather deep rocky reefs. Body is fusiform. In young body has 3 brown bands which become indistinct in adult. Dorsal fin has 13–14 spines and 17–19 rays, anal fin has 3 spines and 8 rays. 55–56 lateral scales. Nocturnal. Feeds on small fish, crustacea and squid. Spawns between June and September. Edible. Up to 45cm in total length.

Distribution — Japan, Ryukyu, Formosa, South China Sea, India

▲ Adult

▲Young Adult

Pomadasyidae

Plectorhynchus lineatus (Linnaeus)

[*Gaterin lineatus* (Linnaeus)]

Yellow-banded Sweetlip

Diagnosis — Inhabits shallow coral reefs. Body is somewhat deep, has 6 black stripes. Lips and fins are yellow. Dorsal fin has 13 spines and 19 rays, anal fin has 3 spines and 7 rays. 60 lateral scales. Edible. Up to 35cm in total length.

Distribution — Ryukyu, Indo-Pacific, South Africa

Pomadasyidae

Plectorhynchus albovittatus (Rüppell)

[*Diagramma albovittatum* Rüppell]

[*Gaterin albovittatum* (Rüppell)]

Yellow-lined Sweetlip

Diagnosis — Body is somewhat deep, has irregular dark blotches. In young, body has 2 black bands. Dorsal fin has 13 spines and 18–20 rays, anal fin has 3 spines and 6–7 rays. Edible. Up to 20cm in total length.

Distribution — Indo-Pacific, Red Sea, Persian Gulf, East Africa

▲Young Adult

▼Adult

Pomadasyidae

⇦ ***Plectorhynchus flavomaculatus*** (Ehrenberg) ⇩

[*Gaterin flavomaculatus* (Ehrenberg)]

Lemonfish

Diagnosis — Inhabits shoreline reefs and coral reefs. Body is oblong, closely covered with orange stripes and blotches. Dorsal fin has 12–13 spines and 21–22 rays, anal fin has 3 spines and 7 rays. 56–59 lateral scales. Edible. Up to 60cm in total length.

Distribution — Japan, Ryukyu, Indo-West Pacific, Red Sea, South Africa

▲Adult

Pomadasyidae

Plectorhynchus picus (Valenciennes) ⇧⇨

[*P. orientalis* (Bloch)]

[*Gaterin orientalis* (Bloch)]

Oriental Sweetlip

Diagnosis — Inhabits outer slopes of coral reefs. Body is somewhat deep, closely covered with small black spots. Dorsal, caudal, anal margins are black. Lips are thick. Dorsal fin has 12–14 spines and 17–20 rays, anal fin has 3 spines and 7–8 rays. 70–75 lateral scales. Edible. Up to 60cm in total length.

Distribution — Japan, Ryukyu, Indo-West Pacific, East Africa

▲Young Adult

▲Young Adult

Pomadasyidae

Plectorhynchus pictus (Thunberg) ⇧⇨⇩

[*Spilotichthys pictus* (Thunberg)]

Blackall, Bluey, Common Thick-lipped Grunt, Painted Grunt, Morwong, Slate Bream, Thicklip, Thick Lip Bream, Sweetlip

Diagnosis — Inhabits shoreline muddy-sand bottoms. Body is deep, profusely dotted with brown. Lips are not thick, mouth is not red inside. Dorsal fin has 9–10 spines and 22–23 rays, anal fin has 3 spines and 7–8 rays. 69–72 lateral scales. Edible. Up to 90cm in total length.

Distribution — Japan, Ryukyu, Indo-Pacific, Persian Gulf, Red Sea, South Africa

▼Young

▲Adult

Pomadasyidae

Plectorhynchus celebicus Bleeker ⇩

Celebes Sweetlips

Diagnosis — Body is somewhat deep with some 10 orange stripes. Fins are yellow. Dorsal fin has 13 spines and 20 rays, anal fin has 3 spines and 7 rays. 56–59 lateral scales. Edible. Up to 50cm in total length.

Distribution — Ryukyu, West Pacific

▲Adult

▲Young

Pomadasyidae

⇧ ***Plectorhynchus chaetodontoides*** Lacépède ⇧⇨

Harlequin Sweetlip, Many-spotted Sweetlip, Sweetlip

Diagnosis — Inhabits outer slopes of coral reefs. Body is deep, profusely covered with black blotches. Lips are thick, mouth is red inside. Dorsal fin has 11–12 spines and 19–20 rays, anal fin has 3 spines and 7–8 rays. 55–58 lateral scales. Edible. Up to 60cm in total length.

Distribution — Southern Japan, Ryukyu, Indo-West Pacific

▼Young Adult

▲Adult

Pomadasyidae

Plectorhynchus diagrammus (Linnaeus) ⇧⇨⇩

[*Gaterin diagrammus* (Linnaeus)]

Silver-banded Sweetlip, Yellow-banded Sweetlip

Diagnosis — Inhabits shallow coral reefs. Body is somewhat deep, has 4 dark bands. Head is roundish, has 6 dark bands. Dorsal, anal, caudal fins are dotted with black. Dorsal fin has 12–13 spines and 19–20 rays, anal fin has 3 spines and 7–8 rays. 53–56 lateral scales. Edible. Up to 50cm in total length.

Distribution — Southern Japan, Ryukyu, Indo-West Pacific, South Africa

▼Adult

▲Young

Pomadasyidae

Plectorhynchus goldmanni (Bleeker) ⇨

Diagonal-banded Sweetlips, Goldmann's Grunt, Many-lined Sweetlips

Diagnosis — Inhabits coral reefs. Body is somewhat deep; head is roundish; back is covered with black oblique bands. Fins are yellow. Pectoral fin bases are red. Dorsal fin has 13 spines and 19–20 rays, anal fin has 3 spines and 7 rays. 56–58 lateral scales. Edible. Up to 60cm in total length.

Distribution — Southern Japan, Ryukyu, West Pacific

Pomadasyidae

⇦ ***Plectorhynchus nigrus*** (Cuvier)

[*P. niger* (Cuvier)]

[*Pseudopristipoma nigra* (Cuvier)]

[*Gaterin niger* (Cuvier et Valenciennes)]

Black Sweetlip, Brown Sweetlips, Blubber-lip Bream

Diagnosis — Body is deep and dark. Back has indistinct dark blotches. Caudal margin is truncated. Lips are thick. Dorsal fin has 14 spines and 15–16 rays, anal fin has 3 spines and 7–8 rays. 51–55 lateral scales. Edible. Up to 60cm in total length.

Distribution — Ryukyu, Indo-West Pacific, Red Sea, South Africa

Pomadasyidae

⇦ ***Plectorhynchus cinctus*** (Temminck et Schlegel)

[*Gaterin cinctus* (Schlegel)]

Grunt, Three-banded Sweetlip

Diagnosis — Inhabits shoreline reefs. Body is deep with 3 dark bands. Posterior dorsal sides, dorsal fin, caudal fin are dotted with black. In young, body is black. Dorsal fin has 12 spines and 15–17 rays, anal fin has 3 spines and 7–8 rays. 53–57 lateral scales. Spawns between May and June. Nocturnal. Edible. Up to 60cm in total length.

Distribution — Japan, Ryukyu, Korea, Indo-Pacific, Persian Gulf

Pomadasyidae

Pomadasys hasta (Bloch) ⇩⇨

Banded Grunter, Common Javelinfish, Grunter Bream, Head-grunt, Javelin-fish, Lined Silver Grunter, Queensland Trumpeter, Silver Grunter

Diagnosis — Body is somewhat deep, silvery-white; back is profusely dotted with black. Caudal fin is truncated. Dorsal fin has 12 spines and 13–15 rays, has 3 rows of dark spots. Anal fin has 3 spines and 7 rays. 47–49 lateral scales. Edible. Up to 45cm in total length.

Distribution — Southern Japan, Ryukyu, Formosa, Indo-West Pacific, Red Sea, South Africa

▲Adult

Pomadasyidae

⇩ ***Hapalogenys nitens*** Richardson

Grunt, Skewband Grunt

Diagnosis — Inhabits rather deep rocky reefs. Body is compressed and deep with 2 broad dark bands. Pelvic fins are black. Cheeks have scales. Dorsal fin has 10–11 spines and 15–16 rays, anal fin has 3 spines and 9 rays. 44–47 lateral scales. Edible. Up to 40cm in total length.

Distribution — Japan, Korea, East China Sea

Pomadasyidae

Hapalogenys nigripinnis (Temminck et Schlegel) ⇧⇩

Beard Grunt, Grunt

Diagnosis — Inhabits rather deep rocky reefs. Body is remarkably deep, has 2 indistinct dark bands; snout is projected; lower jaw is closely covered with barbels. Dorsal fin has 11 spines and 17 rays, anal fin has 3 spines and 10 rays. 45 lateral scales. Edible. Up to 60cm in total length.

Distribution — Japan, Korea, China, South China Sea

▲Young

Pomadasyidae

⇦ ***Hapalogenys mucronatus*** (Eydoux et Sauleyet)

Belted Beard Grunt, Grunt

Diagnosis — Inhabits shoreline reefs. Body is compressed and deep with 6 dark bands. Marginal parts of fins are black. Dorsal fin has 11 spines and 15 rays, anal fin has 3 spines and 9 rays. 48–49 lateral scales. Edible. Up to 30cm in total length.

Distribution — Southern Japan, Formosa, Korea, China, Indonesia

Teraponidae

⇖ ***Terapon jarbua*** (Forskål) ⇧

[*Therapon jarbua* (Forskål)]

Spiky Trumpeter, Striped Therapon, Arrow Bass, Crescent Perch, Jarbua, Jarbua Grunt, Pest of St. Lucia, Threestriped Tigerfish, Thornfish

Diagnosis — Found in schools in sandy bottoms of bays also river mouth. Body is fusiform, has 3 black bow-shaped bands. Snout is rounded; jaws have canines. Carnivorous. Spawns between spring and summer. Grown in air-bladder. Dorsal fin has 12 spines and 10 rays, anal fin has 3 spines and 8 rays. 90–93 lateral scales. Edible. Up to 30cm in total length.

Distribution — Japan, Ryukyu, Korea, Indo-Pacific, Red Sea, South Africa

Teraponidae

Rhyncopelates oxyrhynchus (Temminck et Schlegel) ⇧ ⇨

[*Terapon oxyrhynchus* Temminck et Schlegel]

[*Therapon oxyrhynchus* Temminck et Schlegel]

Sharpnose Tigerfish, Five-striped Pig-face, Fourstriped Grunter

Diagnosis — Adult inhabits shoreline reefs, young grow in bays, are also found in freshwater areas. Body is fusiform, has 7 dark-brown bands. Snout is pointed; mouth is small. Pectoral fins are short. Dorsal fin has 12 spines and 10 rays, anal fin has 3 spines and 8 rays. 51–52 lateral scales. Spawns between spring and summer. Edible. Up to 30cm in total length.

Distribution — Japan, Ryukyu, Formosa, Korea, Indo-Pacific

Teraponidae

⇦ ***Pelates quadrilineatus*** (Bloch)

Trumpeter Perch, Croaker, Four-barred Grunt, Four-lined Theraponid, Red-mouthed Tigerfish, Tigerfish Croaker, Trumpeter

Diagnosis — Inhabits shallow shoreline waters and river mouths. Body is fusiform, has 4–5 indistinct dark stripes. Snout is slightly pointed. Fin membrane and anterior part of dorsal fin are black. Caudal fin has no blotches. Dorsal fin has 12 spines and 10 rays, anal fin has 3 spines and 10–11 rays. 63–73 lateral scales. Edible. Up to 25cm in total length.

Distribution — Japan, Ryukyu, Formosa, Indo-West Pacific, Persian Gulf, Red Sea, South Africa

Cirrhitidae

⇦ ***Cirrhitichthys aprinus*** (Cuvier)

Blotched Hawkfish, Boar Hawkfish, Spotted Hawkfish

Diagnosis — Inhabits shoreline reefs. Body is somewhat deep, covered with irregualr brown blotches. Upper part of opercles have an eye-like blotch. Dorsal fin has 10 spines and 12 rays, tips of spines are fringed. Anal fin has 3 spines and 6 rays. Up to 12cm in total length.

Distribution — Japan, Ryukyu, Indo-West Pacific, South Africa

Cirrhitidae

Neocirrhites armatus Castelnau ⇩

Red Hawkfish (New Common Name)

Diagnosis — Inhabits shallow coral reefs. Body is deep and red. Surface around eyes and dorsal margins are black. Dorsal fin has 10 spines and 13 rays, anal fin has 3 spines and 6 rays. 42−44 lateral scales. Up to 9cm in total length.

Distribution — Ryukyu, Bonin Islands, West Pacific

Cirrhitidae

⇩ ***Cirrhitus pinnulatus*** (Bloch et Schneider)

Stocky Hawkfish

Diagnosis — Inhabits outer slopes of coral reefs. Body is somewhat deep, has irregular brown blotches. Dorsal fin has 10 spines and 11 rays, tips of spines are fringed. Anal fin has 3 spines and 6 rays. 39−44 lateral scales. Up to 28cm in total length.

Distribution — Southern Japan, Indo-Pacific, Hawaii, Red Sea, South Africa

Cirrhitidae

⇩ ***Oxycirrhites typus*** Bleeker

Long-nosed Hawkfish

Diagnosis — Inhabits rather deep rocky reefs. Body is elongated, has reddish-brown checker blotches. Snout is projected. Dorsal fin has 10 spines and 13 rays, anal fin has 3 spines and 7 rays. 51−53 lateral scales. Up to 13cm in total length.

Distribution — Japan, Indo-Pacific, Hawaii, California, Red Sea

Cirrhitidae

⇦⇧ ***Paracirrhites forsteri*** (Bloch et Schneider)

Blacksided Hawkfish, Forster's Curlyfin, Freckled Hawkfish

Diagnosis — Inhabits shallow coral reefs. Body is somewhat deep; back is dark-brown, abdominal side is pale. Head is profusely dotted with dark-red. Dorsal fin has 10 spines and 11 rays, anal fin has 3 spines and 6 rays. 45−49 lateral scales. Up to 45cm in total length.

Distribution — Japan, Ryukyu, Indo-Pacific, Hawaii, Red Sea, South Africa

Cheilodactylidae

Goniistius zebra (Döderlein)

Redlip Morwong

Diagnosis — Inhabits rather deep rocky reefs. Resembles *G. zonatus*, but distinguished by slender tail and red lips. Black oblique bands on body. Pelvic fins and lower lobe of caudal fin are black. Dorsal fin has 17 spines and 32–33 rays, anal fin has 3 spines and 8 rays. 62–65 lateral scales. Up to 40cm in total length.

Distribution — Japan

Cheilodactylidae

Goniistius zonatus (Cuvier)

Flag-fish, White-spot-tail Morwong

Diagnosis — Inhabits shoreline reefs. Body is compressed and somewhat deep, has 8 yellowish-brown oblique bands. Caudal fin is yellow, profusely dotted with white. Mouth is small, directed downward. Dorsal fin has 17 spines and 32 rays, anal fin has 3 spines and 8 rays. 62 lateral scales. Omnivorous. Feeds mostly on seaweed. Up to 45cm in total length.

Distribution — Japan, Ryukyu, Formosa, Korea, China, East China Sea

Cheilodactylidae

Goniistius vittatus (Garrett)

Striped Hawk Fish

Diagnosis — Inhabits rather deep rocky reefs and coral reefs. Body is somewhat deep, has 3 black bands. Tail is extended. Lips are thick, mouth is directed downward. Dorsal fin has 17 spines and 29 rays. Up to 30cm in total length.

Distribution — Hawaii

Labracoglossidae

Labracoglossa argentiventris (Lacépède)

Yellowstriped Butterfish

Diagnosis — Found in schools about rather deep rocky reefs. Body is fusiform, back has a yellow band. Dorsal, anal, caudal fins are yellow. Snout is roundish. Dorsal fin has 10 spines and 26–29 rays, anal fin has 3 spines and 22–24 rays. 62–75 lateral scales. Feeds on zooplankton. Edible. Up to 25cm in total length.

Distribution — Japan

Opisthognathidae
Opisthognathus castelnaui Bleeker

Bigmouth Jawfish
Diagnosis — Inhabits waters more than 100m deep. Found in holes dug in bottoms. Mouth is large, upper jaw is remarkably long. 7–9 black splotches on dorsal fin base. Dorsal fin has 11 spines and 14 rays, anal fin has 2 spines and 14 rays. Edible. Up to 28cm in total length.
Distribution — Ryukyu, Indo-West Pacific

▲Young Adult

Carangidae
Alectis indica (Rüppell)
[*A. indicus* (Rüppell)]

Diamond-fish, Diamond Trevally, Indian Thread-finned Trevally, Indian Threadfish, Mirror Fish, Plumed Trevally
Diagnosis — Body is deep, almost pentagonal. No adipose eyelids on eyes. Surfaces between eyes and mouth are wide. Snout is depressed. First dorsal fin is degenerated, second has 1 spine and 18–19 rays. Anal fin has 1 spine and 16–17 rays. 10–13 small scutes. In young, dorsal, pelvic, anal fin rays extend like thread. Edible. Up to 1.5m in total length.
Distribution — Japan, Ryukyu, Formosa, Indo-Pacific, Hawaii, Red Sea, South Africa

▲Adult

▲Young Adult

Carangidae
Alectis ciliaris (Bloch)
[*A. breviventralis* Wakiya]

Ciliated Threadfish, Gobbler Fish, Gobblefish, Hair Fish, Pennant Fish, Pennant Trevally, Sunfish, Thread Fin, Threadfin Mirrorfish, Threadfish
Diagnosis — Body is compressed and deep. Scales are small, sink under skin. Surfaces between eyes and mouth are narrow. First dorsal fin is degenerated, second has 1 spine and 18–19 rays, has a black blotch. Anal fin has 1 spine and 15–16 rays. 12–19 small scutes. In young, dorsal and anal fin rays extend like thread. Edible. Up to 90cm in total length.
Distribution — Japan, Ryukyu, Korea, Indo-Pacific-Atlantic, Hawaii, Red Sea, South Africa

▼Young Adult

Carangidae

Atropus atropus (Bloch et Schneider) ⇨

Cleftbelly Kingfish, Kuweh, Thin Crevalle, Kuweh Trevally

Diagnosis — Body is compressed and deep. Surfaces between lower jaw and pelvic fins are straight. Pelvic fins are black, can be folded back into groove. Pectoral fins have no scales. Dorsal fin ray extends long. First dorsal fin has 7–8 spines, second has 1 spine and 21–22 rays. Anal fin has 1 spine and 17–18 rays. 30–35 scutes. Edible. Up to 37cm in total length.

Distribution — Japan, Ryukyu, Bonin Islands, Formosa, China, Indo-West Pacific, South Africa

Carangidae

Caranx ignobilis (Forskål)

White Ulua, Forskal's Indo-Pacific Jackfish, Giant Kingfish, Horse Mackerel, Lowly Trevally, Yellow-fin Trevally

Diagnosis — Body is somewhat deep and blackish; head is curved. Abdominal margin is straight. Pectoral sides have nc scales. First dorsal fin has 8 spines, second has 1 spine and 18–20 rays. Anal fin has 1 spine and 15–17 rays. 28–32 large scutes. In young, anal fin and lower lobe of caudal fin are yellow. Upper part of opercles have an indistinct black spot. Edible but has records of ciguatera poisoning. Up to 1.5m in total length.

Distribution — Southern Japan, Ryukyu, Indo-Pacific, Hawaii, Red Sea, South Africa

Carangidae

Caranx leptolepis Cuvier

[*Selaroides leptolepis* (Cuvier et Valenciennes)]

Crevalle, Slender-scaled Scad, Smooth-tailed Trevally, Thin-scaled Trevally, Yellow-banded Scad

Diagnosis — Body is compressed and fusiform with a broad yellow band on dorsal sides. Upper part of opercles have a large black blotch. Adipose eyelids are developed, covers posterior part of eyes. Upper jaw is toothless, lower jaw has 1 row of teeth. First dorsal fin has 8 spines, second has 1 spine and 24–26 rays. Anal fin has 1 spine and 20–22 rays. 23–33 scutes. Edible. Up to 25cm in total length.

Distribution — Japan, Ryukyu, Formosa, Indo-West Pacific

Carangidae

Caranx equula Temminck et Schlegel

[*Kaiwarinus equula* (Temminck et Schlegel)]

Horse Kingfish, Kingfish, Whitefin Crevalle, White-finned Crevalle

Diagnosis — Inhabits rather deep rocky reefs. Body is oval and silvery-white. Adipose eyelids are not developed. Pectoral sides have scales. First dorsal fin has 8 spines, second has 1 spine and 23–25 rays. Anal fin has 1 spine and 21–23 rays. Dorsal and anal fin rays have a dark brown band. 22–29 scutes on entire straight lateral lines. Edible. Up to 40cm in total length.

Distribution — Japan, Ryukyu, Formosa, Korea, Indo-Pacific, Hawaii, South Africa

Carangidae

⇦ ***Caranx elacate*** (Jordan et Evermann) ⇩

Large-mouth Trevally

Diagnosis — Body is fusiform. Resembles *C. sexfasciatus*, but distinguished by shorter snout, a black spot on upper surface of opercles, larger mouth. Pectoral sides are covered with scales. Pectoral fins are long. First dorsal fin has 7–8 spines, second has 1 spine and 19–20 rays. Anal fin has 1 spine and 16 rays. 27–31 black scutes. Edible. Up to 1m in total length.

Distribution — Ryukyu, Indo-Pacific

Carangidae

Caranx speciosus (Forskål) ⇧ ⇨

[*Gnathanodon speciosus* (Forskål)]

Banded Trevally, Golden Crevalle, Golden Kingfish, Golden Trevally, Horse Mackerel, Kingfish, King Trevally, Golden Toothless Trevally

Diagnosis — Body is oval with some 12 black bands. Fins are yellowish. Adipose eyelids are absent on eyes. Jaws are toothless. First dorsal fin has 8 spines, second has 1 spine and 18–20 rays. Anal fin has 1 spine and 15–17 rays. 15–20 scutes. Edible. Up to 1m in total length.

Distribution — Japan, Ryukyu, Indo-Pacific, Hawaii, Red Sea, South Africa

Carangidae

⇦ ***Caranx ferdau*** (Forskål) ⇩

[*Carangoides ferdau* (Forskål)]

[*C. orthogrammus* (Jordan et Gilbert)]

Ferdau's Trevally, Horse Mackerel, Forskal's Jack Fish, Yellow-spotted Crevalle

Diagnosis — Body is oblong, profusely dotted with yellow. Resembles *C. fulvoguttatus*, but distinguished by longer dorsal fin rays, gently curved lateral lines, slightly black pectoral fin bases. Scaleless areas of pectoral sides are narrow. First dorsal fin has 6–8 spines, second has 1 spine and 29–32 rays. Anal fin has 1 spine and 24–26 rays. 20–27 scutes. Edible but has records of ciguatera poisoning. Up to 75cm in total length.

Distribution — Japan, Ryukyu, Indo-Pacific, Hawaii, Red Sea, South Africa

▲Young

Adult▶

Carangidae

Caranx ciliarius (Rüppell) ⇧ ⇨

[*Carangoides ciliarius* (Rüppell)]

[*C. plumbeus* (Quoy et Gaimard)]

Armed Trevally, Kingfish, Long-finned Crevalle, Long-finned Trevally, Longfin Trevally

Diagnosis — Body is compressed and deep. Snout is depressed. Resembles *C. hedlandensis*, but distinguished by shorter snout, fewer scutes. In adult, dorsal and anal fin rays extend long. First dorsal fin has 8 spines, second has 1 spine and 21–22 rays. Anal fin has 1 spine and 17 rays. 19–23 scutes. Edible. Up to 35cm in total length.

Distribution — Japan, Ryukyu, Indo-West Pacific

Carangidae
Caranx delicatissimus
(Döderlein)

[*C. mertensii* Cuvier et Valenciennes]
[*Pseudocaranx dentex*
(Bloch et Schneider)]
[*Longirostrum delicatissimus*
(Döderlein)]

Striped Jack

Diagnosis — Migrates in coastal waters. Young inhabit bays and river mouths. Body is oblong, has a yellow band on middle surface. Snout is long. Back is bluish-green. Upper part of opercles have a black spot. First dorsal fin has 8 spines, second has 1 spine and 23–26 rays. First dorsal fin is higher than second. Anal fin has 1 spine and 21–23 rays. 24–28 scutes. Good for breeding. Edible. Up to 1.5m in total length.

Distribution — Japan, Ryukyu, Indo-Pacific-Atlantic

▼Young Adult

Carangidae
Caranx melampygus Cuvier

[*C. bixanthopterus* Rüppell]

Blue Jack Fish, Black Ulua, Black-tipped Trevally, Blue Crevalle, Blue Kingfish, Blue-finned Trevally, Malauli, Spotted Trevally

Diagnosis — Body is somewhat deep, greenish-brown, profusely dotted with small black. In young, pectoral fins are yellow. Pectoral sides have scales. Eyes are small. Snout is long. First dorsal fin has 8 spines, second has 1 spine and 22–24 rays. Anal fin has 1 spine and 18–20 rays. 34–38 scutes. Edible but has records of ciguatera poisoning. Up to 1.2m in total length.

Distribution — Japan, Ryukyu, Indo-Pacific, Hawaii, Red Sea, South Africa

▲ Adult

▲ Young Adult

Carangidae
Caranx hemigymnostethus (Bleeker) ⇩
[*Carangoides hemigymnostethus* Bleeker]

Black-banded Trevally (New Common Name)

Diagnosis — Body is oblong, has indistinct dark bands. Dorsal fin ray does not extend long. Scaleless areas of pectoral sides are wide. Snout is short. First dorsal fin has 8 spines, second has 1 spine and 30−33 rays. Anal fin has 1 spine and 25−27 rays. 25−28 scutes. Edible. Up to 50cm in total length.

Distribution — Japan, Ryukyu, Indo-West Pacific

Carangidae
Caranx fulvoguttatus (Forskål) ⇧
[*Carangoides fulvoguttatus* (Forskål)]

Dark Gold-spotted Jack, Yellowspotted Kingfish

Diagnosis — Body is oblong with indistinct broad dark bands, profusely dotted with yellow. Pectoral sides are scaleless. Pectoral fins are long, bases are black. First dorsal fin has 7−8 spines, second has 1 spine and 27−29 rays. Anal fin has 1 spine and 22−24 rays. 14−20 scutes. Edible. Up to 1m in total length.

Distribution — Ryukyu, Indo-West Pacific, Red Sea, South Africa

Carangidae
⇦ *Caranx plagiotaenia* (Bleeker)
[*Carangoides compressus* (Day)]

[*C. plagiotaenia* Bleeker]

Oblique-banded Trevally, Shortridge Kingfish

Diagnosis — Body is compressed, fusiform with 5 indistinct dark bands. Pectoral sides have scales. Pectoral fins are long. Lower jaw is long, mouth is directed upward. Preopercular margins are black. First dorsal fin has 8 spines, second has 1 spine and 22−24 rays. Anal fin has 1 spine and 18−19 rays. 12−19 weak scutes. Edible. Up to 45cm in total length.

Distribution — Ryukyu, Indo-West Pacific, Red Sea, South Africa

Carangidae
Caranx helvolus (Forster) ⇨
[*Uraspis helvolus* (Forster)]

[*U. helvola* (Forster)]

White-tongued Crevalle

Diagnosis — Inhabits rather deep waters. Body is oval and compressed with broad dark bands. Jaws and gill slits are black. Pectoral sides are scaleless. First dorsal fin has 6−8 spines, second has 1 spine and 26−30 rays. Anal fin has 1 spine and 20−22 rays. 33−37 scutes. Edible. Up to 75cm in total length.

Distribution — Japan, Ryukyu, Formosa, Indo-Pacific, Hawaii, Red Sea

Carangidae
⇦ *Caranx hedlandensis* (Whitley)
[*Carangoides hedlandensis* (Whitley)]

Long-finned Trevally

Diagnosis — Body is deep and silvery-white. Dorsal and anal fin rays extend long. Pectoral sides are scaleless. Upper part of opercles have a black spot. First dorsal fin has 8 spines, second has 1 spine and 22 rays. Anal fin has 1 spine and 18 rays. 28−30 weak scutes. Edible. Up to 35cm in total length.

Distribution — Southern Japan, Ryukyu, South China Sea, Indian Ocean

Carangidae
Caranx caeruleopinnatus Rüppell ⇨
[*C. chrysophrys* (Valenciennes)]

[*Carangoides caeruleopinnatus* (Rüppell)]

Bluefin Kingfish

Diagnosis — Body is somewhat deep and blackish. First dorsal fin has 7−8 spines, second has 1 spine and 21−23 rays. Anal fin has 1 spine and 17−19 rays. Anterior parts of dorsal and anal fins are curved. 17−25 small scutes. Edible. Up to 35cm in total length.

Distribution — Japan, Ryukyu, Formosa, Indo-West Pacific, South Africa

Carangidae

Caranx mate (Cuvier) ⇩

[*Atule mate* (Cuvier)]

[*Selar mate* (Cuvier)]

Yellow Tail Scad, One-finlet Scad, Kingfish

Diagnosis — Body is fusiform with indistinct dark bands. Upper part of opercles have a black blotch. Adipose eyelids on eyes are developed. Upper jaw has 2 rows, lower has 1 row of teeth. First dorsal fin has 8 spines, second has 1 spine and 23–24 rays. Anal fin has 1 spine and 19–21 rays. 38–50 scutes. Edible. Up to 30cm in total length.

Distribution — Japan, China, Indo-Pacific, South Africa

Carangidae

⇧ *Decapterus akaadsi* Abe

[*D. korroides* Bleeker]

Red-tail Scad

Diagnosis — Body is compressed, somewhat deep; back is bluish-brown. Adipose eyelids on eyes are well-developed. Caudal fin is reddish. Pectoral fins are long. First dorsal fin has 7–8 spines, second has 1 spine and 27–30 rays. Anal fin has 1 spine and 20–24 rays. 32–35 scutes on entire straight lateral lines. Edible. Up to 40cm in total length.

Distribution — Southern Japan, Ryukyu, Formosa, South China Sea

▼ Body is cylindrical.

Carangidae

Caranx lugubris Poey ⇨

Black Jack

Diagnosis — Body is deep and black. Head bulges out. Caudal lobes are short. Teeth are small. Pectoral sides have scales. Dorsal and anal fin rays extend long. First dorsal fin has 8 spines, second has 1 spine and 21–22 rays. Anal fin has 1 spine and 17–19 rays. 24–33 black scutes. Edible but has records of ciguatera poisoning. Up to 1.2m in total length.

Distribution — Japan, Ryukyu, Bonin Islands, Formosa, Indo-Pacific-Atlantic, Hawaii

Carangidae

Decapterus russellii (Rüppell) ⇧ ⇨

[*D. tabl* Berry]

Mackerel Scad, Russell's Scad, Northern Mackerel Scad, Russell's Mackerel Scad

Diagnosis — Body is cylindrical; back is bluish-brown. Caudal fin is reddish. Pectoral fins are long. Upper part of opercles have a black spot. First dorsal fin has 7–8 spines, second has 1 spine and 29–32 rays. Anal fin has 1 spine and 22–26 rays. 35–40 scutes on entire straight lateral lines. Edible. Up to 45cm in total length.

Distribution — Japan, Ryukyu, Indo-Pacific-Atlantic, Red Sea. South Africa

Carangidae

⇦ *Selar crumenophthalmus* (Bloch)

[*Trachurops crumenophthalmus* (Bloch)]

[*Caranx macrophthalmus* Rüppell]

[*C. mauritianus* Quoy et Gaimard]

Bigeye Scad, Horse-eyed Jack, Goggle-eye Jack, Goggler, Large-eyed Scad, Purse-eyed Scad

Diagnosis — Body is fusiform, has a yellow band on middle surface. Back is bluish-green. Eyes are remarkably large, adipose eyelids are developed. First dorsal fin has 8 spines, second has 1 spine and 24–28 rays. Anal fin has 1 spine and 20–23 rays. 33–38 scutes. Edible but has records of ciguatera poisoning. Up to 30cm in total length.

Distribution — Japan, Ryukyu, Bonin Islands, Formosa, Indo-Pacific-Atlantic, Red Sea, South Africa

Carangidae

⇦ *Caranx dinema* (Bleeker)

[*Carangoides dinema* Bleeker]

Shadow Kingfish, Two Thread Trevally

Diagnosis — Body is oval and silvery-white. Pectoral sides are scaleless. Dorsal and anal fin rays extend long. First dorsal fin has 8 spines, second has 1 spine and 18 rays, second dorsal fin base has black spots. Anal fin has 1 spine and 16–17 rays. 25–27 scutes. Edible. Up to 75cm in total length.

Distribution — Japan, Ryukyu, Indo-West Pacific, South Africa

Carangidae

⇙ *Carangoides armatus* (Rüppell)

[*Caranx armatus* (Forskål)]

Longfinned Trevally, Armed Trevally, Longfinned Crevalle

Diagnosis — Body is compressed, has 6 brown bands. Head is roundish, snout length is equal to eye diameters. Second dorsal and anal fins extend long. First dorsal fin has 8 spines, second has 1 spine and 21–22 rays. Anal fin has 1 spine and 17 rays. 17–20 scutes. Edible. Up to 30cm in total length.

Distribution — Japan, Ryukyu, Formosa, Indo-West Pacific

Carangidae

⇩ *Carangoides chrysophrys* (Cuvier)

[*Caranx chrysophrys* (Valenciennes)]

Longnosed Trevally, Grunting Trevally, Longnose Kingfish, Dusky Trevally, Dusky Crevalle

Diagnosis — Body is rhombic; head is straight; snout is long, longer than eye diameters. Upper part of opercles have a black blotch. First dorsal fin has 8 spines, second has 1 spine and 19–20 rays. Anal fin has 1 spine and 16 rays. 24–26 scutes. Edible. Up to 55cm in total length.

Distribution — Ryukyu, Indo-West Pacific

Carangidae

Caranx sexfasciatus Quoy et Gaimard ⇧

[*C. lessonii* Cuvier et Valenciennes]

Jack, Six-banded Jack, Sixband Trevally, Kingfish, Crevalle Jack, Banded Cavalla, Bigeye Kingfish, Forster's Jack, Great Trevally

Diagnosis — Body is compressed and somewhat deep. Eyes are small, adipose eyelids are developed. Resembles *C. elacate*, but distinguished by slightly smaller mouth, more rays of dorsal and anal fins, more scutes. Pectoral sides are covered with scales. First dorsal fin has 8 spines, second has 1 spine and 20–22 scales. Anal fin has 1 spine and 16–18 rays. 30–33 scutes. Edible but has records of ciguatera poisoning. Up to 1.5m in total length.

Distribution — Japan, Ryukyu, Indo-Pacific, Hawaii, Red Sea, South Africa

Carangidae

⇦ *Trachurus japonicus* (Temminck et Schlegel) ⇧

[*T. trachurus* (Linnaeus)]

[*T. argenteus* Wakiya]

[*Caranx trachurus japonica* Temminck et Schlegel]

Atlantic Horse Mackerel, Horse Mackerel, Jack Mackerel, Rough Scad, Yellowfin Horse Mackerel

Diagnosis — Body is fusiform, somewhat deep. In shoreline type, body is light-brown, in offshore type, bluish-green. First dorsal fin has 8 spines, second has 1 spine and 30–35 rays. Anal fin has 1 spine and 26–30 rays. Scales on lateral lines are all scute and 69–73 in number. Edible but may cause histamine poisoning. Up to 50cm in total length.

Distribution — Japan, Ryukyu, Formosa, Korea, Pacific-Atlantic

Carangidae

⇦ ***Decapterus muroadsi*** (Temminck et Schlegel)

Brownstriped Mackerel Scad, Horse-scad Mackerel, Scad

Diagnosis — Body is elongated; back is bluish-green with a yellow band on middle surface. Upper lobe of caudal fin is yellow, lower lobe is dark-black. First dorsal fin has 8 spines, second has 1 spine and 29−33 rays. Anal fin has 1 spine and 24−28 rays. 33−38 scutes cover ¾ of straight part of lateral lines. Edible. Up to 45cm in total length.

Distribution — Japan, Ryukyu, Korea, China, Indo-Pacific, Hawaii, South Africa

Carangidae

Decapterus lajang Bleeker ⇨

[*D. macrosoma* Bleeker]

Horse-mackerel, Lajang, Mackerel-scad, Slender Mackerel Scad

Diagnosis — Body is cylindrical; back is bluish-green. Scales are small, easily removable. Upper part of opercles have a black spot. Pectoral fins are short. First dorsal fin has 7−8 spines, second has 1 spine and 32−37 rays. Anal fin has 1 spine and 26−31 rays. 27−36 scutes cover ¾ of straight part of lateral lines. Edible. Up to 30cm in total length.

Distribution — Japan, Ryukyu, Bonin Islands, Indo-West Pacific, South Africa

Carangidae

⇦ ***Decapterus macarellus*** (Cuvier)

[*D. macrosoma* Bleeker]

Blue Mackerel Scad, Cherootfish, Mackerel Scad, Sear

Diagnosis — Body is cylindrical; back is bluish-green. Adipose eyelids on eyes are developed. Pectoral fins are short. Posterior end of upper jaw does not reach surface below anterior margin of eyes. 21−35 scutes cover ⅔ of straight part of lateral lines. First dorsal fin has 8 spines, second has 1 spine and 32−36 rays. Anal fin has 1 spine and 27−30 rays. Edible. Up to 40cm in total length.

Distribution — Ryukyu, Izu Islands, Indo-Pacific-Atlantic

Carangidae

Decapterus maruadsi (Temminck et Schlegel) ⇨ ⇩

Amberfish, Round Scad, Whitetipped Mackerel Scad

Diagnosis — Body is fusiform, somewhat deep; back is bluish-green. Caudal fin is yellowish. Pectoral fins are long. Upper part of opercles have a black spot. Adipose eyelids on eyes are developed. First dorsal fin has 7−8 spines, second has 1 spine and 27−36 rays. Anal fin has 1 spine and 23−30 rays. 30−37 scutes on entire straight lateral lines. Edible. Up to 35cm in total length.

Distribution — Japan, Ryukyu, Indo-Pacific, Hawaii

Carangidae

Megalaspis cordyla (Linnaeus) ⇨

[*Scomber cordyla* Linnaeus]

Cordyla Scad, Finletted Mackerel Scad, Finny Scad, Horse Mackerel, Hardtail, Mackerel Scad, Torpedo, Torpedo Kingfish, Torpedo Trevally

Diagnosis — Body is fusiform; tail extends long. Eyes are large, adipose eyelids are developed. First dorsal fin has 8 spines, second has 1 spine and 10−11 rays. Anal fin has 1 spine and 8−11 rays. Caudal peduncles are rounded and thick. 53−58 large scutes on straight long lateral lines. Edible. Up to 1.5m in total length.

Distribution — Japan, Ryukyu, Formosa, Indo-West Pacific, Hawaii, South Africa

Carangidae

⇦ *Trachinotus baillonii* (Lacépède)

[*Caesiomorus quadripunctatus* Rüppell]

Baillon's Dart, Moonfish, Black Spotted Dart, Ladyfish, Black-spotted Swallowtail Dart, Pompano, Smallspot Pompano

Diagnosis — Body is oblong. Caudal fin is deeply falcated, lobes extend long. Second dorsal and anal fin rays extend long. Lateral lines are marked by several black spots. Scutes are absent. First dorsal fin has 5–6 remarkably small spines, second has 1 spine and 21–24 rays. Anal fin has 1 spine and 21–24 rays. Edible. Up to 60cm in total length.

Distribution — Japan, Ryukyu, Indo-Pacific, Red Sea, South Africa

Carangidae

⇩ *Trachinotus blochii* (Lacépède) ⇨

Tropical Pompano, Round Pompano, Snubnose Dart, Oyster-eater, Pompano, Pamphlet, Bloch's Dart, Ladyfish, Moonfish, Buck-nosed Trevally, Derbio

Diagnosis — Body is oval and compressed. Snout is rounded; mouth is small and toothless. Scales are small. Dorsal and anal fin rays extend long. Scutes are absent. First dorsal fin has 6 remarkably small spines, second has 1 spine and 18–20 rays. Anal fin has 1 spine and 16–18 rays. Edible. Up to 80cm in total length.

Distribution — Southern Japan, Ryukyu, Indo-West Pacific, South Africa

Carangidae

Zonichthys nigrofasciatus (Rüppell) ⇩

[*Z. nigrofasciata* (Rüppell)]

[*Seriolina nigrofasciata* (Rüppell)]

[*S. intermedia* (Temminck et Schlegel)]

Black-banded Amberjack, Bowstriped Jack, Butter Yellowtail, Black-banded Kingfish

Diagnosis — Body is fusiform with 5–7 black bands. Snout is short. Fins are dark. Scales are small, scutes are absent. First dorsal fin has 5–6 spines, second has 1 spine and 26–34 rays. Anal fin has 1 spine and 15–16 rays. Edible. Up to 90cm in total length.

Distribution — Southern Japan, Ryukyu, Indo-Pacific-Atlantic, Red Sea, South Africa

Carangidae

Naucrates ductor (Linnaeus) ⇧ ⇨

[*N. indicus* Cuvier et Valenciennes]

Lootsman, Pilot Fish, Romero, Rudderfish, Shark Pilot

Diagnosis — Body is fusiform, has 6 dark bands. Snout is rounded. Scales are small, scutes are absent. Caudal peduncles have a keel. Fins are black. First dorsal fin has 4–5 remarkably short spines, second has 1 spine and 26–29 rays. Anal fin has 1 spine and 15–17 rays. Moves with large fishes. Edible. Up to 60cm in total length.

Distribution — Japan, Ryukyu, Indo-Pacific-Atlantic, South Africa

Carangidae

⇦ ***Scomberoides commersonianus*** Lacépède

Talang, Leatherskin, Queen Fish, Spotted Leatherskin, Whitefish, Skinnyfish, Giant Dart

Diagnosis — Resembles *S. lysan*, but distinguished by somewhat deeper body, larger mouth, posterior end of upper jaw extending far beyond surface below posterior margin of eyes. Lateral lines have dark round blotches. First dorsal fin has 7 separated spines, second has 1 spine and 20 rays. Anal fin has 1 spine and 18 rays. Edible. Up to 1.2m in total length.

Distribution — Southern Japan, Formosa, Indo-West Pacific, South Africa

Carangidae

⇩ ***Scomberoides lysan*** (Forskål) ⇨

[*S. orientalis* (Temminck et Schlegel)]

[*Chorinemus orientalis* (Temminck et Schlegel)]

[*C. lysan* (Forskål)]

Black-tipped Leatherskin, Doubledotted Queenfish, Queenfish, Oriental Leatherskin, Saint Peter's Leatherskin

Diagnosis — Body is compressed and fusiform; tail extends long. Scales are needle-shaped, sink under skin, scutes are absent. In adult, above and below lateral lines are marked by 2 rows of black spots. First dorsal fin has 6–7 separated spines, second has 1 spine and 19–21 rays. Anal fin has 1 spine and 17–19 rays. Edible. Up to 1.2m in total length.

Distribution — Japah, Ryukyu, Formosa, Indo-Pacific, Persian Gulf

Carangidae

Scomberoides tol (Cuvier) ⇩

[*Chorinemus tol* Cuvier et Valenciennes]

Leatherskin, Queen Fish, Slender Leatherskin, Slender Queenfish

Diagnosis — Body is fusiform, remarkably compressed. Lateral lines have several black spots. Scales are needle-shaped, sink under skin, scutes are absent. First dorsal fin has 7 short separated spines, second has 1 spine and 19–21 rays. Anal fin has 1 spine and 18–20 rays. Edible. Up to 60cm in total length.

Distribution — Japan, Ryukyu, Formosa, Indo-West Pacific, South Africa

Carangidae

Elagatis bipinnulata (Quoy et Gaimard) ⇧⇨

Rainbow Runner, Skipjack, Yellowtail, Blue-striped Runner, Hawaiian Salmon, Prodigal Son, Runner, Shoemaker

Diagnosis — Found in schools in upper layers of ocean, also found near shoreline. Body is fusiform with 3 dark bands. Caudal fin is deeply falcated, lobes are extended. Scales are round and small, scutes are absent. First dorsal fin has 6 spines, second has 1 spine and 24–27 rays. Anal fin has 1 spine and 15–17 rays. Edible but has records of ciguatera poisoning. Up to 1.5m in total length.

Distribution — Japan, Ryukyu, Indo-Pacific-Atlantic, Hawaii, Red Sea, South Africa

Carangidae
Seriola lalandi Valenciennes ⇨
[*S. aureovittata* Temminck et Schlegel]

Yellow-tail, Amberjack, Goldstriped Amberjack

Diagnosis — Body is fusiform. A yellow band runs from eyes to tail. Fins are yellow. Pectoral fins are shorter than pelvic fins. Posterior end of upper jaw is rounded. First dorsal fin has 6–8 spines, second has 1 spine and 33–36 rays. Anal fin has 1 spine and 20–22 rays. Edible but has records of ciguatera poisoning. Up to 2m in total length.

Distribution — Japan, Ryukyu, Korea, China, Pacific, Hawaii, Brazil, South Africa

Carangidae
⇦⇩ *Seriola quinqueradiata* Temminck et Schlegel

Japanese Amberjack, Yellow Tail, Yellowtails, Amberjacks

Diagnosis — Migratory fish of temperate zone. Body is fusiform and roundish. Back is bluish-green, has a yellow band. Caudal fin is yellow, pectoral fins are equal in length to pelvic fins. Posterior end of upper jaw is angular. First dorsal fin has 5–6 spines, second has 1 spine and 29–36 rays. Anal fin has 1 spine and 17–22 rays. Good for breeding. Edible but has records of histamine poisoning. Up to 1.5m in total length.

Distribution — Japan, Ryukyu, Formosa, Korea, East China Sea

Carangidae
Seriola dumerili (Risso) ⇧⇨
[*S. purpurascens* Temminck et Schlegel]

[*Caranx dumerili* Risso]

Purplish Amberjack, Rudder Fish, Allied Kingfish, Amberjack, Greater Amberjack, Greater Yellowtail

Diagnosis — Body is fusiform, somewhat deep. Back is purplish-brown. A yellow band runs between eyes and tail. Head has a black band crossing eyes. Posterior end of lower jaw is roundish. First dorsal fin has 7 spines, second has 1 spine and 26–31 rays. Anal fin has 1 spine and 19–21 rays. Edible but has records of ciguatera poisoning. Up to 1.5m in total length.

Distribution — Japan, Ryukyu, Formosa, Korea, China, Hawaii, Indo-Pacific-Atlantic, Red Sea

Carangidae
Seriola rivoliana Valenciennes ⇨
[*S. songoro* Smith]

Longfin Amberjack, Almaco Jack, Longfin Yellowtail

Diagnosis — Resembles *S. dumerili*, but distinguished by somewhat dark body, fewer fin rays, extended anterior rays. Lower caudal tip does not turn white. First dorsal fin has 7 spines, second has 1 spine and 26–29 rays. Anal fin has 1 spine and 19–21 rays. Edible. Up to 1.5m in total length.

Distribution — Southern Japan, Ryukyu, Indo-Pacific-Atlantic

Formionidae

⇧ ***Parastromateus niger*** (Bloch)

[*Formio niger* (Bloch)]

Jack-pomfret, False Butterfish, Black-pomfret, Black Butterfish, German Fish

Diagnosis — Inhabits rather deep shoreline reefs. Body is deep and black. Dorsal fin has 1 spine and 40−44 rays, anal fin has 1 spine and 35−38 rays, correspond to both fins. Pectoral fins are long, pelvic fins are absent. Mouth is small. Caudal peduncles have a keel. 15−17 scutes. Edible. Up to 75cm in total length.

Distribution — Southern Japan, Ryukyu, Formosa, China, Indo-Pacific, Hawaii, Persian Gulf, South Africa

Menidae

⇧ ***Mene maculata*** (Bloch et Schneider)

Moon Fish, Ponyfish, Razor Trevally, Spotted Moonfish

Diagnosis — Inhabits rather deep sea bottoms. Disk-like body is remarkably compressed, abdominal side bulges out. Mouth is small, directed upward, and extensile. Scaleless. Dorsal fin has 3−4 spines and 40−41 rays, anal fin has 30−32 rays. Pelvic fins extend long. Edible. Up to 30cm in total length.

Distribution — Japan, Ryukyu, Formosa, Indo-Pacific, Hawaii, Persian Gulf, South Africa

Lactariidae

⇦ ***Lactarius lactarius*** (Bloch et Schneider)

False Trevally, Milk Trevally

Diagnosis — Inhabits rather deep waters. Resembles *Caranx*, but distinguished by absence of scutes, anal fin base longer than second dorsal fin base, 1 pair of canine teeth in jaws. Scales are small, easily removable. Body depth is equal to head length. First dorsal fin has 8 spines, second has 1 spine and 21 rays. Anal fin has 3 spines and 24 rays. Edible. Up to 40cm in total length.

Distribution — China, South China Sea, Indo-West Pacific

Leiognathidae

⇩ ***Leiognathus bindus*** (Valenciennes)

Orange-finned Pony Fish

Diagnosis — Found in schools in muddy-sand bottoms of coastal waters. Body is somewhat deep. Abdominal margin bulges out. Back has irregular dark blotches. Dorsal fin has 8 spines and 16 rays, is yellowish. Anal fin has 3 spines and 14 rays. Edible. Up to 10cm in total length.

Distribution — Ryukyu, Formosa, China, Indo-West Pacific, East Africa

Leiognathidae

⇩ ***Leiognathus fasciatus*** Lacépède

Banded Pony Fish

Diagnosis — Found in schools in muddy-sand bottoms of bays. Body is deep. Resembles *L. equulus*, but distinguished by extended spines of dorsal fin. Back has irregular black stripes. Dorsal fin has 8 spines and 16 rays, anal fin has 3 spines and 14 rays, spines are not stiff. Edible. Up to 20cm in total length.

Distribution — Southern Japan, Ryukyu, Indo-Pacific, East Africa

Leiognathidae

⇩ ***Leiognathus splendens*** (Cuvier)

Splendid Pony Fish

Diagnosis — Found in schools in coastal waters. Body is oval, back bulges out. Scales are somewhat large. Upper part of eyes have 2 spines. In young, body has irregular black bands. Dorsal fin has 8 spines and 15−17 rays, anal fin has 3 spines and 13−14 rays, anal spines are black. Edible. Up to 13cm in total length.

Distribution — Ryukyu, Formosa, China, Indo-Pacific, Red Sea

Leiognathidae
⇦ ***Leiognathus leuciscus*** (Günther)

Ponyfish, Slipmouth, Miralla

Diagnosis — Found in schools in coastal waters. Body is oval, stripe-patterned on back. Upper part of eyes have 2 black blotches and 2 spines. Second spine of dorsal fin is long and filiform. Dorsal fin has 8 spines and 16 rays, anal fin has 3 spines and 14 rays. Edible. Up to 15cm in total length.

Distribution — Ryukyu, Formosa, Indo-West Pacific

Leiognathidae
⇧ ***Secutor insidiator*** (Bloch)

[*Zeus insidiator* Bloch]

[*Leiognathus insidiatrix* (Day)]

[*Equula insidiatrix* Day]

Slender Barred Pony Fish, Pug-nosed Ponyfish

Diagnosis — Found in schools in muddy-sand bottoms of coastal waters. Body is somewhat shallow, has dark stripes on back. Mouth is directed upward, lower jaw is vertical. Dorsal fin has 8 spines and 16 rays, anal fin has 3 spines and 14 rays. Spines do not extend long. Feeds on zooplankton. Edible. Up to 10cm in total length.

Distribution — Formosa, Indo-Pacific, South Africa

Leiognathidae
⇦⇩ ***Leiognathus equulus*** (Forskål)

[*L. equula* (Forskål)]

Common Slipmouth, Greater Pony Fish, Slimy, Soapy

Diagnosis — Found in schools in coastal waters. Body is remarkably deep. Posterior head bulges out. Back has thin, dark stripes. Upper part of eyes have 2 spines. Dorsal fin has 8 spines and 16 rays, spines do not extend long. Anal fin has 3 spines and 14 rays. Edible. Up to 30cm in total length.

Distribution — Ryukyu, Indo-West Pacific, Red Sea, South Africa

Leiognathidae
⇧ ***Leiognathus rivulatus*** (Temminck et Schlegel)

Offshore Ponyfish, Slimy, Soapy

Diagnosis — Found in schools in offshore waters. Body is shallow; back has dark blotches. Dorsal fin has 8 spines and 16 rays, anal fin has 3 spines and 14 rays. Edible. Up to 10cm in total length.

Distribution — Southern Japan, Ryukyu, Formosa

Leiognathidae
Gazza minuta (Bloch) ⇩

[*Scomber minuta* Bloch]

Toothed Ponyfish

Diagnosis — Found in schools in rather deep muddy-sand bottoms of coastal waters. Body is oval, has dark blotches on back. Mouth is large and extensible. Jaws have large canines. Eyes have adipose eyelids. Dorsal fin has 8 spines and 16 rays, anal fin has 3 spines and 14 rays, spines are thin and soft. Edible. Up to 15cm in total length.

Distribution — Ryukyu, Indo-West Pacific, South Africa

▲ Mouth is extensible.

Leiognathidae
⇦⇧ ***Leiognathus nuchalis*** (Temminck et Schlegel)

Spotnape Ponyfish, Slimy, Soapy

Diagnosis — Found in schools in muddy-sand bottoms of bays. Body is oval and silvery. Mouth is extensile. Scales are small, present only on posterior half of body. Dorsal fin has 8 spines and 16 rays, spines are sharp-pointed. Anal fin has 3 spines and 14 rays. Posterior head and anterior part of dorsal fin have a black blotch. Produces sound or emits light. Spawns in seaweed bed between April and May. Edible. Up to 17cm in total length.

Distribution — Japan, Ryukyu, Korea, Pacific

▼ Head is well-developed in male adult.

Coryphaenidae

⇧ ***Coryphaena hippurus*** Linnaeus ⇧⇨

Dolphin Fish, Common Dolphin, Common Dolphin Fish, Dorado, Dourado

Diagnosis — Migratory fish found in schools in open seas. Body is compressed and extended. Caudal fin is remarkably falcated. In male adult, head is well-developed. Scales are remarkably small. Mouth is large, teeth are present on tongue. Lacks air bladder. Bones are bluish. Dorsal fin has 55−65 rays, anal fin has 26−29 rays. 240−300 lateral scales. Spawns in Japanese waters in summer. Edible but may cause ciguatera or histamine poisoning. Up to 2m in total length.

Distribution — Japan, Ryukyu, Formosa, Indo-Pacific-Atlantic, Hawaii, Persian Gulf, Mediterranean

Coryphaenidae

⇦ ***Coryphaena equiselis*** Linnaeus

[*C. equisetis* Günther]

Lesser Dolphin-fish, Little Dolphin, Pompano Dolphin, Small Dolphin, Small Pompano Dolphin

Diagnosis — Migratory fish found in open seas. Caught among schools of *C. hippurus*. Body is fusiform; dorsal margin is rounded. Caudal fin is thin and long. Dorsal fin has 44 −56 rays, fewer than in *C. hippurus*. Anal fin has 24−28 rays. 160−200 lateral scales. Pelvic fins are situated slightly posterior. Edible. Up to 1m in total length.

Distribution — Japan, Ryukyu, Indo-Pacific-Atlantic, Mediterranean, Hawaii, Galapagos Islands

Rachycentridae

⇦⇧ ***Rachycentron canadum*** (Linnaeus)

[*R. canadus* (Linnaeus)]

Black Bonito, Black King, Black Kingfish, Cabio, Cobia, Crab Eater, Prodigal Son, Sergent Fish, Snook, West Indian Kingfish, Ling, Runner

Diagnosis — Inhabits coastal waters. Body is elongated; head is flat and broad. Dorsal fin has 8 spines and 28−35 rays, spines are remarkably short and separate. Anal fin has 25−27 rays. Scales are remarkably small. Abdominal side is pale. Like shark sucker, has a habit of moving with large fishes. Up to 2m in total length.

Distribution — Japan, Ryukyu, Indo-Pacific-Atlantic, Persian Gulf, South Africa

Embiotocidae

Ditrema temmincki Bleeker

Japanese Sea-perch, Sea Chub, Surf-fish, Temminck's Surfperch

Diagnosis — Found in schools among shoreline reefs and seaweed beds. Body is oval and compressed, blackish-brown or reddish-brown. Lower part of eyes have 2 black spots. Dorsal fin has 9–11 spines and 19–22 rays, anal fin has 3 spines and 25–28 rays. In male, rays are extended. Ovoviviparous. Produces 10–30 juveniles between May and June. Edible. Up to 30cm in total length.

Distribution — Japan, Korea, East China Sea

Pomacentridae

Chromis atripectoralis Welander et Schultz

Black-axil Chromis

Diagnosis — Found in schools about shallow coral reefs. Resembles *C. caeruleus*, but distinguished by slightly larger body, black pectoral fin bases. Canines of lower jaw are projected. Feeds on zooplankton and seaweed. Dorsal fin has 12 spines and 9–10 rays, anal fin has 2 spines and 9–10 rays. 15–16 pored lateral scales. Aquarium fish. Up to 10cm in total length.

Distribution — Ryukyu, Pacific, Hawaii

Pomacentridae

Teixeirichthys jordani (Rutter)

[*T. mossanbicus* Smith]

Bluespot Damselfish

(New Common Name)

Diagnosis — Inhabits rather deep waters. Body is compressed and deep. Pectoral fin bases have a black blotch. Scales on body and opercles have a blue spot. Mouth has 1 row of teeth. Dorsal fin has 13 spines and 12–14 rays, anal fin has 2 spines and 14–15 rays. Aquarium fish. Up to 15cm in total length.

Distribution — Southern Japan, Formosa, China, South Africa

Pomacentridae

Chromis analis (Cuvier)

Gold Reeffish, Puller, Yellow Chromis

Diagnosis — Inhabits rather deep rocky reefs and coral reefs. Body is somewhat deep and dark-yellow. Pectoral fin bases have a small black spot. Dorsal fin has 13 spines and 11–13 rays, anal fin has 2 spines and 11–12 rays, spines are long. 16–19 pored lateral scales. Spawns in summer. Up to 17cm in total length.

Distribution — Japan, Ryukyu, China, Pacific, East Africa

Pomacentridae

Chromis caeruleus (Cuvier)

[*C. caerulea* Cuvier]

[*Heliases caeruleus* (Cuvier)]

Blue-green Puller, Blue Puller, Bluegreen Chromis

Diagnosis — Found in schools about shallow coral reefs. Body is somewhat shallow and bluish-green. Resembles *C. atripectoralis*, but distinguished by slightly smaller body, no black pectoral fin bases. Dorsal fin has 12 spines and 9–10 rays, anal fin has 2 spines and 10 rays. 15–16 pored lateral scales. Feeds on zooplankton. Spawns between spring and summer. Aquarium fish. Up to 10cm in total length.

Distribution — Ryukyu, Indo-Pacific, Red Sea, East Africa

Pomacentridae

Chromis chrysurus (Bliss)

[*C. chrysura* (Bliss)]
[*C. isharai* (Schmidt)]
[*Shiphonochromis lepidostethicus* Fowler]

Whitetail Reeffish, Yellowfin Reeffish, Stout-body Chromis

Diagnosis — Found in schools among rather deep rocky reefs and coral reefs. Body is deep; tail is white. Pelvic fins are black. Dorsal fin has 13 spines and 14−15 rays, anal fin has 2 spines and 10 rays. 17−19 pored lateral scales. Spawns in summer. Edible. Up to 17cm in total length.

Distribution — Japan, Ryukyu, Formosa, Indo-Pacific

Pomacentridae

Chromis xanthurus (Bleeker)

[*C. xanthura* (Bleeker)]

Pale-tail Chromis

Diagnosis — Inhabits rather deep rocky reefs and coral reefs. Body is black; tail is white. Caudal lobes extend long. Dorsal fin has 13 spines and 10−11 rays, anal fin has 2 spines and 9−10 rays. 16−19 pored lateral scales. Up to 15cm in total length.

Distribution — Japan, Ryukyu, Indo-West Pacific

Pomacentridae

Chromis flavomaculatus Kamohara

Yellowfreckled Reeffish

Diagnosis — Found in schools among rather deep coral reefs and rocky reefs. Body is deep. Caudal fin is yellowish, deeply falcated. Pectoral fin bases are black. Dorsal fin has 13−14 spines and 11−13 rays, anal fin has 2 spines and 10−11 rays. 17−19 pored lateral scales. Feeds on zooplankton. Spawns between spring and fall. Up to 15cm in total length.

Distribution — Japan, Ryukyu, Bonin Islands, Formosa, West Pacific

Pomacentridae

Chromis notatus notatus (Temminck et Schlegel)

[*C. notata* (Temminck et Schlegel)]

Coral Fish, Demoiselles, Coral Fish Demoiselles, Whitesaddled Reeffish

Diagnosis — Found in schools about rather deep rocky reefs. Highly cold resistant, also found in the Sea of Japan. Body is somewhat deep. Pectoral fin bases are black. Anterior part of caudal peduncles have a white spot. Dorsal fin has 13−14 spines and 12−14 rays, anal fin has 2 spines and 10−11 rays. 16−19 pored lateral scales. Feeds on zooplankton. Spawns in summer. Edible. Up to 18cm in total length.

Distribution — Japan, Ryukyu, Korea, China, Philippines

Pomacentridae

Chromis margaritifer Fowler

[*C. bicolor* (Macleay)]

Bicolor Chromis, Bicolor Reeffish

Diagnosis — Inhabits outer slopes of coral reefs. Body is somewhat shallow, tail is white. Posterior parts of dorsal and anal fins are white. Pectoral fin bases are black. Dorsal fin has 12 spines and 12−13 rays, anal fin has 2 spines and 11−12 rays. 16−18 pored lateral scales. Feeds on zooplankton and fish-egg. Up to 12cm in total length.

Distribution — Japan, Ryukyu, Formosa, Indo-Pacific, Hawaii, Red Sea, East Africa

Pomacentridae

Chromis lepidolepis Bleeker ⇧

Scaly Chromis

Diagnosis — Inhabits shallow coral reefs. Body is shallow and pale. Caudal lobes are black. Margin of preoperculum is indented. Dorsal fin has 12 spines and 11−13 rays, anal fin has 2 spines and 11−12 rays. 15−18 pored lateral scales. Up to 7cm in total length.

Distribution — Japan, Ryukyu, Formosa, Indo-Pacific

Pomacentridae

Chromis ternatensis (Bleeker) ⇧

Ternate Chromis

Diagnosis — Inhabits outer slope of coral reefs. Body is deep. Upper and lower parts of caudal fin are black. Dorsal fin has 12−13 spines and 10−12 rays, anal fin has 2 spines and 11 rays. 14−17 pored lateral scales. Feeds on zooplankton. Up to 11cm in total length.

Distribution — Ryukyu, Indo-West Pacific, Red Sea, East Africa

Pomacentridae

⇦⇧ ***Pomacentrus lepidogenys*** Fowler et Bean

Scaly Damsel, Scalycheek Damselfish

Diagnosis — Inhabits shallow coral reefs. Preopercular margins are serrated. Dorsal fin has 12 spines and 13−14 rays, margin is black. Anal fin has 2 spines and 13−14 rays. Lateral lines are not complete. Feeds on zooplankton. Up to 7cm in total length.

Distribution — Ryukyu, Pacific

▲Young

Pomacentridae

⇦ ***Pomacentrus flavicauda*** Whitley ⇧

[*P. dorsalis* Gill]

[*P. rhodonotus* Bleeker]

Spotless Yellowtail Damselfish, Yellowtailed Damsel-fish, White-tail Damsel

Diagnosis — Inhabits singly shallow coral reefs. Body is somewhat deep. Caudal fin is white with dark margin. In young, dorsal fin has a black blotch. Feeds on seaweed. Spawns between spring and summer. Dorsal fin has 13 spines and 14−17 rays, anal fin has 2 spines and 14−17 rays. Lateral lines are not complete. Up to 8cm in total length.

Distribution — Japan, Ryukyu, Indo-West Pacific

Pomacentridae

Pomacentrus melanopterus Bleeker ⇨

[*P. alexanderae* Evermann et Seale]

Charcoal Damsel, Bar-finned Demoiselle, Dark-finned Damsel-fish

Diagnosis — Found in schools about shallow coral reefs. Body is pale. Pectoral fin bases are black. Has scales below eyes. Dorsal fin has 13 spines and 13−14 rays, anal fin has 2 spines and 14−15 rays. Lateral lines are not complete. Feeds on zooplankton. Up to 10cm in total length.

Distribution — Ryukyu, Formosa, Pacific

Pomacentridae
Pomacentrus vaiuli
Jordan et Seale ⇨

Princess Damsel

Diagnosis — Inhabits outer slope of coral reefs. Body is somewhat deep and light-brown. Upper part of opercles and dorsal fin ray have a black blotch. Dorsal fin has 13 spines and 15–16 rays, anal fin has 2 spines and 15–16 rays. Lateral lines are not complete. Up to 9cm in total length.

Distribution — Ryukyu, Izu Islands, Pacific

Pomacentridae
Pomacentrus amboinensis Bleeker ⇧

Amboina Demoiselle, Pallid Damselfish, Ambon Damsel

Diagnosis — Inhabits outer slope of coral reefs. Resembles *P. moluccensis*, but distinguished by brownish-yellow body, a large black blotch on pectoral fin bases. In young, dorsal fin has a black blotch. Dorsal fin has 13 spines and 14–16 rays, anal fin has 2 spines and 14–16 rays. Feeds on zooplankton and seaweed. Up to 10cm in total length.

Distribution — Ryukyu, Pacific

Pomacentridae
Pomacentrus moluccensis Bleeker ⇧

Molucca Sea Damselfish, Molucca Damsel

Diagnosis — Inhabits shallow coral reefs. Body is somewhat deep and yellow. Back is dark. Pectoral fin bases and lateral line origins have a small black spot. Anus is black. In young, dorsal fin has no black blotch. Dorsal fin has 13 spines and 13–16 rays, anal fin has 2 spines and 14–15 rays. Up to 8cm in total length.

Distribution — Ryukyu, Formosa, Pacific

Pomacentridae
Pomachromis richardsoni (Snyder) ⇧

Richardson's Damselfish

Diagnosis — Found in schools about outer slope of coral reefs. Body is shallow. Upper and lower parts of caudal fin are black, lobes are extended. Jaws have a row of teeth. Dorsal fin has 14 spines and 13–14 rays, anal fin has 2 spines and 11–14 rays. Lateral lines do not run over full length. Feeds on zooplankton. Up to 10cm in total length.

Distribution — Ryukyu, Indo-Pacific

Pomacentridae
Pomacentrus bankanensis Bleeker ⇨

Monocle Damselfish, Spotfin Damselfish, Speckled Damsel

Diagnosis — Inhabits shallow coral reefs. Body is deep and dark. Caudal fin is white. Fore head is orange. Dorsal fin has 13 spines, 14–16 rays, a black blotch. Anal fin has 2 spines and 14–16 rays. Lateral lines are not complete. Up to 10cm in total length.

Distribution — Ryukyu, Formosa, Pacific

Pomacentridae
⇦ *Pomacentrus philippinus* Evermann et Seale

Philippine Damselfish

Diagnosis — Inhabits singly or in small schools among coral reefs. Body is somewhat shallow. Dorsal and anal fin rays are yellow. Caudal fin is also yellow. Pectoral fin bases are black. Dorsal fin has 13 spines and 13–16 rays, anal fin has 2 spines and 14–16 rays. Up to 13cm in total length.

Distribution — Ryukyu, Pacific

Pomacentridae
⇦ *Pomacentrus coelestis*
Jordan et Starks

Heavenly Damselfish

Diagnosis — Found in schools about shoreline reefs and coral reefs. Body is elongated and blue. Caudal fin is pale. Dorsal fin has 13 spines and 14−15 rays, anal fin has 2 spines and 14−15 rays. Spawns in summer. Aquarium fish. Up to 9cm in total length.

Distribution — Japan, Ryukyu, Pacific

Pomacentridae
Neopomacentrus taeniurus
(Bleeker) ⇩

Fresh-water Demoiselle, Blue Ribboned Demoiselle, Crescent Damsel

Diagnosis — Inhabits shallow coral reefs. Body is dark. Caudal lobes extend long. Margins of fins are blue. Dorsal fin has 12 spines and 10−12 rays, anal fin has 2 spines and 11−12 rays. Lateral lines are not complete. Up to 8cm in total length.

Distribution — Ryukyu, Indo-West Pacific

Pomacentridae
Neopomacentrus violascens
(Bleeker) ⇨

[*Pomacentrus violascens* (Bleeker)]

Purple Damselfish, Yellow-tailed Demoiselle, Violet Demoiselle

Diagnosis — Inhabits shallow rocky reefs and coral reefs. Body is slightly elongated and dark-purple. Caudal fin is yellow. Upper part of opercles have a black spot. Dorsal fin has 13 spines and 11−12 rays, anal fin has 2 spines and 10−11 rays. Up to 8cm in total length.

Distribution — Japan, Ryukyu, China, West Pacific

Pomacentridae
⇦ *Chrysiptera rex* (Snyder)

[*Glyphidodontops rex* (Snyder)]

Lemon Damselfish

Diagnosis — Inhabits singly shallow coral reefs. Body is shallow and yellowish-brown. Head is bluish. Opercles have a small black spot on upper surface. Dorsal fin has 13 spines and 12−15 rays, anal fin has 2 spines and 11−14 rays. Aquarium fish. Up to 8cm in total length.

Distribution — Ryukyu, West Pacific

Pomacentridae
Eupomacentrus nigricans (Lacépède) ⇩

[*Pomacentrus nigricans* (Lacépède)]

[*Parapomacentrus nigricans* (Lacépède)]

Black Damsel, Dusky Demoiselle, Olive Damselfish, Dusky Gregory

Diagnosis — Inhabits shallow coral reefs. Body is oval, snout is slightly pointed. Dorsal fin has 12 spines and 14−17 rays, a black spot present on posterior end. Anal fin has 2 spines and 12−13 rays. Pored lateral scales are not complete. Feeds on zooplankton. Aquarium fish. Up to 15cm in total length.

Distribution — Ryukyu, Formosa, China, Indo-Pacific, East Africa

Pomacentridae
Neopomacentrus azysron (Bleeker) ⇩

Yellow-tail Demoiselle

Diagnosis — Found in schools in outer slope of coral reefs. Body is elongated, tail is extended. Upper surface of opercles have a black splotch. Pectoral fin base is black, tail is yellow. Caudal fin is truncated, lobes are extended. Dorsal fin has 13 spines and 11−12 rays, anal fin has 2 spines and 11−12 rays. 17−19 pored lateral scales. Aquarium fish. Up to 8cm in total length.

Distribution — West Pacific

Pomacentridae

Dischistodus prosopotaenia (Bleeker) ⇨

[*D. notophthalmus* (Bleeker)]

[*Pomacentrus prosopotaenia* Bleeker]

Honey-head Damsel, Spotted Yellow Damselfish

Diagnosis — Inhabits shallow coral reefs. Body is oval, back is brown, abdominal side is white. In young, body has 2 brown bands, dorsal fin has a large black blotch. Dorsal fin has 13 spines and 14–15 rays, anal fin has 2 spines and 13–14 rays, ray part is triangular. Jaws have 2 rows of teeth. Up to 18cm in total length.

Distribution — Ryukyu, West Pacific

▼Female (Variety)

▼Male (above) Female (below)

Pomacentridae

⇧ ***Chrysiptera hemicyanea*** (Weber)

[*C. parasema* (Fowler)]

[*Glyphidodontops parasema* (Fowler)]

[*G. hemicyaneus* (Weber)]

Azure Demoiselle

Diagnosis — Inhabits shallow coral reefs. Body is oblong, of brilliant cobalt blue. Tail is yellow. Caudal margin is lunated. Pectoral fin bases have a black spot. Dorsal fin has 13 spines and 10–12 rays, anal fin has 2 spines and 11–13 rays. Aquarium fish. Up to 7cm in total length.

Distribution — Ryukyu, West Pacific

Pomacentridae

⇧ ***Chrysiptera cyanea*** (Quoy et Gaimard) ⇧

[*Abudefduf assimilis* (Günther)]

[*A. cyaneus* (Quoy et Gaimard)]

[*Glyphidodontops cyaneus* (Quoy et Gaimard)]

One-spot Sergent-major, Blue-devil, Blue Damselfish, Blue Duskytail, Cornflower Sergent-major, Emerald Damselfish, One-spot Demoiselle

Diagnosis — Found in schools about shoreline reefs and coral reefs. Body is slightly elongated, of beautiful cobalt blue. In female, dorsal and anal fin rays, caudal fin are transparent. In some individuals, caudal fin is orange. Dorsal fin has 13 spines and 12–13 rays, anal fin has 2 spines and 12–14 rays. Spawns between spring and summer. Aquarium fish. Up to 8cm in total length.

Distribution — Japan, Ryukyu, Formosa, Pacific

◀Adult

▼Young

Pomacentridae

⇦ ***Chrysiptera leucopoma*** (Lesson) ⇧

[*C. biocellata* (Cuvier)]

[*Glyphidodontops leucopomus* (Lesson)]

White-banded Damsel-fish

Diagnosis — Inhabits singly shallow coral reefs. Body is shallow, has 2 white bands. Cheeks have 3 rows of scales. Pectoral fin bases, opercles and caudal fin are white. In young, a blue line runs from eyes to back. Dorsal fin has 13 spines and 12–15 rays, anal fin has 2 spines and 11–14 rays. Aquarium fish. Up to 8cm in total length.

Distribution — Japan, Ryukyu, Indo-West Pacific

▲ Young

Pomacentridae

Paraglyphidodon behni (Bleeker) ⇦⇩

[*P. nigroris* (Cuvier)]

[*Chromis bitaeniatus* Fowler et Bean]

Behn's Damsel, Chocolate Damselfish, Scar-face Damsel-fish

Diagnosis — Inhabits shallow coral reefs. Body is oval. Fins are extended. Head has 2 black bands. In young, body is yellow with 2 longitudinal black bands. Dorsal fin has 13 spines and 14−16 rays, anal fin has 2 spines and 13−15 rays. Up to 12cm in total length.

Distribution — Ryukyu, Formosa, Indo-West Pacific

Pomacentridae

Plectroglyphidodon dickii (Liénard) ⇩

[*Abudefduf dickii* (Liénard)]

Blackband Damselfish, Narrowbar Damsel, Dick's Damsel

Diagnosis — Inhabits singly shallow coral reefs. Body is somewhat deep and light-brown. Tail is white. A black band runs from dorsal fin to anal fin. Dorsal fin has 12 spines and 16−18 rays, anterior surface has a black spot. Anal fin has 2 spines and 14−16 rays. Feeds on seaweed. Up to 9cm in total length.

Distribution — Japan, Ryukyu, Indo-Pacific

▲ Adult

Pomacentridae

Plectroglyphidodon johnstonianus Fowler et Ball ⇧

[*P. nitidus* Smith]

Johnston Damsel

Diagnosis — Inhabits shallow coral reefs. Resembles *P. dickii*, but distinguished by sufficiently broad black bands on hind to reach caudal peduncles. Tail is light-yellow. Dorsal fin has 12 spines and 18−19 rays, anal fin has 2 spines and 16−18 rays. Up to 9cm in total length.

Distribution — Ryukyu, Bonin Islands, Indo-Pacific, Hawaii

Pomacentridae

Pristotis jerdoni (Day) ⇩

Gulf Damsel

Diagnosis — Inhabits rather deep waters. Body is oblong, snout is short. Teeth are in 1 row and pointed. Dorsal fin has 13 spines and 12−13 rays, anal fin has 2 spines and 12−14 rays. Edible. Up to 15cm in total length.

Distribution — Ryukyu, Formosa, South China Sea, Indo-West Pacific, Persian Gulf

Pomacentridae

⇦ ***Plectroglyphidodon lacrymatus*** (Quoy et Gaimard)

[*Abudefduf lacrymatus* (Quoy et Gaimard)]

Jewel Damsel, Fleckled Spotfin Damselfish, White-speckled Demoiselle

Diagnosis — Inhabits singly shallow coral reefs. Body is deep, profusely dotted with blue. In young, dorsal fin has a black blotch. Dorsal fin has 12 spines and 16−18 rays, anal fin has 2 spines and 13−15 rays. Lateral lines are not complete. Feeds on seaweed. Aquarium fish. Up to 10cm in total length.

Distribution — Ryukyu, Indo-Pacific, Red Sea, South Africa

▼Adult ▲Young

Pomacentridae
Plectroglyphidodon leucozona (Bleeker) ⇧⇨

[*P. leucozonus leucozonus* (Bleeker)]

Coralfish, Whiteband Damselfish, Whitebar Damselfish

Diagnosis — Inhabits outer slope of coral reefs. Body is somewhat deep, has an indistinct white band on middle surface. In young, dorsal fin has a large black blotch. Dorsal fin has 12 spines and 14–16 rays, anal fin has 2 spines and 11–13 rays. Up to 12cm in total length.

Distribution — Japan, Ryukyu, Indo-Pacific, Red Sea, East Africa

Pomacentridae
Abudefduf septemfasciatus (Cuvier) ⇧

Seven-band Demoiselle, Seven-banded Sergent-major, Sevenbar Damsel, Banded Sergent

Diagnosis — Inhabits outer slope of coral reefs. Body is somewhat deep, has 7 dark bands. Dorsal fin has 13 spines and 12–14 rays, anal fin has 2 spines and 12–13 rays. Lateral lines are not complete. Aquarium fish. Up to 18cm in total length.

Distribution — Japan, Ryukyu, Indo-Pacific

▲Young

▲Adult

Pomacentridae
⇧ ***Paraglyphidodon melas*** (Cuvier) ⇧

[*Glyphidodon melanopus* Bleeker]

[*Abudefduf melas* (Cuvier et Valenciennes)]

Black Damsel-fish, Zulu Damsel, Black Sergentmajor, Royal Damselfish, Yellowback Damselfish

Diagnosis — Inhabits shallow coral reefs. Body is deep, black over full length. In young, body is white, part from above eyes to dorsal fin is yellow. Anterior parts of pelvic and anal fins are dark blue. Dorsal fin has 13 spines and 14–16 rays, anal fin has 2 spines and 13–14 rays. Aquarium fish. Up to 20cm in total length.

Distribution — Southern Japan, Ryukyu, Formosa, Indo-West Pacific, Red Sea, East Africa

Pomacentridae
Dascyllus reticulatus (Richardson) ⇧⇨

Reticulated Humbug, Reticulated Puller, Twobar Humbug, Two-striped Damselfish, Reticulated Dascyllus

Diagnosis — Found in schools about rather deep coral reefs. Body is deep, has 1–2 black bands. Dorsal fin has 12 spines and 14–16 rays, margin is black. Anal fin has 2 spines and 13–14 rays. 18–19 pored lateral scales. Aquarium fish. Up to 6.5cm in total length.

Distribution — Japan, Ryukyu, Indo-Pacific, Hawaii, Red Sea, South Africa

Pomacentridae

⇦ ***Dascyllus aruanus*** (Linnaeus) ⇩

[*Tetradrachmum aruanum* (Linnaeus)]

Banded Humbug, Banded Dascyllus, Indo-Pacific Threestripe Damsel, Threeband Puller, Whitespotted Puller, White-tailed Footballer, Zebra Humbug

Diagnosis — Found in schools about rather shallow coral reefs. Body is deep, has 3 black bands. Caudal fin is white. Dorsal fin has 12 spines and 12–13 rays, anal fin has 2 spines and 12–13 rays. 16–18 pored lateral scales. Feeds on zooplankton and seaweed. Spawns between spring and summer. Aquarium fish. Up to 8cm in total length.

Distribution — Ryukyu, Indo-Pacific, Hawaii, Red Sea, South Africa

Pomacentridae

Dascyllus melanurus Bleeker ⇨

Four-striped Damsel, Black-tailed Dascyllus

Diagnosis — Inhabits shallow coral reefs. Resembles *D. aruanus*, but distinguished by black caudal margin. Dorsal fin has 12 spines and 12–13 rays, anal fin has 2 spines and 12–13 rays. Feeds on zooplankton and seaweed. Aquarium fish. Up to 9cm in total length.

Distribution — Ryukyu, West Pacific

▲ Young

Pomacentridae

⇦ ***Dascyllus trimaculatus*** (Rüppell) ⇩

Domino, Three-spot Humbug, White-spot Puller, White-spotted Puller, Three-spot Dascyllus

Diagnosis — Lives together with sea anemone inhabiting sandy bottoms of shallow coral reefs. Body is deep; snout is short. Back and head crown have white spots which disappear in adult. Dorsal fin has 11–12 spines and 14–16 rays, anal fin has 2 spines and 13–15 rays. Feeds on zooplankton and seaweed. Spawns in summer. Aquarium fish. Up to 15cm in total length.

Distribution — Japan, Ryukyu, Indo-Pacific, Hawaii, Red Sea, South Africa

▲ Adult

Pomacentridae

Amblyglyphidodon leucogaster (Bleeker) ⇩

[*Abudefduf leucogaster* (Bleeker)]

Golden-scaled Damsel-fish, Whitebreasted Sergent-major, Whitebell Damselfish

Diagnosis — Inhabits shallow coral reefs. Body is deep, dark on upper half, abdominal side is yellow. Pectoral fin bases have a black spot. Dorsal fin has 13 spines and 12–13 rays, anal fin has 2 spines and 13–14 rays. Lateral lines are not complete. Up to 18cm in total length.

Distribution — Southern Japan, Ryukyu, Indo-Pacific, Red Sea, East Africa

Pomacentridae

⇦ ***Dascyllus albisella*** Gill

Hawaiian Dascyllus, White-spotted Damsel Fish, Single Spot Dascyllus

Diagnosis — Found in schools about rather deep rocky reefs. Body is deep, has a large white blotch on back. Fish that has been identified as *D. albisella* in Japan is actually *D. trimaculatus*, this species is not found in waters of Japan. Dorsal fin has 12 spines and 15–16 rays, anal fin has 2 spines and 15–16 rays. 19–20 pored lateral scales. Aquarium fish. Up to 11cm in total length.

Distribution — Hawaii

Pomacentridae

⇦ ***Abudefduf sordidus*** (Forskål) ⇨

Coralfish, Spot Damsel, Spot Pom, Yellow-banded Sergent-major, Black-spot Sergent

Diagnosis — Inhabits outer slope of coral reefs. Body is deep, has broad dark bands. Caudal peduncles have a black blotch on upper part. Dorsal fin has 13 spines and 14–16 rays, anterior part is black. Anal fin has 2 spines and 14–16 rays. Aquarium fish. Up to 22cm in total length.

Distribution — Japan, Ryukyu, China, Indo-Pacific, Hawaii, Red Sea, South Africa

Pomacentridae

⇦ ***Abudefduf sexfasciatus*** (Lacépède) ⇩

[*A. coelestinus* (Cuvier et Valenciennes)]

Striped-tail Damselfish, Scissor-tail Sergent, Six- banded Sergent-major, Stripetail Damsel, Convict Damselfish

Diagnosis — Found in schools about shallow coral reefs. Body is somewhat deep, has 6 black bands. Upper and lower surfaces of caudal fin are black. Dorsal fin has 13 spines and 12–14 rays, anal fin has 2 spines and 12–14 rays. A row of scales below eyes. Feeds on zooplankton and seaweed. Spawns between spring and summer. Edible but has records of ciguatera poisoning. Aquarium fish. Up to 20cm in total length.

Distribution — Japan, Ryukyu, Formosa, China, Indo-Pacific-Atlantic, Hawaii, Red Sea, South Africa

Pomacentridae

⇦ ***Amblyglyphidodon curacao*** (Bloch)

[*Abudefduf curacao* (Bloch)]

Black-snouted Sergent-major, Clouded Damsel-fish, Demoiselle

Diagnosis — Found in schools about shallow coral reefs. Body is deep, has 5 dark bands. 1–2 rows of scales below eyes. Dorsal fin has 13 spines and 11–14 rays, anal fin has 2 spines and 13–14 rays. Feeds on zooplankton and seaweed. Aquarium fish. Up to 10cm in total length.

Distribution — Ryukyu, China, Pacific

Pomacentridae

⇦ ***Abudefduf vaigiensis*** (Quoy et Gaimard)

Five-banded Damsel-fish, Fivebanded Sergent-major, Sergent Major, Waigieu Demoiselle

Diagnosis — Found in schools about shallow coral reefs and rocky reefs. Body is oval, has 5 black bands. Back is yellowish. A row of scales below eyes. Dorsal fin has 13 spines and 11–13 rays, anal fin has 2 spines and 11–13 rays. Aquarium fish. Up to 21cm in total length.

Distribution — Japan, Ryukyu, Formosa, Indo-Pacific, Galapagos Islands, Red Sea, South Africa

Pomacentridae

Hemiglyphidodon plagiometopon (Bleeker) ⇨

Lagoon Damsel, Blackfin Damselfish, Purple-spotted Sergent-major

Diagnosis — Inhabits singly shallow coral reefs. Body is oval. Fins are well-developed. Margin of eyes are orange. Dorsal fin has 13 spines and 13–15 rays, anal fin has 2 spines and 13–15 rays. Lateral lines are not complete. Up to 20cm in total length.

Distribution — Ryukyu, China, West Pacific

Pomacentridae

⇦ *Amphiprion clarkii* (Bennett) ⇩

[*A. xanthurus* Cuvier et Valenciennes]

[*A. polymnus* (Linnaeus)]

[*A. bicinctus* Rüppell]

Anemone Fish, Goldbelly Anemonefish, Three-banded Anemonefish, Two-banded Anemonefish, Yellow-tailed Anemone Fish, Clark's Anemonefish

Diagnosis — Lives together with large sea anemone inhabiting coral reefs. Body has 3 white bands. Caudal fin is white, yellow, yellow-margined. In young, snout is white, white bands on body extend to dorsal fin. Dorsal fin has 10–11 spines and 14–17 rays, anal fin has 2 spines and 12–15 rays. Spawns between summer and fall. Aquarium fish. Up to 15cm in total length.

Distribution — Japan, Ryukyu, Bonin Islands, China, Indo-Pacific, Persian Gulf, Red Sea, East Africa

▲ Young Adult

Pomacentridae

Amphiprion perideraion Bleeker ⇩

Pink Anemonefish

Diagnosis — Lives together with sea anemone inhabiting coral reefs. Body is somewhat deep and orange. Head has a white band. Dorsal fin has 9–10 spines and 16–17 rays, anal fin has 2 spines and 12–13 rays. Feeds on zooplankton and seaweed. Up to 20cm in total length.

Distribution — Ryukyu, Pacific

Pomacentridae

Amphiprion melanopus Bleeker ⇨

Red and Black Anemonefish

Diagnosis — Lives together with sea anemone inhabiting coral reefs. Resembles *A. frenatus*, but distinguished by fewer rays on dorsal fin. In adult, pelvic and anal fins turn blackish. In young, body has 2 white bands. Dorsal fin has 10 spines and 15–18 rays, anal fin has 2 spines and 13–15 rays. Aquarium fish. Up to 9cm in total length.

Distribution — Ryukyu, Pacific

Pomacentridae

⇧ *Amphiprion biaculeatus* (Bloch)

[*Chaetodon biaculeatus* Bloch]

[*Premnas biaculeatus* (Bloch)]

Spine-cheek Anemonefish, Tomato Clownfish

Diagnosis — Lives together with sea anemone inhabiting coral reefs. Resembles *A. ocellaris*, but distinguished by narrower white bands on body. Margins of fins are not black. Cheeks have 2 spines. Dorsal fin has 10 spines and 16–19 rays, anal fin has 2 spines and 13–15 rays. Aquarium fish. Up to 15cm in total length.

Distribution — Indo-West Pacific

Pomacentridae

Amphiprion frenatus Brevoort ⇩

Tomato Anemonefish, Bridled Anemonefish, Oneband Anemonefish, Onebar Anemonefish

Diagnosis — Lives together with sea anemone inhabiting coral reefs. Body is reddish; head has a white band fringed in black. In young, body has 2–3 white bands. Dorsal fin has 9–10 spines and 16–18 rays, anal fin has 2 spines and 13–15 rays. Transsexual. Spawns between spring and fall. Aquarium fish. Up to 15cm in total length.

Distribution — Japan, Ryukyu, Formosa, China, Indo-Pacific, Red Sea, South Africa

Pomacentridae

Amphiprion akindynos Allen ⇨

White-tailed Clownfish, Barrier Reef Anemonefish

Diagnosis — Lives together with sea anemone inhabiting coral reefs. Resembles *A. frenatus*, but distinguished by lighter body color, narrower white bands on middle of body. Dorsal fin has 10 spines and 14–17 rays, anal fin has 2 spines and 13–14 rays. Aquarium fish. Up to 12cm in total length.

Distribution — New Caledonia, Australia

Pomacentridae

Amphiprion sandracinos Allen ⇧

Orange Anemonefish, White-backed Anemonefish, Allen's Clownfish

Diagnosis — Lives together with sea anemone inhabiting coral reefs. Body is shallow; back has a white longitudinal band. Dorsal fin has 8–10 spines and 16–18 rays, anal fin has 2 spines and 12 rays. Aquarium fish. Up to 8cm in total length.

Distribution — Ryukyu, Formosa, West Pacific

Pomacentridae

⇦ *Amphiprion ephippium* (Bloch)

[*A. calliops* Schultz]

Black-backed Clownfish, Red Saddleback Anemonefish

Diagnosis — Lives together with sea anemone inhabiting coral reefs. Resembles *A. frenatus*, but white bands on body disappear in adult, posterior part of dorsal is dark. In young, head has a white band. Dorsal fin has 10 spines and 16–18 rays, anal fin has 2 spines and 13–14 rays. Aquarium fish. Up to 9cm in total length.

Distribution — Indonesia, Malaysia, Thailand, Nicobar Islands, Andaman Islands

Pomacentridae

Amphiprion polymnus (Linnaeus) ⇨

Twoband Anemonefish, White-tipped Anemonefish, Saddleback Anemonefish

Diagnosis — Lives together with sea anemone inhabiting sandy bottoms of bays. Body is somewhat deep. Head, dorsal side, caudal margin are white. In adult, head turns reddish. Dorsal fin has 10–11 spines and 12–16 rays, anal fin has 2–3 spines and 12–16 rays. Aquarium fish. Up to 12 cm in total length.

Distribution — Ryukyu, Formosa, China, Indo-West Pacific, South Africa

Pomacentridae

⇦ *Amphiprion ocellaris* Cuvier

[*Actinicola percula* (Lacépède)]

Amphiprion, Anemone Demoiselle, Clown Anemone Fish, Clown Fish, Harlequin Fish, Orange Anemone Fish, False Clown Anemonefish

Diagnosis — Lives together with sea anemone inhabiting coral reefs. Body is shallow, has 3 white bands. Margins of fins are black. Dorsal fin has 10–11 spines and 13–17 rays, anal fin has 2 spines ans 11–13 rays. Feeds on zooplankton and seaweed. Aquarium fish. Up to 15cm in total length.

Distribution — Ryukyu, Formosa, Indo-Pacific

Cichlidae

Tilapia nilotica (Linnaeus)

Tilapia, Nile Mouthbreeder

Diagnosis — Freshwater fish of African origin, also found in brackish areas. Resembles *T. mossambica*, but distinguished by slightly deeper body. Also grows to larger size, has higher cold-resistance. Omnivorous. Good for breeding. Dorsal fin has 16–17 spines and 12–13 rays, anal fin has 3 spines and 10–11 rays. Lateral lines are not complete. Edible. Up to 57cm in total length.

Distribution — Japan, Ryukyu, Indo-Pacific, Africa

▲ Adult

▼Young

▲ Young

Cichlidae

Tilapia mossambica (Peters)

Tilapia, Leather Jacket, Mozambique Mouthbreeder, Mozambique Mouthbrooder, Mozambique Tilapia

Diagnosis — Inhabits freshwater and brackish areas. Resembles *T. nilotica*, but distinguished by lump-like protuberance on back of snout. In adult, body turns blackish, pectoral, dorsal, caudal margins turn reddish. Grows in 3–4 months, spawns in sand bottoms. Omnivorous. Dorsal fin has 15–17 spines and 12–13 rays, anal fin has 3 spines and 9–10 rays. Edible. Up to 30cm in total length.

Distribution — Japan, Ryukyu, Indo-Pacific, Africa

▼Young

▲ Adult

Cichlidae

Tilapia galilaea (Linnaeus)

Galilee Ciclid, Mouth-breeder, Galilee Mouthbrooder

Diagnosis — Resembles *T. mossambica*, but distinguished by blackish anterior half of body. Edible. Aquarium fish. Up to 30cm in total length.

Distribution — Africa

Labridae

⇖⇧ ***Choerodon azurio*** (Jordan et Snyder)

Scarbreast Tuskfish

Diagnosis — Inhabits rather deep rocky reefs. Head is developed and angular. A dark oblique band runs from dorsal fin to pectoral fins. Dorsal fin has 12 spines and 7 rays, anal fin has 3 spines and 10 rays. 24–28 lateral scales. Spawns in June. Edible. Up to 40cm in total length.

Distribution — Japan, Ryukyu, Formosa, Korea, China, East China Sea

Labridae

⇦⇙ ***Choerodon schoenleinii*** (Valenciennes) ⇨

Black Spot Tusk Fish, Green Wrasse, Parrot-fish, Purple Grouper, Rainbow Fish

Diagnosis — Inhabits rather deep rocky reefs. Body is somewhat deep, profusely dotted with blue. Snout is roundish. Dorsal fin has 12–13 spines, 7–8 rays and a black spot. Anal fin has 3 spines and 9–10 rays. 26–29 lateral scales. Edible. Up to 1m in total length.

Distribution — Ryukyu, China, West Pacific

Green Wrasse has blue canines.▲

Labridae

Choerodon anchorago (Bloch) ⇨

Orange Dotted Tusk Fish, Pigtoothed Wrasse, Yellow-cheeked Tuskfish

Diagnosis — Inhabits rather deep coral reefs. Body is compressed and deep. Pectoral and anal fins are yellow. Dorsal fin has 12–13 spines and 7 rays, anal fin has 3 spines and 7–9 rays. 26–30 lateral scales. Edible. Up to 32cm in total length.

Distribution — Ryukyu, Indo-Pacific

▼Female

Labridae

⇦ ***Choerodon robustus*** (Günther)

[*Xiphocheilus robustus* Günther]

Sword-lipped Wrasse

Diagnosis — Inhabits bottoms more than 100m deep. Body is deep; snout is rounded. In male, body has a yellow band, in female, lower part is white. Lower jaw is blue. Dorsal fin has 12 spines and 8 rays, anal fin has 3 spines and 10 rays. 29 lateral scales. Edible. Up to 35cm in total length.

Distribution — Ryukyu, Indo-Pacific, Persian Gulf, East Africa

Labridae

⇧ ***Choerodon jordani*** (Snyder)

Black-banded Wrasse

(New Common Name)

Diagnosis — Inhabits rather deep rocky reefs. Body is slightly elongated, has a black oblique band. Snout is rounded. Anterior part of caudal peduncles have a large white blotch. Dorsal fin has 13 spines and 6–7 rays, anal fin has 3 spines and 9–10 rays. 29 lateral scales. Up to 20cm in total length.

Distribution — Southern Japan, Ryukyu, Formosa

Labridae

Lienardella fasciata (Günther) ⇧

Harlequin Tusk-fish, Harlequin Wrasse

Diagnosis — Inhabits shallow coral reefs. Teeth are bluish and projected. Body has 7–8 orange bands. Caudal fin is white. In young, dorsal and anal fins have a black blotch. Dorsal fin has 12 spines and 7–8 rays, anal fin has 3 spines and 10 rays. 28–29 lateral scales. Edible. Up to 30cm in total length.

Distribution — Ryukyu, West Pacific

Labridae

⇦ ***Bodianus perditio*** (Quoy et Gaimard) ⇧

[*Lepidaplois perditio* (Quoy et Gaimard)]

[*Labrus perditio* Quoy et Gaimard]

Freckleface Hogfish, Reef Parrot, Golden-spot Pig Fish, Golden-spot Wrasse, Goldsaddle Hogfish

Diagnosis — Inhabits singly rather deep rocky reefs. Back has a white cross band. Dorsal fin has 12–13 spines and 10 rays, anterior part is black. Anal fin has 3 spines and 12 rays. 33–34 lateral scales. Carnivorous. Edible but may cause ciguatera poisoning. Up to 90cm in total length.

Distribution — Japan, Ryukyu, Bonin Islands, Indo-Pacific, Red Sea, South Africa

Labridae

Bodianus diana (Lacépède) ⇨

[*Lepidaplois diana* (Lacèpède)]

[*L. aldabrensis* Smith]

Diana's Wrasse, Three-eyed Hogfish, Diana's Hogfish

Diagnosis — Inhabits shoreline reefs and coral reefs. Resembles *B. axillaris*, but distinguished by entirely reddish-brown body, bright-white spots on back. Pectoral fin bases have no black blotch in adult. Dorsal fin has 12 spines and 9–11 rays, anal fin has 3 spines and 10–12 rays. 31–32 lateral scales. Aquarium fish. Up to 25cm in total length.

Distribution — Japan, Ryukyu, Indo-West Pacific, Red Sea, South Africa

Labridae

Bodianus hirsutus (Lacépède)

[*B. loxozonus* (Snyder)]

[*Lepidaplois hirsutus* (Lacépède)]

[*Labrus hirsutus* Lacépède]

Blackfin Hogfish, Tarry Hogfish

Diagnosis — Inhabits singly rather deep rocky reefs. Resembles *B. bilineatus*, but distinguished by black pelvic and anal fins. A black broad band runs between dorsal ray and lower part of caudal fin. Dorsal fin has 12 spines and 10–11 rays, anal fin has 3 spines and 11–12 rays. 28–36 lateral scales. Edible. Up to 90cm in total length.

Distribution — Japan, Ryukyu, Bonin Islands, Indo-Pacific, South Africa

Labridae

Bodianus anthioides (Günther)

[*Lepidaplois anthioides* (Bennett)]

Bronzehead Hogfish, Lyretail Hogfish

Diagnosis — Inhabits rather deep rocky reefs and coral reefs. Head is roundish. Upper and lower parts of caudal fin are black, extended long. Posterior part of body is profusely dotted with black. Dorsal fin has 12 spines and 10 rays, anal fin has 3 spines and 12 rays. 31 lateral scales. Aquarium fish. Up to 30cm in total length.

Distribution — Japan, Ryukyu, Indo-West Pacific, Red Sea, South Africa

Labridae

Bodianus oxycephalus (Bleeker)

[*Cheilinus oxycephalus* Bleeker]

Banded Pigfish, Blackspot Hogfish, Black-spot Pig Fish, Fox Berd, Pigfish, Rainbow Fish

Diagnosis — Inhabits rather deep rocky reefs. Body is somewhat deep, has red blotches. Snout is projected. Dorsal fin has 12 spines, 10–11 rays and a black blotch. Anal fin has 3 spines and 11–12 rays. 33–35 lateral scales. Edible. Up to 40cm in total length.

Distribution — Japan, Ryukyu, Indo-Pacific, East Africa

Labridae

Bodianus luteopunctatus (Smith)

[*Lepidaplois luteopunctatus* Smith]

Whitespotted Hogfish

Diagnosis — Inhabits rather deep rocky reefs. Resembles *B. oxycephalus*, but distinguished by absence of a black blotch on dorsal fin. Body has red broken lines. Dorsal fin has 12 spines and 10 rays, anal fin has 3 spines and 12 rays. 29 lateral scales. Aquarium fish. Up to 27cm in total length.

Distribution — Japan, Ryukyu, Indian Ocean, South Africa

Labridae

Bodianus bilunulatus (Lacépède)

[*Lepidaplois bilunulatus* (Lacépède)]

Crescent Banded Wrasse, Crescent Band Wrasse

Diagnosis — Inhabits rather deep rocky reefs. Resembles *B. perditio*, but distinguished by presence of a large black blotch on caudal peduncles. In young, posterior part of body is black. Dorsal fin has 12 spines and 9–10 rays, anal fin has 3 spines and 12 rays. 28–32 lateral scales. Aquarium fish. Up to 60cm in total length.

Distribution — Japan, Ryukyu, Bonin Islands, Indo-Pacific, Hawaii

▲Young Adult

▲Adult

Labridae

⇧ ***Semicossyphus reticulatus*** (Valenciennes) ⇗ ⇨

Bulgyhead Wrasse, Cold Porgy

Diagnosis — Inhabits rather deep rocky reefs. In adult male, head is swollen, lower jaw is dilated. In juvenile, body has a white band, fins are black. Dorsal fin has 12 spines and 9−10 rays, anal fin has 3 spines and 12 rays. 40−50 lateral scales. Ferocious. Feeds on gastropoda and crustacea. Spawns in spring. Edible but may cause ciguatera poisoning. Up to 1m in total length.

Distribution — Japan, Korea, South Chian Sea

▲Large Adult

Labridae

⇦ ***Bodianus axillaris*** (Bennett)

[*Lepidaplois axillaris* (Bennett)]

[*L. albomaculatus* Smith]

[*Stethojulis axillaris* (Quoy et Gaimard)]

Blackspotted Hogfish, Turncoat Hogfish, Axilspot Hogfish, Red-spot Rainbow Fish

Diagnosis — Inhabits shallow coral reefs. Snout is projected. Anterior half of body is dark. Pectoral fin bases, dorsal and anal fins have a black blotch. Dorsal fin has 12 spines and 10 rays, anal fin has 3 spines and 12 rays. 30−31 lateral scales. Aquarium fish. Up to 20cm in total length.

Distribution — Japan, Ryukyu, Korea, Indo-Pacific, Red Sea, South Africa

Labridae

Bodianus mesothorax (Schneider) ⇨

[*Lepidaplois mesothorax* (Schneider)]

Eclipse Pig-fish

Diagnosis — Inhabits rather deep rocky reefs. Snout is projected. Anterior half of body is black. Pectoral fin bases have a black blotch; anal fin has no black blotches. Dorsal fin has 12 spines and 9−10 rays, anal fin has 3 spines and 11−12 rays. 30−32 lateral scales. Aquarium fish. Up to 21cm in total length.

Distribution — Japan, Ryukyu, China, West Pacific

Labridae

Bodianus sp. ⇧

Red-finned Pigfish (New Common Name)

Diagnosis — Inhabits rather deep rocky reefs. Resembles *B. diana*, but distinguished by absence of black blotches on body. Dorsal fin and anal fin has a red blotch. Up to 20cm in total length.

Distribution — Pacific

Labridae

⇦ ***Bodianus trilineatus*** (Fowler)

[*Lepidaplois trilineatus* Fowler]

Threeline Hogfish

Diagnosis — Resembles *B. luteopunctatus*, but distinguished by black pectoral fin bases. Body has 4 red stripes. Anal fin is yellow. Dorsal fin has 12 spines and 10 rays, anal fin has 3 spines and 12 rays. 31−32 lateral scales. Edible. Up to 25cm in total length.

Distribution — Japan, Ryukyu, Indo-West Pacific, Red Sea, South Africa

▲ Male

▲ Female

Labridae

⇧ ***Anampses caeruleopunctatus*** Rüppell ⇨

[*A. diadematus* Rüppell]

Bluebreast Wrasse, Bluefreckeled Wrasse, Bluespotted Tamarin, Spotted Chiseltooth-wrasse, Bluespotted Wrasse

Diagnosis — Inhabits shoreline reefs and coral reefs. Body is somewhat deep. In male, body is remarkably bluish. In female, body is dark-brown, profusely dotted with blue. Surface around eyes has bluish radial stripes. Dorsal fin has 9 spines and 12 rays, anal fin has 3 spines and 12 rays. 27 lateral scales. Edible. Up to 30cm in total length.

Distribution — Japan, Ryukyu, Formosa, Indo-Pacific, Red Sea, South Africa

▲ Female

Labridae

Anampses meleagrides Valenciennes ⇩

[*A. amboinensis* Bleeker] ⇩

Archbar Wrasse, Speckled Wrasse, Yellowtail Tamarin, Yellowtail Wrasse, Guinea-fowl Wrasse

Diagnosis — Inhabits coral reefs. In male, body is dark, margins of dorsal and anal fins are white, caudal lobes extend long. In female, body is profusely dotted with white. Caudal fin is yellow. Dorsal fin has 9 spines and 12 rays, anal fin has 3 spines and 12 rays. 26–28 lateral scales. Up to 25cm in total length.

Distribution — Japan, Ryukyu, Indo-Pacific, Red Sea, South Africa

▲ Male

Labridae

Anampses geographicus Valenciennes ⇧

Blackspotted Wrasse, Cheeklined Wrasse, Speckled Wrasse, Lined Chiseltooth-wrasse, Scribbled Wrasse, Ocellated Chiseltooth-wrasse

Diagnosis — Inhabits coral reefs. In male, body is dark, caudal margin is white and emarginated. In female, dorsal and anal fins have a black blotch. Dorsal fin has 9 spines and 12 rays, anal fin has 3 spines and 12 rays. 48–50 lateral scales. Up to 25cm in total length.

Distribution — Ryukyu, Indo-Pacific, Hawaii

▲ Female

▲ Male

Labridae

Pseudodax moluccanus (Valenciennes) ⇧

Molucca Sea Wrasse, Chiseltooth Wrasse

Diagnosis — Inhabits coral reefs. Body is compressed, somewhat deep. Pectoral and caudal fins are small. Dorsal fin has 11 spines and 12–13 rays, anal fin has 2–3 spines and 11–14 rays. 31–33 lateral scales. Edible. Up to 27cm in total length.

Distribution — Southern Japan, Ryukyu, Indo-Pacific, Red Sea, East Africa

◄ Male ▼

▲ Female

Labridae

Cheilio inermis (Forskål)

Cigar Wrasse, Quaker, Scooper, Sharp-nosed Rainbow Fish, Sharp-nosed Wrasse, Slender Wrasse

Diagnosis — Inhabits shoreline reefs and coral reefs. Body is elongated and cylindrical. Snout is pointed. In male, body is somewhat deep and dark. In female, body is elongated, back is dark. Digs in sand to sleep. Dorsal fin has 9 spines and 13 rays, anal fin has 3 spines and 12 rays. 46−48 lateral scales. Edible. Up to 50cm in total length.

Distribution — Japan, Ryukyu, Indo-Pacific, Hawaii, Red Sea, South Africa

Labridae

Gomphosus varius Lacépède

Blue Long-nose, Long-nose, Olive Club-nosed Wrasse, Purple Clubnosed Wrasse, Yellow Club-nosed Wrasse, Olive Clubnose, Bird Wrasse

Diagnosis — Inhabits shallow coral reefs. Swims very swiftly. In adult, snout is pointed like fountain pen filler, preys on small animals. In male, body is blue. In female, body is profusely dotted with black; snout is orange. In young, snout is short, body has 2 black stripes. Dorsal fin has 8 spines and 13−14 rays, anal fin has 2−3 spines and 10−13 rays. 26−30 lateral scales. Aquarium fish. Up to 30cm in total length.

Distribution — Japan, Ryukyu, Indo-Pacific, Hawaii, Red Sea, South Africa

▲ Large Adult (Male)

◄Female

▼Male

Labridae

Thalassoma hardwickii (Bennett)

[*T. hardwickei* (Bennett)]

[*T. schwanefeldii* (Bleeker)]

[*Julis dorsale* Quoy et Gaimard]

[*J. schwanenfeldii* Bleeker]

Hardwicke Wrasse, Sixbar Wrasse, Hardwick's Wrasse, Six-barred Wrasse

Diagnosis — Inhabits shallow coral reefs. Upper part of body has 6 black bands. Caudal lobes are brown and extended. Opercles have 3−4 rows of scales. Dorsal fin has 8 spines and 12−14 rays, anal fin has 3 spines and 11 rays. 25−28 lateral scales. Aquarium fish. Up to 45cm in total length.

Distribution — Japan, Ryukyu, Bonin Islands, China, Indo-Pacific, Hawaii, Red Sea, South Africa

▲Female▼ ▲ Male

Labridae
⇦ *Thalassoma lutescens* (Lay et Bennett)

Green Moon Wrasse, Yellow Wrasse

Diagnosis — Inhabits shallow rocky reefs and coral reefs. In male, body is bluish-green, caudal lobes are extended. In female, body is yellow. Margins of pectoral fins are black. Head has brown stripes. Transsexual. Dorsal fin has 8 spines and 13 rays, anal fin has 2 spines and 11 rays. Carnivorous. Aquarium fish. Up to 25cm in total length.

Distribution — Japan, Ryukyu, Bonin Islands, Indo-Pacific, Hawaii

Labridae
⇩ *Thalassoma purpureum* (Forskål)

[*T. purpurea* (Forskål)]

[*T. umbrostigma* (Rüppell)]

Freckleback Wrasse, Purple Wrasse, Rainbow Fish, Rainbow Wrasse

Diagnosis — Inhabits shallow coral reefs. Body has 3 brown bands. Caudal margin is truncated. In male, body is remarkably bluish. In female, reddish-brown. Dorsal fin has 8 spines and 13 rays, anal fin has 2–3 spines and 11 rays. Aquarium fish. Up to 45cm in total length.

Distribution — Japan, Ryukyu, Bonin Islands, Indo-Pacific, Hawaii, Red Sea, South Africa

▼Male

Labridae
⇦ *Thalassoma lunare* (Linnaeus)

Crescent-tail Wrasse, Moon Wrasse, Rainbow Fish, Green Wrasse

Diagnosis — Inhabits shoreline reefs and coral reefs. Head has red stripes. Pectoral fins are blue, have a reddish stripe. Caudal lobes are extended like thread. Opercles have 1 row of scales. Dorsal fin has 8 spines and 13 rays, anal fin has 3 spines and 11 rays. Aquarium fish. Up to 30cm in total length.

Distribution — Japan, Ryukyu, Indo-Pacific, Red Sea, South Africa

▲ Male

Labridae
⇦ *Thalassoma trilobatum* (Lacépède)

[*T. fuscum* (Lacépède)]

Christmas Wrasse, Duskyred Wrasse, Gray Wrasse, Green-blocked Wrasse, Ladder Wrasse, Rainbow Fish, Brown Wrasse Fish

Diagnosis — Inhabits coral reefs. Resembles *T. quinquevittatum* and *T. klunzingeri*, but distinguished by absence of red stripes on head, shorter caudal lobes. Dorsal fin has 8 spines and 12–14 rays, anal fin has 3 spines and 11–13 rays. Aquarium fish. Up to 30cm in total length.

Distribution — Ryukyu, Indo-Pacific, Hawaii, East Africa

Labridae
Thalassoma quinquevittatum (Lay et Bennett) ⇨

[*T. albolineum* Nichols]

Blackspot Wrasse, Ladderstripe Wrasse, Red-banded Wrasse

Diagnosis — Inhabits coral reefs. Abdominal side is yellow. Upper lobe of caudal fin is extended. Dorsal fin has 8 spines and 12–14 rays, anal fin has 3 spines and 10–13 rays. Aquarium fish. Up to 15cm in total length.

Distribution — Ryukyu, Indo-West Pacific

Labridae
Thalassoma cupido
(Temminck et Schlegel) ⇨

Cupid Wrasse, Rainbow Fish

Diagnosis — Inhabits shoreline reefs and coral reefs. Body is elongated. Caudal margin is roundish. In male, body is blue. In female, reddish-brown. Dorsal fin has 8 spines and 13 rays, anal fin has 3 spines and 11 rays. Aquarium fish. Up to 20cm in total length.

Distribution — Japan, Ryukyu, Indo-Pacific

Labridae
Thalassoma klunzingeri
Fowler et Steinitz ⇧

[*T. rüppellii* Klunzinger]

Klunzinger's Wrasse

Diagnosis — Inhabits coral reefs. Head has broad, reddish-brown stripes. Upper and lower parts of caudal fin are reddish-brown and extended. Dorsal fin has 8 spines and 13 rays, anal fin has 3 spines and 11 rays. 26 lateral scales. Aquarium fish. Up to 20cm in total length.

Distribution — Red Sea

Labridae
Thalassoma duperrey
(Quoy et Gaimard) ⇨

Saddle Wrasse Fish, Duperrey's Wrasse, Saddle-back Wrasse

Diagnosis — Inhabits shallow coral reefs. Head is blue. Posterior part of opercles are reddish-brown. Caudal lobes are extended. In young, back is dark, abdominal side is pale. Aquarium fish. Up to 30cm in total length.

Distribution — Philippines, Hawaii

▲Female

▲Female

▲Male

Labridae
Pseudolabrus japonicus
(Houttuyn) ⇧

Bambooleaf Wrasse

Diagnosis — Inhabits offshore rocky reefs and bays. Body is somewhat deep, varies in color. Digs in sand to sleep. Dorsal fin has 9 spines and 10−11 rays, anal fin has 3 spines and 10 rays. 23−26 lateral scales. Edible. Aquarium fish. Up to 25cm in total length.

Distribution — Japan, Ryukyu

Labridae
Hemigymnus fasciatus (Bloch) ⇩

Barred Thicklip, Five-banded Wrasse

Diagnosis — Inhabits coral reefs. Resembles *H. melapterus*, but distinguished by 5 black bands on body. Lips are thick. Mouth has 2 canines. Dorsal fin has 9 spines and 11 rays, anal fin has 3 spines and 11 rays. 26−30 lateral scales. Edible. Aquarium fish. Up to 80cm in total length.

Distribution — Japan, Ryukyu, Indo-Pacific, Red Sea, South Africa

Labridae
Pteragogus flagellifera
(Valenciennes) ⇨

[*Duymaeria flagellifera* (Valenciennes)]

Cocktail Fish, Cocktail Wrasse

Diagnosis — Inhabits singly shallow rocky reefs. Body is deep; snout is pointed; cheeks have scales. Dorsal fin has 9 spines and 11−12 rays, dorsal spines are extended in male. Anal fin has 3 spines and 9−10 rays. 22−25 lateral scales. Edible. Up to 20cm in total length.

Distribution — Japan, Ryukyu, Formosa, China, Indo-West Pacific, South Africa

Labridae

⇦ ***Hemigymnus melapterus***

(Bloch) ⇩

Black-eyed Thick-lip, Blackeye Thicklip, Thick-lipped Wrasse

Diagnosis — Inhabits shallow coral reefs. Body is deep. Lips are thick. Anterior half of body is pale. Caudal fin is yellow. Dorsal fin has 9 spines and 10−11 rays, anal fin has 3 spines and 11 rays. 27−28 lateral scales. Edible. Up to 1m in total length.

Distribution — Ryukyu, China, Indo-Pacific, Red Sea, South Africa

Labridae

⇧ ***Labroides phthirophagus*** Randall

Louse-eating Wrasse, Cleaner Wrasse Fish

Diagnosis — Inhabits shallow coral reefs and rocky reefs. Body is elongated; snout is pointed; lips are thick. Feeds on parasitic copepoda on large fishes. Aquarium fish. Up to 8cm in total length.

Distribution — Hawaii

Labridae

⇦ ***Labroides bicolor*** Fowler et Bean

Black-and-yellow Wrasse

Diagnosis — Snout is slender and pointed. Upper jaw has 2 canines. Anterior half of body is dark, posterior half is white. Feeds on parasitic copepoda on other kinds of fish. Aquarium fish. Up to 12cm in total length.

Distribution — Japan, Ryukyu, Indo-Pacific, South Africa

Labridae

Labroides dimidiatus

(Valenciennes) ⇨

[*L. paradiseus* Bleeker]

[*Fissilabrus dimidiatus* (Valenciennes)]

Blueback Black Wrasse, Blue Streak, Bluestreak Cleaner Wrasse, Bridled Beauty, Cleaner-fish, Paradise-fish, Sea Swallow, Doctorfish, Barbarfish, Cleaner Wrasse

Diagnosis — Inhabits shallow coral reefs and rocky reefs. Body is elongated; snout is pointed. Preys on parasitic copepoda on other kinds of fish. Transsexual. Aquarium fish. Up to 12cm in total length.

Distribution — Japan, Ryukyu, Indo-Pacific, Hawaii, Red Sea, South Africa

Labridae

Macropharyngodon meleagris (Valenciennes)

[*M. pardalis* (Kner)]

Guineafowl Wrasse, Reticulated Wrasse

Diagnosis — Inhabits shallow coral reefs. Body is somewhat deep. Jaws have 2 pairs of canines. In male, body is dark. In female, body has reticulate patterns. Dorsal fin has 9 spines and 11 rays, anal fin has 3 spines and 10–11 rays. 25–27 lateral scales. Digs in sand to sleep. Aquarium fish. Up to 15cm in total length.

Distribution — Japan, Ryukyu, Pacific, Hawaii

▲ Male

▲ Female

Labridae

Stethojulis interrupta terina Jordan et Snyder

[*S. interrupta* (Bleeker)]

[*S. trossula* Jordan et Snyder]

[*S. kalosoma* (Bleeker)]

[*S. terina* Jordan et Snyder]

[*S. phekadopleura* (Bleeker)]

One-line Rainbowfish, Spotted-bellied Rainbowfish, Spotted-belly Rainbowfish

Diagnosis — Inhabits shallow sandy bottoms. In male, surface between eyes and pectoral fins are reddish. In female, abdominal side has black spots, a black stripe runs from pectoral fin bases. Dorsal fin has 9 spines and 11–12 rays, anal fin has 3 spines and 11–12 rays. 25–27 lateral scales. Digs in sand to sleep. Aquarium fish. Up to 15cm in total length.

Distribution — Japan, Ryukyu, Korea, Indo-Pacific, Red Sea, South Africa

Labridae

Halichoeres gymnocephalus (Schneider)

Lightgreen Wrasse

(New Common Name)

Diagnosis — Inhabits shallow coral reefs. Body is light-green; head has irregular rosy blotches. Dorsal fin has 9 spines and 12 rays, anal fin has 3 spines and 11 rays. Aquarium fish. Up to 16cm in total length.

Distribution — West Pacific

▲ Female

▲ Male

Labridae

Stethojulis strigiventer (Bennett)

[*S. renardi* (Bleeker)]

Lined Rainbow-fish, Renard's Rainbowfish, Silver-streaked Rainbowfish, Threeribbon Wrasse

Diagnosis — Inhabits shallow coral reefs. In male, a white stripe runs from snout to caudal fin. Upper surfaces of pectoral fins are reddish. In female, surface below eyes are striped. Caudal peduncles have a small black spot. Dorsal fin has 9 spines and 11 rays, anal fin has 3 spines and 11 rays. 26 lateral scales. Up to 14cm in total length.

Distribution — Japan, Ryukyu, Indo-West Pacific, Hawaii, South Africa

Labridae

Halichoeres miniatus (Kuhl et Van Hasselt)

[*H. miniatus* (Cuvier et Valenciennes)]

[*H. binotopsis* (Bleeker)]

Spotted-finned Rainbowfish

Diagnosis — Inhabits shallow coral reefs. Radial stripes on head are remarkably curved. Middle of dorsal fin has a black spot. In female, upper part of pectoral fins have a black blotch. Dorsal fin has 9 spines and 11 rays, anal fin has 3 spines and 11 rays. 27 lateral scales. Aquarium fish. Up to 10cm in total length.

Distribution — Ryukyu, China, Indo-Pacific

◀Male

▼Sleeping Rainbowfish in sandy bottom.

▲Female

Labridae
Halichoeres poecilopterus
(Temminck et Schlegel)

Multicolorfin Rainbowfish

Diagnosis — Inhabits sandy bottoms of bays. In male, head is striped and bluish. In female, a black band runs from snout to caudal fin. Dorsal fin has 9 spines and 14 rays, anal fin has 3 spines and 14 rays. 26 lateral scales. Digs in sand to sleep. Edible. Aquarium fish. Up to 34cm in total length.

Distribution — Japan, Korea, China, East China Sea, Philippines

▲Female

Labridae
Halichoeres chrysus Randall

Golden Rainbowfish
(New Common Name)

Diagnosis — Inhabits shallow coral reefs and rocky reefs. Body is yellow. In male, head has brown stripes. In female, dorsal and caudal fins have small black spots. Dorsal fin has 9 spines and 12 rays, anal fin has 3 spines and 12 rays. 28 lateral scales. Up to 10cm in total length.

Distribution — Japan, Ryukyu, West Pacific

▼Female

Labridae
Halichoeres melanurus (Bleeker)
[*H. kallochroma* (Bleeker)]

Orange-tipped Rainbow-fish, Three-eyed Wrasse

Diagnosis — Inhabits shallow coral reefs. In male, head has crimson stripes, back has broad dark transverse bands. Caudal fin also has crimson stripes. In female, body is covered with yellowish-brown longitudinal stripes with black spots on dorsal fin and caudal peduncles. Dorsal fin has 9 spines and 12 rays, anal fin has 3 spines and 11−12 rays. 28 lateral scales. Aquarium fish. Up to 12cm in total length.

Distribution — Ryukyu, West Pacific

▲Male

Labridae
Halichoeres melanochir
Fowler et Bean
[*H. prosopeion* (Bleeker)]

Darkfront Wrasse, Masked Rainbowfish

Diagnosis — Inhabits shallow coral reefs. Body is dark, profusely dotted with small black. Pelvic fins are yellow, pectoral fin bases are black. In young, body has 3 white bands. Dorsal fin has 9 spines and 12 rays, anal fin has 3 spines and 12 rays. 27−28 lateral scales. Up to 18cm in total length.

Distribution — Japan, Ryukyu, West Pacific

▼Female

Labridae
Halichoeres trimaculatus
(Quoy et Gaimard)

Spot-tail Rainbow-fish, Three-spot Wrasse

Diagnosis — Inhabits shallow coral reefs. In male, head has red stripes, caudal peduncles and upper part of pectoral fins have a black blotch. In female, body is yellowish-white, caudal peduncles have a black blotch. Dorsal fin has 9 spines and 11 rays, anal fin has 3 spines and 10 rays. 27−28 lateral scales. Aquarium fish. Up to 18cm in total length.

Distribution — Japan, Ryukyu, Pacific

▲ Male

Labridae

⇧ *Halichoeres hortulanus* (Lacépède)

[*H. nigrescens* Fowler]

[*H. centiquadrus* (Lacépède)]

[*Labrus centiquadrus* Lacépède]

Checkerboard Wrasse

Diagnosis — Inhabits shallow coral reefs. Body is slightly deep; snout is pointed. In male, head has radial red lines. In female, body is slightly elongated, pale, profusely dotted with black. Dorsal fin has 9 spines and 10–11 rays, anal fin has 3 spines and 11 rays. 26–28 lateral scales. Edible. Up to 30cm in total length.

Distribution — Japan, Ryukyu, Indo-Pacific, Red Sea, South Africa

Labridae

Halichoeres tenuispinis (Günther) ⇩

Motleystripe Rainbowfish

Diagnosis — Inhabits bays. In male, body is reddish-brown, anterior part of dorsal fin is black. In female and young, body is greenish-brown. Dorsal fin has 9 spines and 12 rays, anal fin has 3 spines and 12 rays. 25–26 lateral scales. Up to 18cm in total length.

Distribution — Japan, Korea, China, East China Sea, Philippines

▲ Male

◀Young

▲Young

Labridae

Coris gaimard (Quoy et Gaimard)

[*C. gaimardi* (Quoy et Gaimard)]

[*Julis gaimardi* (Quoy et Gaimard)]

[*J. greenovii* Bennett]

Rainbow Wrasse, African Coris, Gaimard's Wrasse, Gaimard's Rainbowfish

Diagnosis — Inhabits shallow coral reefs. Body is profusely dotted with blue. Head has blue lines. In juvenile, body has white blotches. Dorsal fin has 9 spines and 12–13 rays, anterior spines are extended. Anal fin has 3 spines and 12–13 rays. 70–85 lateral scales. Edible but has records of ciguatera poisoning. Aquarium fish. Up to 45cm in total length.

Distribution — Japan, Ryukyu, Indo-Pacific, Hawaii, Red Sea, South Africa

▲▼Young Adult

▲ Adult

◀Adult ▲Young Adult

▲Young

▲Young Adult

Labridae

Coris aygula Lacépède ⇧

[*C. angulata* Lacépède]

Clown Coris, Redthroat Wrasse, Redthroated Rainbow Fish, Twinspot Wrasse, Red-blotched Rainbowfish

Diagnosis — Inhabits rather deep coral reefs and rocky reefs. Body is somewhat deep. In large adult, head is projected. In young, body is white with reddish blotches, black spots over head. Dorsal fin has 8–9 spines and 12–13 rays, anal fin has 3 spines and 12 rays. 61–66 lateral scales. Edible. Aquarium fish. Up to 1.2m in total length.

Distribution — Japan, Ryukyu, Indo-Pacific, Hawaii, Red Sea, South Africa

Labridae

Coris pica (Schneider) ⇩

[*C. musume* (Jordan et Snyder)]

Combfish

Diagnosis — Inhabits rather deep rocky reefs. In male, body is somewhat deep; back is black; pectoral fins are reddish. In female, body has a broad black band, abdominal side is white. Dorsal fin has 9 spines and 12 rays, anal fin has 3 spines and 12 rays. 76–92 lateral scales. Digs in sand to sleep. Edible. Aquarium fish. Up to 25cm in total length.

Distribution — Japan, Ryukyu, Formosa, Australia, New Zealand

▲ Female

Labridae

Coris sp. ⇩

Long-finned Crown Coris

(New Common Name)

Diagnosis — Inhabits rather deep muddy-sand bottoms. Resembles *C. aygula*, but fins are extended. Edible. Up to 70cm in total length.

Distribution — Gulf of California

Labridae

Coris variegata (Rüppell) ⇩

Sevenbar Wrasse, Variegated Rainbowfish, Dapple Coris

Diagnosis — Inhabits shallow coral reefs. Back has 7 dark bands. Dorsal fin has a large black blotch, 9 spines and 11 rays. Anal fin has 3 spines and 11–12 rays. 50–58 lateral scales. Aquarium fish. Up to 20cm in total length.

Distribution — Japan, Ryukyu, Indo-Pacific, Hawaii, Red Sea, East Africa

Labridae

Coris flavovittata (Bennett) ⇩

Yellowstripe Coris, Black-striped Wrasse

Diagnosis — Inhabits coral reefs. Body has 5 black stripes. In young, back has black bands. Up to 45cm in total length.

Distribution — Philippines, Guam, Hawaii

▲ Female

Male

Female

Labridae

Pseudocoris yamashiroi (Schmidt)

[*Coris yamashiroi* (Schmidt)]

Yamashiro's Rainbowfish

Diagnosis — Inhabits coral reefs. Body is slightly elongated. In male, body is dark, caudal lobes are pointed, dorsal fin spines are extended. In female, body is reddish-brown. Dorsal fin has 9 spines and 12 rays, anal fin has 3 spines and 12 rays. 70–72 lateral scales. Aquarium fish. Up to 15cm in total length.

Distribution — Southern Japan, Ryukyu

▲ Male

Labridae

Cirrhilabrus cyanopleura (Bleeker)

Rimmed-scale Wrasse, Thread-finned Tuskfish

Diagnosis — Inhabits shallow coral reefs. Cheeks have 2 rows of scales. In male, caudal fin is pointed, pelvic fins are extended. In female, caudal peduncles have a black spot. Dorsal fin has 11–12 spines and 8–9 rays, anal fin has 3 spines and 8–10 rays. Lateral lines are not complete. Aquarium fish. Up to 15cm in total length.

Distribution — Ryukyu, West Pacific

▲ Female

Labridae

Cirrhilabrus temminckii Bleeker

Threadfin Wrasse

Diagnosis — Inhabits shallow coral reefs. In male, back is reddish, pelvic fins are extended. In female, body is somewhat pale. Dorsal fin has 11–12 spines and 8–9 rays, anal fin has 3 spines and 9 rays. Lateral lines are not complete. Up to 20cm in total length.

Distribution — Japan, Ryukyu, West Pacific

▲ Male

Labridae

Cirrhilabrus jordani Snyder

Flame Wrasse, Jordan's Wrasse Fish

Diagnosis — Inhabits rather deep rocky reefs. Back is dark-red, abdominal side is pale. Dorsal fin has 11 spines and 8–9 somewhat high rays. Caudal margin is rounded. Aquarium fish. Up to 10cm in total length.

Distribution — Hawaii

▼Male

▲ Female

Labridae

Hologymnosus annulatus (Lacépède) ⇨ ⇩

[*H. semidiscus* (Lacépède)]

[*H. doriatus* (Lacépède)]

Finescale Wrasse, Ringed Wrasse, Narrow-banded Rainbowfish, Ring Wrasse

Diagnosis — Inhabits shallow coral reefs and rocky reefs. Body is compressed, slightly elongated. Caudal margin is white and crescent-shaped. In male, body is bluish, in female, has brown stripes. Dorsal fin has 9 spines and 12 rays, anal fin has 3 spines and 12 rays. 100−124 lateral scales. Feeds on small fish. Digs in sand to sleep. Edible. Aquarium fish. Up to 60cm in total length.

Distribution — Japan, Ryukyu, Indo-West Pacific, Red Sea, South Africa

▲ Male

Labridae

Xyrichtys aneitensis (Günther) ⇩

[*Hemipteronotus aneitensis* (Günther)]

White-marked Razorfish (New Common Name)

Diagnosis — Inhabits rather deep muddy-sand bottoms. Body is compressed; has a large white blotch; head is developed. Surface below eyes is covered with 3 rows of scales. Dorsal fin base has a black blotch, dorsal fin has 9 spines and 12 rays, anal fin has 3 spines and 12 rays. Lateral lines are not complete. Edible. Up to 25cm in total length.

Distribution — Ryukyu, Pacific, Hawaii

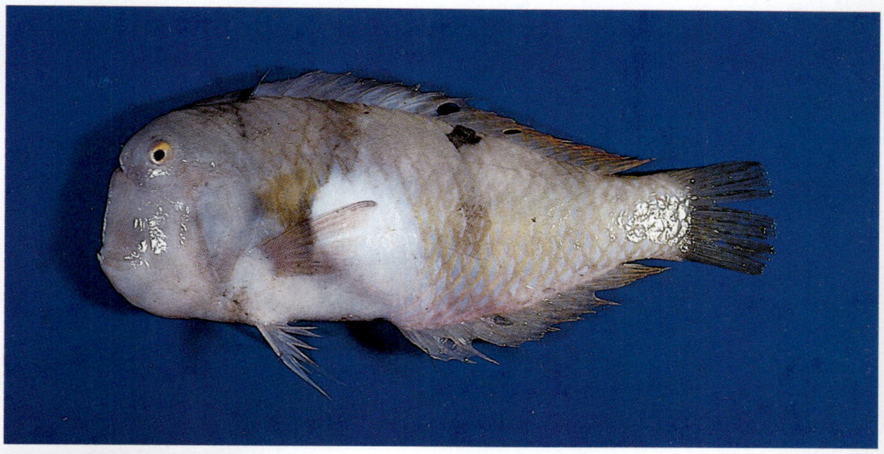

▲Male

Labridae

Xyrichtys verrens (Jordan et Evermann) ⇩

Red-spotted Razorfish, White-cheeked Razorfish (New Common Names)

Diagnosis — Inhabits rather deep muddy-sand bottoms. Body is compressed; head is developed. In male, back has a reddish flower-like blotch. In female, opercles have a large bluish-white blotch. Dorsal fin has 9 spines and 12 rays, anal fin has 3 spines and 12 rays. Lateral lines are not complete. Edible. Up to 30cm in total length.

Distribution — Japan, Formosa

▼Male

Labridae

Xyrichtys evides (Jordan et Richardson) ⇩

Black-spotted Razorfish (New Common Name)

Diagnosis — Inhabits rather deep muddy-sand bottoms. Body is compressed; head is developed. Abdominal side is white; back has a distinct black spot. Dorsal fin has 9 spines and 12 rays, anal fin has 3 spines and 12 rays. Lateral lines are not complete. Edible. Up to 20cm in total length.

Distribution — Ryukyu, Formosa

◀Female

▲Male

Labridae

⇦ ***Xyrichtys dea*** Temminck et Schlegel

[*Hemipteronotus dea* (Temminck et Schlegel)]

Goddess Razorfish

Diagnosis — Inhabits rather deep muddy-sand bottoms. Body is compressed and deep. Head is developed. Surface below eyes are covered with 1−2 rows of scales. Lower part of dorsal fin has a black spot. Dorsal fin has 9 spines and 11−13 rays, anterior spines are extended. Anal fin has 3 spines and 11−12 rays. Lateral lines are not complete. Edible. Up to 30cm in total length.

Distribution — Japan, Formosa, East China Sea, Indonesia

Labridae

⇩ ***Novaculichthys taeniourus*** (Lacépède) ⇩⇨

[*Hemipteronotus taeniurus* (Lacépède)]

[*Xyrichtys taeniurus* (Lacépède)]

[*Julis bifer* Lay et Bennett]

Dragon Wrasse, Olive-scribbled Wrasse, Bar-cheeked Wrasse, Redbelly Wrasse, Rockmover Wrasse

Diagnosis — Inhabits sandy bottoms among coral reefs. Body is compressed. Caudal fin is rounded, has a white band. Surface around eyes has dark stripes. Dorsal fin has 9 spines and 13−14 rays with 2 black spots on anterior surface. In young, dorsal fin spines are extended. Anal fin has 3 spines and 13 well-developed rays. Lateral lines are not complete. Aquarium fish. Up to 30cm in total length.

Distribution — Ryukyu, Indo-West Pacific, Red Sea, South Africa

▲Adult

▲ Young

Labridae

Xyrichtys twistii (Bleeker) ⇩

Red-marked Razorfish (New Common Name)

Diagnosis — Inhabits rather deep muddy-sand bottoms. Body is compressed, has a red blotch. Head is developed. Dorsal fin has 9 spines and 12 rays, spines are not extended. Anal fin has 3 spines and 12 rays. Lateral lines are not complete. Edible. Up to 30cm in total length.

Distribution — Japan, Ryukyu, Indonesia

Labridae

⇧ ***Cheilinus fasciatus*** (Bloch) ⇨

Barred Wrasse, Scarlet-breasted Maori Wrasse

Diagnosis — Inhabits coral reefs. Body is compressed, somewhat deep, has 6−7 blackish bands. Caudal lobes are extended. Dorsal fin has 9 spines and 10 rays, anal fin has 3 spines and 8 rays. Lateral lines are not complete. Edible but has records of ciguatera poisoning. Up to 35cm in total length.

Distribution — Ryukyu, Indo-Pacific, Red Sea, South Africa

▲Young Adult

Labridae

Epibulus insidiator (Pallas)

Longjaw Wrasse, Sling Jaw, Telescopefish

Diagnosis — Inhabits shallow coral reefs. Body is compressed and deep. Mouth is long and extensile. Lower jaw is projected backward. Body is yellow or black. Young has white bands. Dorsal fin has 10–11 spines and 9–11 rays, anal fin has 3 spines and 8–9 rays. Edible but has records of ciguatera poisoning. Up to 35cm in total length.

Distribution — Ryukyu, China, Indo-Pacific, Hawaii, Red Sea, South Africa

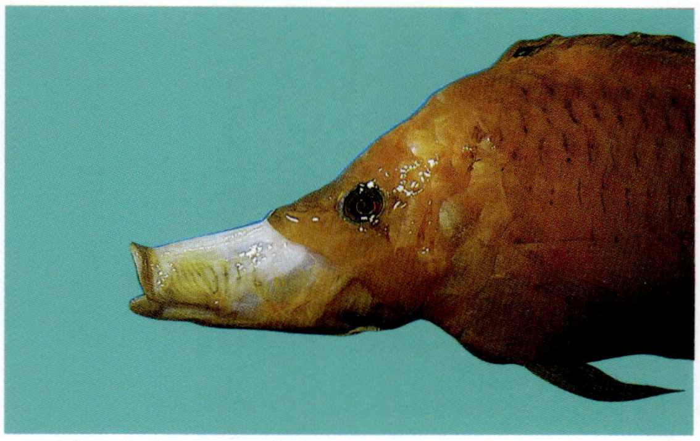

▲ Mouth is extensible.

▼Adult

Labridae

Cheilinus undulatus Rüppell

Napoleonfish, Blue-tooth Grouper, Maori-wrasse, Double-headed Maori-wrasse, Double-headed Parrot-fish, Hump-headed Maori Wrasse, Hump-headed Wrasse, Giant Wrasse, Wavylined Wrasse

Diagnosis — Body is deep. Lips are thick. In large adult, head is projected. Scales are large. Dorsal fin has 9 spines and 9–10 rays, anal fin has 3 spines and 8 rays. Lateral lines are not complete. World-largest wrasse measuring 2.3m in total length, 190kg in weight. Edible but has records of ciguatera poisoning.

Distribution — Ryukyu, Indo-Pacific, Red Sea, South Africa

▲ Head is projected in large adult.

Labridae

⇦ ***Cheilinus chlorurus*** (Bloch)

[*Thalliurus chlorurus* (Bloch)]

Floral Wrasse, Rainbow Fish, Redspot Green Wrasse, Yellow-dotted Maori-wrasse

Diagnosis — Inhabits coral reefs. Body is compressed, somewhat deep. Head has red spots and stripes. Caudal fin is profusely dotted with white. In adult, caudal fin is double emarginated. Dorsal fin has 10 spines and 8–9 rays, anal fin has 3 spines and 8 rays. Lateral lines are not complete. Edible. Up to 30cm in total length.

Distribution — Ryukyu, Indo-Pacific, South Africa

Labridae

Cheilinus trilobatus Lacépède ⇨

Maori Wrasse, Trilobed Maori Wrasse, Triple-tailed Wrasse, Tripletail Maori Wrasse

Diagnosis — Inhabits coral reefs. Body is slightly deep. Scales are large. Head is profusely dotted with red. Caudal fin is double emarginated. Dorsal fin has 9 spines and 10 rays, anal fin has 3 spines and 8 rays. Lateral lines are not complete. Edible but has records of ciguatera poisoning. Up to 50cm in total length.

Distribution — Ryukyu, Indo-Pacific, Red Sea, South Africa

Labridae

⇦ ***Cheilinus unifasciatus*** Streets ⇩

Rose-colored Wrasse Fish, Ring-tail Maori Wrasse, Ringtail Wrasse

Diagnosis — Inhabits rather deep coral reefs. *C. rhodochrous* is misidentified name. Body is elongated. Lips are thick. Head has red stripes. Caudal peduncles have a pale band. Dorsal fin has 9 spines and 10 rays, anal fin has 3 spines and 8 rays. Lateral lines are not complete. Edible but has records of ciguatera poisoning. Up to 30cm in total length.

Distribution — Ryukyu, Pacific, Hawaii,

Labridae

Cheilinus diagrammus (Lacépède) ⇩⇘

[*C. diagramma* (Lacépède)]

Cheeklined Wrasse, Rainbow Fish, Scribbled Wrasse, Violet Lined Maori Wrasse, Bandcheek Wrasse

Diagnosis — Inhabits rather deep coral reefs. Body is slightly elongated. Lips are thick. Head has many brown stripes. Upper and lower parts of caudal fin are dark with yellow margin. Dorsal fin has 9 spines and 10 rays, anal fin has 3 spines and 8 rays. Lateral lines are not complete. Edible. Up to 30cm in total length.

Distribution — Ryukyu, Indo-Pacific, Red Sea, South Africa

▼Adult

▲Young Adult

Labridae

Cheilinus celebicus Bleeker ⇨

[*C. kitlitzii* Valenciennes]

Long-nosed Wrasse

Diagnosis — Inhabits rather deep coral reefs. Body is elongated; snout is pointed. In male, surface around eyes has radial stripes. Dorsal fin has 9 spines and 10 rays, anal fin has 3 spines and 8 rays. Lateral lines are not complete. Up to 20cm in total length.

Distribution — Japan, Ryukyu, West Pacific

Labridae
⇦ *Pseudocheilinus octotaenia* Jenkins

Eightline Wrasse

Diagnosis — Body is slightly elongated; snout is pointed. Body has 8 black lines. Fins are transparent. Dorsal fin has 9 spines and 11–12 rays, anal fin has 3 spines and 9–10 rays. Lateral lines are not complete. Up to 10cm in total length.

Distribution — Ryukyu, Pacific, Hawaii

▼Young Adult (Female)

Labridae
Pseudocheilinus hexataenia (Bleeker) ⇧

Sixstripe Wrasse, Six-lined Wrasse

Diagnosis — Inhabits coral reefs. Body has 6 brown stripes. Upper part of caudal peduncles have a black spot. Dorsal fin has 9 spines and 11–12 rays, anal fin has 3 spines and 9 rays. Lateral lines are not complete. Up to 45cm in total length.

Distribution — Ryukyu, Indo-Pacific, Hawaii, Red Sea, South Africa

▲Male

Scaridae
⇧ *Bolbometopon bicolor* (Rüppell) ⇨⇩

[*Scarus pulchellus* Rüppell]

[*Callyodon bicolor* (Rüppell)]

[*Cetoscarus bicolor* (Rüppell)]

Two-colored Parrotfish, Red-spotted Parrotfish, Spotted Parrotfish

Diagnosis — Inhabits coral reefs. Body is slightly elongated; snout is pointed. Cheeks have 3 rows of scales. Teeth are white. 5–7 predorsal scales. Edible. Up to 80cm in total length.

Distribution — Ryukyu, Indo-Pacific, Red Sea, East Africa

▲Young

▲Female

▼Male

Scaridae
⇦ *Scarus* sp.

Striped-belly Parrotfish

(New Common Name)

Diagnosis — Inhabits coral reefs. Snout is pointed. A yellow line runs from eyes to abdominal side. Upper and lower parts of caudal fin are yellow. Edible. Up to 30cm in total length.

Distribution — Ryukyu, Indo-Pacific

▲Male

Scaridae

Scarus bowersi (Snyder)

Bower's Parrotfish

Diagnosis — Inhabits coral reefs. Caudal margin is truncated. Anterior half of body has a large orange blotch. Teeth are white. 4 predorsal scales. Edible. Up to 30cm in total length.

Distribution — Ryukyu, Pacific

▲Male

▼Female

Scaridae

Scarus globiceps Valenciennes

[*S. pronus* Fowler]

[*Callyodon globiceps* (Valenciennes)]

[*C. spilonotus* Jordan et Seale]

Brown Parrotfish, Violet-lined Parrotfish

Diagnosis — Inhabits coral reefs. A brown line runs from forehead to abdominal side. Cheeks have 2 rows of scales. Mouth has canines. 5–6 predorsal scales. Dorsal fin has a small blackish spot. Edible but may cause ciguatera poisoning. Up to 30cm in total length.

Distribution — Ryukyu, Indo-Pacific, East Africa

Scaridae

Scarus rivulatus Valenciennes

[*S. fasciatus* Valenciennes]

[*Callyodon fasciatus* (Valenciennes)]

Surf Parrotfish, Rivulated Parrotfish, Tattooed Parrotfish

Diagnosis — Inhabits coral reefs. Surface around eyes has radial blotches. Cheeks have 3 rows of scales. 7 predorsal scales. Edible but may cause ciguatera poisoning. Up to 40cm in total length.

Distribution — Ryukyu, Indo-Pacific

Male

Female

▲Female▶

Scaridae

Scarus ghobban Forskål

[*S. fehlmanni* Schultz]
[*S. mus* (Smith)]
[*S. dussumieri* Cuvier et Valenciennes]
[*Callyodon ghobban* (Forskål)]
[*Pseudoscarus cantori* Macleay]
[*P. pyrrhostethus* Günther]

Blue-barred Orange Parrotfish, Blue Parrotfish, Dussumier's Parrotfish, Flame-breasted Parrotfish, Cream Parrotfish, Mouse Parrotfish

Diagnosis — Inhabits coral reefs. Teeth are white. Cheeks have 3 rows of scales. Fins are orange. 5–6 predorsal scales. In male, body has no blue bands, is greenish over full length. Edible but may cause ciguatera poisoning. Up to 80cm in total length.

Distribution — Japan, Ryukyu, China, Indo-Pacific, Red Sea, East Africa

▼Male

▼Male

Scaridae

Scarus javanicus Bleeker

Yellow-finned Parrotfish

Diagnosis — Inhabits coral reefs. Posterior parts of pectoral and dorsal fins are yellow. Radial stripes run from eyes. Cheeks have 3 rows of scales. 4 predorsal scales. Edible. Up to 30cm in total length.

Distribution — Ryukyu, Indo-Pacific

Scaridae

Scarus quoyi Valenciennes

[*S. blochi* (Cuvier et Valenciennes)]
[*Callyodon blochi* (Cuvier et Valenciennes)]

Bloch's Parrotfish

Diagnosis — Inhabits coral reefs. Caudal margin is truncated. Pectoral fins are blue. 6 predorsal scales. Edible but has records of ciguatera poisoning. Up to 35cm in total length.

Distribution — Ryukyu, Indo-Pacific

▼Male

Male

Scaridae

⇦ ***Scarus festivus*** Valenciennes

[*S. lunula* (Snyder)]

Paumotu Parrotfish, Newmoon Parrotfish

Diagnosis — Inhabits coral reefs. Surface around eyes has 2 radial blue lines. Upper part of pectoral fins are blue. Caudal lobes are extended. Cheeks have 3 rows of scales. 4−5 predorsal scales. Edible. Up to 50cm in total length.

Distribution — Ryukyu, Indo-Pacific, East Africa

Scaridae

Scarus sp. ⇨

Blue-spotted Parrotfish

(New Common Name)

Diagnosis — Inhabits coral reefs. Surface around eyes has blue spots. Caudal lobes are slightly extended. 4 predorsal scales. Edible. Up to 35cm in total length.

Distribution — Ryukyu, Indo-Pacific

Male

▼Male

Scaridae

Scarus prasiognathos Valenciennes ⇧⇨⇘

[*S. janthochir* Bleeker]

[*S. chlorodon* Jenyns]

[*Callyodon janthochir* (Bleeker)]

Green-beak Parrotfish, Ocean Parrotfish, Pink-faced Parrotfish

Diagnosis — Inhabits coral reefs. Lower half of head is green. Canines are absent. Cheeks have 3 rows of scales. 6 predorsal scales. Edible. Up to 70cm in total length.

Distribution — Southern Japan, Ryukyu, Indo-Pacific, Hawaii, Persian Gulf

Male

Female

▲ Female

Scaridae

Scarus niger Forskål ⇧⇨

[*Pseudoscarus madagascariensis* Steindachner]

Masked Parrotfish, Dusky Parrotfish, Swarthy Parrotfish

Diagnosis — Inhabits coral reefs. Teeth are blue. Snout is pointed. Opercles have a green blotch on top surface. 6−8 predorsal scales. Edible but may cause ciguatera poisoning. Up to 60cm in total length.

Distribution — Ryukyu, Indo-Pacific, Red Sea, East Africa

Male

▲Male

Scaridae

⇦ ***Scarus rubroviolaceus*** Bleeker ⇩

[*S. paluca* Jenkins]
[*S. ruberrimus* Jordan et Seale]
[*Scarops rubroviolaceus* (Bleeker)]
[*Callyodon lacerta* Seale]
[*C. upolensis* Jordan et Seale]
[*Pseudoscarus jordani* Jenkins]

Meadow Parrotfish, Jordan's Parrotfish, African Giant Parrotfish, Ember Parrotfish, Black-veined Red Parrotfish, Red and Violet Parrotfish

Diagnosis — Inhabits coral reefs. Body is slightly elongated; head is angular. In female and young, teeth are white. In male adult, teeth are green. Cheeks have 3 rows of scales. 5–7 predorsal scales. Edible but may cause ciguatera poisoning. Up to 60cm in toal length.

Distribution — Ryukyu, Indo-Pacific, Hawaii, Persian Gulf, East Africa

▲Male

▲Female

Scaridae

Scarus tricolor Bleeker ⇧⇨

[*S. lepidus* Jenyns]
[*Callyodon lepidus* (Jenyns)]
[*C. pectoralis* (Valenciennes)]
[*C. cyanognathus* (Bleeker)]

Blackstripe Parrotfish, Multicolored Parrotfish, Green Parrotfish, Banana Parrotfish, Yellow-barred Parrotfish

Diagnosis — Inhabits coral reefs. Body is slightly elongated; abdominal side has 1–2 blue lines. Teeth are canines. Cheeks have 3 rows of scales. Anus is yellow. 6–7 predorsal scales. In female, body has an indistinct broad black band on middle surface. Edible. Up to 50cm in total length.

Distribution — Japan, Ryukyu, Indo-Pacific, East Africa

▼Female

Male

▲Male

Scaridae

⇦ ***Scarus japanensis*** (Bloch) ⇧

Japanese Parrotfish (New Common Name)

Diagnosis — Inhabits coral reefs. Resembles *S. sordidus*, but distinguished by white teeth, a broad yellow band on dorsal and anal fins. Abdominal side has no blue lines. 4 predorsal scales. Edible. Up to 35cm in total length.

Distribution — Ryukyu, Indo-West Pacific

Male

Scaridae

⇦ ***Scarus frenatus*** Lacépède

Fall Parrotfish, Bridled Parrotfish

Diagnosis — Inhabits coral reefs. Upper jaw has canines. In both male and female, caudal peduncles are pale, caudal lobes are slightly extended. Cheeks have 3 rows of scales. 6–7 predorsal scales. Edible. Up to 40cm in total length.

Distribution — Southern Japan, Ryukyu, Indo-Pacific, Red Sea, East Africa

Male

▲Female

Female

Scaridae

Scarus dimidiatus Bleeker

[*S. scaber* Valenciennes]
[*Callyodon dimidiatus* (Bleeker)]
[*C. fumifrons* Jordan et Seale]
[*Pseudoscarus dimidiatus* (Bleeker)]

Fivesaddle Parrotfish, Parrot Wrasse, Variable Parrotfish

Diagnosis — Inhabits coral reefs. Caudal fin is truncated. Cheeks have 3 rows of scales. In male, surface around mouth and upper half of body are blue. 5–6 predorsal scales. Edible. Up to 35cm in total length.

Distribution — Ryukyu, Indo-Pacific, South Africa

▼Male

Scaridae

Leptoscarus vaigiensis (Quoy et Gaimard)

Petroleum Parrotfish, Blue-speckled Parrotfish, Blue-spotted Parrotfish, Marbled Parrotfish

Diagnosis — Body is elongated; snout is roundish. Cheeks have 1 row of scales. In male, mouth has canines. Caudal margin is rounded. Body is profusely dotted with blue. 4 predorsal scales. Edible but has records of ciguatera poisoning. Up to 35cm in total length.

Distribution — Southern Japan, Ryukyu, Indo-Pacific, Mediterranean, South Africa

▼Male

Scaridae

Scarus oviceps Valenciennes

[*S. lupus* Fowler]
[*Pseudoscarus zonatus* Macleay]
[*Callyodon oviceps* (Valenciennes)]

Yellow-barred Parrotfish, Parrotfish

Diagnosis — Inhabits coral reefs. Mouth has no canines. Teeth are not exposed. Cheeks have 3 rows of scales. 6–7 predorsal scales. Female, resembles *S. scaber*, but distinguished by more oblique blackish bands on back. Edible. Up to 30cm in total length.

Distribution — Ryukyu, Indo-Pacific, South Africa

Scaridae

Scarus frontalis Valenciennes

[*S. jonesi* (Streets)]

Tattooed Parrotfish, Blue Parrotfish

Diagnosis — Inhabits coral reefs. Male resembles *S. gibbus*, but distinguished by dark body, white teeth. Head is pale. Fins have no spots and lines. 4 predorsal scales. Edible but has records of ciguatera poisoning. Up to 50cm in total length.

Distribution — Ryukyu, Pacific

Female

▲Male

▲Female

Male

Scaridae

Scarus sordidus Forskål

[*Callyodon sordidus* (Forskål)]
[*C. bipallidus* Smith]

Garnet Red Parrotfish, Green-finned Parrotfish, Green Parrotfish, Troschel's Parrotfish, Red-toothed Parrotfish, Burnt Parrotfish, Bullethead Parrotfish

Diagnosis — Inhabits coral reefs. In male, teeth are blue. In female, mouth is red, has canines. Cheeks have 2 rows of scales. 3−4 predorsal scales. Edible but may cause ciguatera poisoning. Up to 50cm in total length.

Distribution — Japan, Ryukyu, China, Indo-Pacific, Hawaii, Red Sea, Persian Gulf, South Africa

▲Male

▲Female

Female (Adult)

▲Male

Scaridae

Scarus microrhinos Bleeker

[*S. strongylocephalus* Bleeker]

[*Callyodon microrhinos* (Bleeker)]

Blunt-headed Parrotfish, Purple-headed Parrotfish, Big Blue Parrotfish, Steepheaded Parrotfish

Diagnosis — Inhabits coral reefs. In adult, head is developed. 4 predorsal scales. Red-bodied, Indian Ocean or Red Sea species distinct from this species (*S. gibbus* Rüppell). Edible but may cause ciguatera poisoning. Up to 80cm in total length.

Distribution — Ryukyu, China, Pacific

Male (Young Adult)

▲ Male

▼ Teeth-marks of Parrotfish.

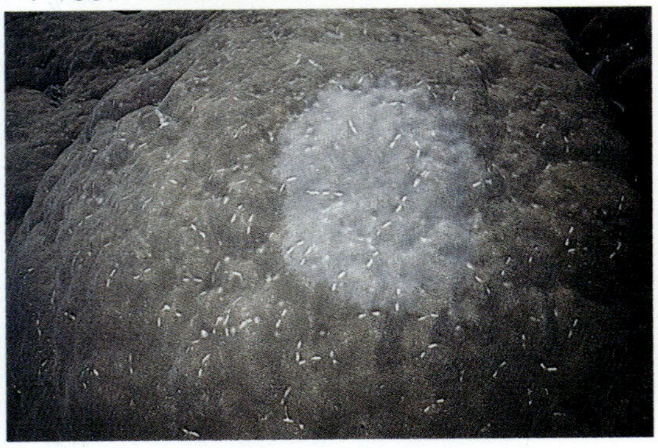

Scaridae

Scarus schlegeli (Bleeker)

[*S. venosus* Valenciennes]

Five-banded Parrotfish, Cut Parrotfish, Banded Parrotfish

Diagnosis — Inhabits coral reefs. In male, posterior half of body has a black band, 5 brown bands in female. Cheeks have 3 rows of scales. 4 predorsal scales. Edible but may cause ciguatera poisoning. Up to 50cm in total length.

Distribution — Ryukyu, Indo-Pacific, East Africa

▲ Female

▲ Male

▲ Male

Scaridae

Scarus psittacus Forskål

[*S. forsteri* Valenciennes]

Forster's Parrot Fish

Diagnosis — Body is purplish-brown; snout is slightly projected. Abdominal side has blue stripes. Caudal lobes are pointed. A blue stripe runs from snout to eyes. 23 lateral scales. Edible. Up to 25cm in total length.

Distribution — Japan, Ryukyu, Indo-Pacific

▲ Female

▼ Female

Scaridae

Scarus sp.

Yellow-tail Parrotfish (New Common Name)

Diagnosis — Inhabits shallow coral reefs. Snout is slightly projected. Margins of dorsal, pelvic and anal fins are blue. Caudal fin is yellow. 22 lateral scales. Edible. Up to 25cm in total length.

Distribution — Ryukyu

▲Female▼

▲Male

Scaridae

Scarus longiceps Valenciennes

[*Scarus harid longiceps* Valenciennes]

Yellow Parrotfish, Long-nosed Parrotfish, Red-lined Parrotfish

Diagnosis — Inhabits coral reefs; snout is long. Teeth are white. Margins of scales are yellow. Very similar between male and female. Cheeks have 3 rows of scales. 4 predorsal scales. Edible but may cause ciguatera poisoning. Up to 60cm in total length.

Distribution — Ryukyu, Pacific

▲Male

▲Female

Scaridae

Scarus ovifrons Temminck et Schlegel

[*Callyodon ovifrons* (Temminck et Schlegel)]

[*Ypsiscarus ovifrons* (Temminck et Schlegel)]

Blue Parrotfish, Blue Humphead Parrotfish

Diagnosis — Inhabits rather deep rocky reefs and coral reefs. Teeth are blue and exposed. Canines are absent. In male adult, head is projected. Cheeks have 3 rows of scales. 6 predorsal scales. Edible but has records of ciguatera or liver poisonings. Up to 90cm in total length.

Distribution — Japan, Ryukyu

▲ Male (Large Adult)

▲Male

▲Female

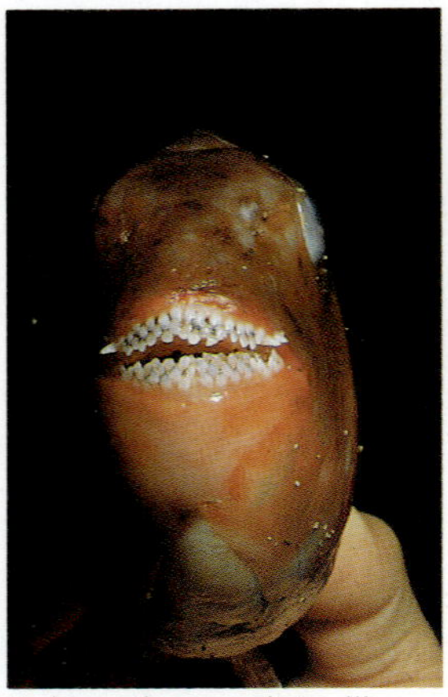

▲Jaws have plate-like incisor teeth.

▲ Cycloid scale of Japanese Parrotfish.

Scaridae
Calotomus japonicus (Valenciennes) ⇧
[*Leptoscarus japonicus* (Cuvier et Valenciennes)]

Japanese Parrotfish, White-spotted Parrotfish

Diagnosis — Inhabits rocky reefs and coral reefs. Jaws have dental plate of incisor-like teeth. In male, body is blue, in female and juvenile, reddish. Cheeks have 1 row of scales. 4 predorsal scales. Omnivorous. Feeds on seaweed and benthos. Edible. Up to 60cm in total length.

Distribution — Japan, Ryukyu, Korea, Indo-Pacific, Hawaii

Scaridae
Calotomus spinidens (Quoy et Gaimard) ⇩
[*Cryptotomus spinidens* (Quoy et Gaimard)]

Halftoothed Parrotfish, Camouflaged Parrotfish, Toothed Parrotfish

Diagnosis — Inhabits rocky reefs and coral reefs. Body is somewhat roundish. In male, surface around eyes has radial stripes. Cheeks have 1 row of scales. Teeth are white. 4 predorsal scales. Edible but has records of ciguatera poisoning. Up to 35cm in total length.

Distribution — Ryukyu, Formosa, Indo-Pacific, Hawaii, South Africa

▼Male

▼Female

Scaridae
Bolbometopon muricatus (Valenciennes) ⇨
[*Scarus muricatus* Valenciennes]
[*Callyodon muricatus* (Valenciennes)]

Double-headed Parrotfish, Humphead Parrotfish, Bumphead Parrotfish

Diagnosis — Inhabits coral reefs. In male adult, head is projected. Male and female resemble in color. Cheeks have 3 rows of scales. Mouth has no canines. 4–5 predorsal scales. Edible. Up to 1.5m in total length.

Distribution — Ryukyu, Indo-Pacific, Red Sea, East Africa

Monodactylidae

Monodactylus argenteus (Linnaeus) ⇨

Diamond-fish, Kite-fish, Leather-jacket, Moon-fish, Moony, Natal Moony, Sea Kite, Silver Batfish, Silvery Batfish, Common Fingerfish, Butter Bream

Diagnosis — Inhabits brackish areas of bays and shallow waters, also descends freshwater rivers. Body is deep. Dorsal and anal fins are developed. In young, head has 2 black bands. Anterior margin of anal fin is black. Dorsal fin has 7–8 spines and 28–31 rays, anal fin has 3 spines and 28–32 rays. Scales are small ctenoid. Aquarium fish. Up to 23cm in total length.

Distribution — Ryukyu, Formosa, China, Indo-Pacific, Red Sea, South Africa

Toxotidae

⇦ *Toxotes jaculator* (Pallas)

[*Sciaena jaculator* Pallas]

Archer Fish

Diagnosis — Inhabits brackish river mouths and mangrove swamps, also found in freshwater lakes and rivers. Body is oblong, silvery with 5–6 black blotches down from back, snout is pointed. Dorsal fin has 4 spines and 12–13 rays, anal fin has 3 spines and 15–17 rays. Fin margins are black. 28–30 lateral scales. Aquarium fish. Up to 30cm in total length.

Distribution — Ryukyu, South East Asia, India, Sri Lanka, New Guinea, Australia

Toxotidae

Toxotes chatareus (Buchanan-Hamilton) ⇨

[*Coius chatareus* Buchanan-Hamilton]

Seven Spot Archer, Seven Spotted Archerfish

Diagnosis — Resembles *T. jaculator*, but distinguished by smaller scales, 6–8 body blotches. Dorsal fin has 5 spines and 12–14 rays, anal fin has 3 spines and 17–18 rays. 33–34 lateral scales. Aquarium fish. Up to 30cm in total length.

Distribution — South East Asia, India, New Guinea, Australia

Scorpididae

⇦⇩ *Microcanthus strigatus* (Cuvier)

Stripey Convict Fish, Stripey, Footballer

Diagnosis — Found in schools among shoreline reefs. Young are found in tidepools. Body is deep, has 5–6 dark bands. Snout is pointed. Dorsal fin has 11 spines and 16–18 rays, anal fin has 3 spines and 13–14 rays. 56–60 lateral scales. Mouth is small. Feeds on zooplankton and small animals. Spawns in early summer. Edible. Aquarium fish. Up to 20cm in total length.

Distribution — Japan, Ryukyu, Formosa, Pacific, Hawaii

▲Young Adult

Ephippididae

⇦ *Platax teira* (Forskål) ⇨

Long-finned Batfish, Round-faced Batfish

Diagnosis — Found in schools among shallow coral reefs. Resembles *P. orbicularis*, but distinguished by rough scales, slender pelvic fins. In young, dorsal and anal fins are remarkably extended. Dorsal fin has 5 spines and 30−37 rays, anal fin has 3 spines and 23−27 rays. 51 lateral scales. Edible. Aquarium fish. Up to 50cm in total length.

Distribution — Ryukyu, Formosa, Indo-West Pacific

▲Young

▼Young Adult

Ephippididae

⇩ *Platax pinnatus* (Linnaeus) ⇨

[*P. melanosoma* Bleeker]

Orange-ring Batfish, Long-finned Batfish

Diagnosis — Found in schools among shallow coral reefs. Resembles *P. orbicularis*, but distinguished by projected snout, somewhat rough scales. Anterior parts of pelvic and anal fins are black. In juvenile, body is leaf-shaped with red margins. Dorsal fin has 5 spines and 36−37 rays, anal fin has 3 spines and 25−27 rays. 67−69 lateral scales. Edible. Aquarium fish. Up to 40cm in total length.

Distribution — Ryukyu, Indo-West Pacific, Red Sea, South Africa

▲Adult

Ephippididae

Platax orbicularis (Forskål) ⇧⇩

Angel-fish, Batfish, Butterflyfish, Leaf-fish, Long-finned Batfish, Sea Bat, Sunfish, Turbot, Circular Spadefish

Diagnosis — Found in schools among shoreline reefs and coral reefs. Leaf-like juvenile move with driftweed. In young, body is dark, dorsal and anal fins are extended, but turns roundish in adult. Body is discoid; snout is not projected. Black bands on body disappear in adult. Dorsal fin has 5 spines and 32−37 rays, anal fin has 3 spines and 25−27 rays. 76−80 lateral scales. Edible. Up to 50cm in total length.

Distribution — Japan, Ryukyu, Bonin Islands, Indo-Pacific, Red Sea, South Africa

Young

Adult

◄Young

Ephippididae

Drepane punctata (Linnaeus) ⇩

Sickle, Concertina Fish, Spotted Batfish

Diagnosis — Inhabits muddy-sand bottoms. Body is remarkably deep, slivery with numerous rows of black spots. Pectoral fins are remarkably long. Dorsal fin has 8–9 spines and 19–22 rays, spine part is compressed. Anal fin has 3 spines and 17–19 rays. 50–55 lateral scales. Feeds on benthos. Edible. Up to 40cm in total length.

Distribution — Ryukyu, Formosa, China, Indo-West Pacific, South Africa

Ephippididae

Ephippus orbis (Bloch) ⇨

[*Tripterodon orbis* Playfair]

Orbfish, Spade Fish

Diagnosis — Inhabits muddy-sand bottoms of shoreline reefs. Body is discoid with 4–5 indistinct dark bands. Dorsal fin has 9 spines and 19–20 rays, spine part is compressed. Anal fin has 3 spines and 15–16 rays. 35–39 lateral scales. Scales are large ctenoid. Edible. Up to 25cm in total length.

Distribution — Southern Japan, Ryukyu, Formosa, China, Indo-Pacific, South Africa

Pomacanthidae

Chaetodontoplus melanosoma (Bleeker) ⇨

Black Velvet Angelfish

Diagnosis — Inhabits shoreline reefs and coral reefs. Body is oval; head and back are pale. In young, pectoral fins are crossed by a yellow band. Caudal margin is yellow. Dorsal fin has 13 spines and 17–18 rays, anal fin has 3 spines and 18 rays. Feeds on sponge and ascidian. Aquarium fish. Up to 20cm in total length.

Distribution — Japan, Ryukyu, Formosa, West Pacific

Pomacanthidae

⇩ *Chaetodontoplus septentrionalis* (Temminck et Schlegel) ⇧

[*Holacanthus septentrionalis* Temminck et Schlegel]

[*Pomacanthus albofasciatus* (Tanaka)]

Bluelined Angelfish, Blue-stripe Angelfish

Diagnosis — Inhabits shoreline reefs and somewhat deeper rocky reefs. Body is oval with many blue stripes. Caudal fin is yellow. Preoperculum has a sharp-pointed spine. In young, pectoral fins are crossed by a yellow band. Dorsal fin has 13 spines and 18–19 rays, anal fin has 3 spines and 17–19 rays. Feeds on sponge and ascidian. Aquarium fish. Up to 22cm in total length.

Distribution — Japan, Ryukyu, Formosa, Korea, China, Hong Kong, Philippines, Indonesia

Pomacanthidae

Chaetodontoplus mesoleucus (Bloch) ⇧⇩

Vermiculated Angelfish

Diagnosis — Inhabits singly shallow coral reefs. Body is oval with a broad black band across the eyes. Posterior part of body is dark with crepy patterns. Caudal fin is yellow. Dorsal fin has 12 spines and 17–18 rays, anal fin has 3 spines and 17–18 rays. Feeds on sponge, ascidian and seaweed. Aquarium fish. Up to 18cm in total length.

Distribution — Ryukyu, Formosa, West Pacific

Pomacanthidae

Pomacanthus sextriatus ⇩ (Kuhl et Van Hasselt) ⇩

[*Arusetta sexstriata* (Cuvier et Valenciennes)]
[*Heteropyge sexstriatus* (Cuvier)]
[*Euxiphipops sexstriatus* (Cuvier et Valenciennes)]

Six-banded Angelfish

Diagnosis — Inhabits singly coral reefs. Body has 6 dark bands; head has a white band. Dorsal and anal fin rays are developed with roundish margin. Young resemble *P. semicirculatus*, but distinguished by straight white stripes on body. Dorsal fin has 13 spines and 18–20 rays, anal fin has 3 spines and 18 rays. Feeds on sponge and ascidian. Aquarium fish. Up to 50cm in total length.

Distribution — Ryukyu, Formosa, Indo-West Pacific

Pomacanthidae

⇩ *Pomacanthus annularis* (Bloch)

[*Pomacanthoides annularis* (Bleeker)]

Ringed Angelfish, Blue-ringed Angelfish

Diagnosis — Inhabits singly coral reefs. Body is deep, has 6 blue curved stripes. Dorsal fin ray is extended. Opercles have a blue ring-like blotch on upper surface. Caudal fin is white. Dorsal fin has 13 spines and 20–21 rays, anal fin has 3 spines and 20 rays. Feeds on sponge and ascidian. Aquarium fish. Up to 45cm in total length.

Distribution — Ryukyu, Formosa, Indo-West Pacific

Pomacanthidae

⇧ *Pomacanthus navarchus* (Cuvier)

[*Euxiphipops navarchus* (Cuvier)]

Blue-girdled Angelfish

Diagnosis — Inhabits singly outer slope of coral reefs. Body is oval. Fins and blotches are margined with blue lines. Opercles have a sharp-pointed spine. Dorsal fin has 13–14 spines and 14–18 rays, anal fin has 3 spines and 16–18 rays. Feeds on sponge and ascidian. Aquarium fish. Up to 25cm in total length.

Distribution — Indo-West Pacific

Pomacanthidae

Pomacanthus xanthometopon (Bleeker) ⇨

[*Euxiphipops xanthometopon* (Bleeker)]
[*Arusetta xanthometopon* (Bleeker)]
[*Holacanthus xanthometopon* (Bleeker)]

Yellow-faced Angelfish

Diagnosis — Inhabits singly outer slope of coral reefs. Body is oval with fine reticulate patterns. Head is blue; surface around eyes is yellow. Dorsal fin ray has a black blotch. In young, white stripes on body are straight. Dorsal fin has 14 spines and 16–17 rays, anal fin has 3 spines and 16–18 rays. Feeds on sponge and ascidian. Aquarium fish. Up to 45cm in total length.

Distribution — Ryukyu, Indo-West Pacific

Pomacanthidae

Pomacanthus imperator (Bloch)

[*Holacanthus imperator* (Bloch)]
[*Chaetodon nicobariensis* Bloch et Schneider]
[*Pomacanthoides imperator* (Bleeker)]

Emperor Angelfish, Imperial Angelfish

Diagnosis — Inhabits singly rather deep coral reefs. Body is deep with yellow stripes. Dorsal and anal fin rays are developed. Preoperculum have a sharp-pointed spine. In young, body has white ring-like blotches. Dorsal fin has 13 spines and 19–21 rays, anal fin has 3 spines and 18–21 rays. Feeds on sponge and ascidian. May cause ciguatera poisoning. Aquarium fish. Up to 45cm in total length.

Distribution — Japan, Ryukyu, China, Indo-Pacific, Hawaii, Persian Gulf, Red Sea, South Africa

▼Young

Pomacanthidae

Pomacanthus semicirculatus (Cuvier)

[*Holacanthus semicirculatus* Cuvier et Valenciennes]
[*H. lepidolepis* Bleeker]
[*Acanthochaetodon nigrolunatus* Tanaka]
[*Pomacanthoides semicirculatus* (Cuvier)]

Zebra Angelfish, Blue Angelfish, Semicircle Angelfish, Koran Butterflyfish

Diagnosis — Inhabits outer slope of shoreline reefs and coral reefs. Body is deep, profusely dotted with dark color. Snout is yellow. Dorsal and anal fin rays are extended. Opercular margins are blue with a sharp-pointed spine. In young, white stripes on body are curved. Dorsal fin has 13 spines and 20–23 rays, anal fin has 3 spines and 18–21 rays. Feeds on sponge and ascidian. Aquarium fish. Up to 45cm in total length.

Distribution — Japan, Ryukyu, Indo-Pacific, Red Sea, South Africa

▲Young

▼Adult

Pomacanthidae

Pygoplites diacanthus (Boddaert)

Regal Angelfish, Blue-banded Angelfish, Royal Angelfish

Diagnosis — Inhabits singly shallow coral reefs. Body has 8–9 blue bands. Pelvic and caudal fins are yellow. Lateral lines are not complete. Preoperculum have a sharp-pointed spine. Dorsal fin has 14 spines and 17–18 rays, anal fin has 3 spines and 17–18 rays. May cause ciguatera poisoning. Aquarium fish. Up to 30cm in total length.

Distribution — Ryukyu, Indo-Pacific, Red Sea, East Africa

Pomacanthidae

Centropyge loriculus (Günther)

Flame Angelfish

Diagnosis — Inhabits singly outer slope of coral reefs. Body is orange-red with 4 dark bands. Posterior part of opercles have a black blotch. Posterior ends of dorsal and anal fins are blue. Margins of eyes are black. Omnivorous. Aquarium fish. Up to 12cm in total length.

Distribution — Pacific, Hawaii

Pomacanthidae
Centropyge bispinosus (Günther) ⇩
[*C. multispinis* (Playfair)]
Dusky Cherub, Spiny Angelfish, Many-spined Angelfish
Diagnosis — Inhabits singly passage or outer slope of coral reefs. Body is oblong, purplish-blue, has many dark stripes; abdominal side is red. Caudal margin is rounded. Dorsal fin has 14 spines and 15 rays, anal fin has 3 spines and 16 rays. Feeds on seaweed. Aquarium fish. Up to 12cm in total length.
Distribution — Ryukyu, Formosa, Indo-Pacific, South Africa

Pomacanthidae
Apolemichthys xanthopunctatus Burgess ⇧
Golden Spotted Angelfish
Diagnosis — Body is oval, profusely dotted with yellow. Snout is blue. Head has a black blotch on back surface. Dorsal, anal, caudal fins are blue-margined. Dorsal fin has 14 spines and 17−19 rays, anal fin has 3 spines and 17−19 rays. Aquarium fish. Up to 25cm in total length.
Distribution — Micronesia, Line Islands

Pomacanthidae
Centropyge acanthops (Norman) ⇩
African Pygmy Angelfish
Diagnosis — Found in schools among rather deep coral reefs. Body is oval; head and back are yellow; abdominal side and anal fin are blue. Opercular spines and surface around eyes are blue. Dorsal fin has 14 spines and 15−17 rays, anal fin has 3 spines and 16−17 rays. Feeds on seaweed. Aquarium fish. Up to 7cm in total length.
Distribution — West Indian Ocean, South Africa

Pomacanthidae
Centropyge ferrugatus Randall et Burgess ⇩
Rusty Pygmy Angelfish
Diagnosis — Inhabits singly shallow coral reefs. Body is oval, profusely dotted with dark color. Dorsal and anal fins are blue-margined. Dorsal fin has 14 spines and 17 rays, anal fin has 3 spines and 17−18 rays. Aquarium fish. Up to 10cm in total length.
Distribution — Japan, Ryukyu, Formosa, Philippines, Indonesia

Pomacanthidae
Centropyge vrolicki (Bleeker) ⇩
Brown Angelfish, Pearl-scaled Angelfish
Diagnosis — Inhabits singly shallow coral reefs. Body is oval, posterior part is dark, has many indistinct brown stripes. Dorsal, anal, caudal margins are blue. Dorsal fin has 14 spines and 15−16 rays, anal fin has 3 spines and 16 rays. Feeds on seaweed. Aquarium fish. Up to 10cm in total length.
Distribution — Japan, Ryukyu, Formosa, West Pacific

▲ Female

Pomacanthidae
Centropyge heraldi
Woods et Schultz ⇧

Herald's Angelfish

Diagnosis — Inhabits singly coral reefs. Resembles *C. flavissimus*, but distinguished by absence of blue margins, dark posterior part of eyes. In male, body is orange, dorsal and anal fin margins are black. Dorsal fin has 15 spines and 15 rays, anal fin has 3 spines and 17 rays. Feeds on seaweed. Aquarium fish. Up to 12cm in total length.

Distribution — Japan, Ryukyu, Formosa, West Pacific

Pomacanthidae
Centropyge bicolor (Bloch) ⇧

Black and Gold Angelfish, Bicolor Angelfish

Diagnosis — Inhabits singly shallow coral reefs. Anterior part of body and caudal fin are yellow, posterior half of body and above eyes are dark-blue. Opercular margins have a sharp-pointed spine. Dorsal fin has 15 spines and 14–15 rays, anal fin has 3 spines and 17–18 rays. Feeds on seaweed. Aquarium fish. Up to 15cm in total length.

Distribution — Japan, Ryukyu, Pacific

▼Male

Pomacanthidae
⇧ *Centropyge interruptus* (Tanaka)

[*C. fisheri* (Snyder)]

[*Holacanthus interruptus* (Tanaka)]

[*Angelichthys interruptus* Tanaka]

Fisher's Angelfish

Diagnosis — Temperate species inhabits shoreline reefs and rather deep rocky reefs. Anterior half of body is yellowish-red, posterior half is bluish-purple, profusely dotted with blue. Caudal fin is yellow. In male, opercles, posterior parts of dorsal and anal fins have blue lines. Dorsal fin has 14 spines and 16 rays, anal fin has 3 spines and 17 rays. Aquarium fish. Up to 16cm in total length.

Distribution — Japan, Izu Islands, Bonin Islands, Kure Atoll (off Midway Islands)

Pomacanthidae
⇧ *Centropyge potteri* Jordan et Metz

Potter's Angel Fish

Diagnosis — Inhabits singly coral reefs and shoreline reefs. Body is oval with many orange stripes. In some individuals, body is dark from lower surface to anal fin. Posterior ends of dorsal, anal, caudal fins are margined with blue. Feeds on seaweed. Aquarium fish. Up to 10cm in total length.

Distribution — Hawaii

Pomacanthidae
Centropyge flavissimus
(Cuvier) ⇧

Lemonpeel Angelfish, Gold Angelfish, Yellow Angel-fish

Diagnosis — Inhabits shallow coral reefs. Body is yellow. Surface around eyes, opercles, fin margins are blue. In young male, body has a black blotch. Dorsal fin has 14 spines and 15 rays, anal fin has 3 spines and 16 rays. Feeds on seaweed. Aquarium fish. Up to 9cm in total length.

Distribution — Ryukyu, Bonin Islands, Pacific, Hawaii

Pomacanthidae
⇦ *Centropyge tibicen* (Cuvier)

Piper Angelfish, Keyhole Angelfish

Diagnosis — Inhabits shallow coral reefs. Body is dark-blue, has a white blotch. Pelvic and anal fin margins are yellow. Transsexual. Dorsal fin has 14 spines and 16–17 rays, anal fin has 3 spines and 16–17 rays. Feeds on seaweed. Aquarium fish. Up to 19cm in total length.

Distribution — Japan, Ryukyu, Formosa, West Pacific

Pomacanthidae

Apolemichthys trimaculatus
(Cuvier) ⇧

[*Holacanthus trimaculatus* Lacépède]

Fourspot Angelfish, Three-spot Angelfish

Diagnosis — Inhabits singly shoreline reefs and coral reefs. Body is oval, posterior head has a black spot. Lips are blue. In young, body has dark stripes with a black blotch on posterior part. Dorsal fin has 14 spines and 17–18 rays, anal fin has 3 spines and 17–18 rays. Anal fin margin is black. Feeds on sponge and ascidian. Aquarium fish. Up to 30cm in total length.

Distribution — Japan, Ryukyu, Indo-Pacific, South Africa

Pomacanthidae

⇧ *Apolemichthys arcuatus* (Gray)

[*Holacanthus arcuatus* Gray]

Bandit Angelfish

Diagnosis — Inhabits singly shoreline reefs. Body is oval with a white-margined broad black band running from eyes to ray part of dorsal fin. Caudal and anal fins also have a black band. Dorsal fin has 13 spines and 17–19 rays, anal fin has 3 spines and 17–19 rays. Feeds on sponge. Aquarium fish. Up to 18cm in total length.

Distribution — Hawaii

▼Male

▲Female

Pomacanthidae

Genicanthus melanospilos
(Bleeker) ⇩ ⇩

Blackspot Angelfish

Diagnosis — Inhabits rather deep coral reefs and rocky reefs. Body is oblong. Caudal lobes are extended. In male, body has many dark bands. In female, top and bottom of caudal fin are black. Transsexual. Dorsal fin has 15 spines and 16 rays, anal fin has 3 spines and 18 rays. Feeds on zooplankton. Aquarium fish. Up to 21cm in total length.

Distribution — Ryukyu, Pacific

Pomacanthidae

Genicanthus personatus
Randall ⇧ ⇨

Masked Angel Fish

Diagnosis — Inhabits rather deep rocky reefs. Body is oval. Anterior half of caudal fin is black. In male, head and fins are yellow except caudal fin. In female, pelvic fins are slightly yellow, head is black. Dorsal fin has 15 spines and 15–16 rays, anal fin has 3 spines and 16–18 rays. Feeds on seaweed and benthos. Aquarium fish. Up to 21cm in total length.

Distribution — Hawaii

▲Male

▼Female

Pomacanthidae

Genicanthus lamarck (Lacépède)

[*Holacanthus lamarck* Lacépède]

Freckletail Angelfish, Lamarck's Angelfish

Diagnosis — Inhabits passage or outer slope of coral reefs. Body is oblong with 4 black stripes extending from eyes. Caudal lobes are extended. In male, pelvic fins are black. In female, top and bottom of caudal fin are black. Dorsal fin has 15 spines and 15–16 rays, anal fin has 3 spines and 16–17 rays. Feeds on zooplankton. Aquarium fish. Up to 24cm in total length.

Distribution — Japan, Ryukyu, Indo-West Pacific, East Africa

▲ Female

▲ Female

Chaetodontidae

Chaetodon ephippium Cuvier

Saddled Butterflyfish, Black-blotched Butterflyfish, Saddled Coralfish, Saddleback Butterflyfish

Diagnosis — Inhabits shallow coral reefs. Snout is projected. Surface from snout to pectoral sides is yellow. Back has a large black blotch. Dorsal fin has 12–14 spines and 21–24 rays, anal fin has 3 spines and 20–22 rays. Dorsal fin rays are extended like thread. Edible but may cause ciguatera poisoning. Feeds on coral polyp and calcareous algae. Aquarium fish. Up to 30cm in total length.

Distribution — Japan, Ryukyu, Formosa, Pacific, Hawaii

Chaetodontidae

Chaetodon vagabundus Linnaeus

[*Linophora vagabundus* (Linnaeus)]

Crisscross Butterflyfish, Vagabond Butterflyfish, Vagabond Coralfish

Diagnosis — Inhabits shoreline reefs and coral reefs. Has a broad black band across the eyes. Body has dark stripes crossing at right angle. Caudal fin has 2 black bands. A black band runs from dorsal fin base to caudal peduncles and anal fin. Dorsal fin has 12–13 spines and 22–25 rays, anal fin has 3 spines and 19–20 rays. Feeds on coral polyp and seaweed. Aquarium fish. Up to 36cm in total length.

Distribution — Japan, Ryukyu, Formosa, Indo-Pacific, Red Sea, South Africa

Pomacanthidae

Holacanthus passer Valenciennes

King Angelfish

Diagnosis — Inhabits shoreline reefs, rather deep rocky reefs. Body is deep, has a white band. Ray parts of dorsal and anal fins are extended. Caudal fin is yellow. In young, body is yellowish, has 8 bluish-white stripes. Feeds on sponge and ascidian. Aquarium fish. Up to 25cm in total length.

Distribution — Gulf of California, Mexico, Panama, East Pacific, Galapagos Islands

Chaetodontidae

Chaetodon lineolatus Cuvier

Lined Butterfly Fish, Newmoon Coral Fish

Diagnosis — Inhabits outer slope of coral reefs. Has a broad black band across the eyes. Snout is projected. Body has many black stripes; back is black. Dorsal fin has 12 spines and 24–28 rays, anal fin has 3 spines and 20–22 rays. Feeds on coral polyp and calcareous algae. Aquarium fish. Up to 36cm in total length.

Distribution — Japan, Ryukyu, Formosa, Indo-Pacific, Hawaii, Red Sea, South Africa

Chaetodontidae
Chaetodon punctatofasciatus Cuvier ⇧

Sevenband Butterflyfish, Spots-and-bands Coralfish, Spot-banded Butterflyfish

Diagnosis — Inhabits shallow coral reefs. Has an orange band bordered with black stripes across the eyes. Body has 7–8 dark bands and spots. Caudal fin is transparent with a black band. Caudal peduncles are orange. Dorsal fin has 13–14 spines and 22–25 rays, anal fin has 3 spines and 17–18 rays. Feeds on coral polyp and seaweed. Aquarium fish. Up to 12cm in total length.

Distribution — Ryukyu, Formosa, Pacific, Hawaii

Chaetodontidae
Chaetodon rafflesi Bennett ⇧

[*Linophora rafflesi* (Bennett)]

Latticed Butterflyfish, Raffles' Coralfish

Diagnosis — Inhabits shallow coral reefs. A distinct black band across the eyes. Snout is pointed. Caudal fin has a black band. Body has reticulate patterns. Dorsal fin is black-margined, has 12–13 spines and 21–23 rays. Anal fin has 3 spines and 18–20 rays. Feeds on coral polyp and benthos. Aquarium fish. Up to 16cm in total length.

Distribution — Japan, Ryukyu, Formosa, Indo-Pacific

Chaetodontidae
Chaetodon miliaris Quoy et Gaimard ⇩

Milletseed Butterflyfish, Lemon Butterflyfish, Crochet Butterflyfish

Diagnosis — Inhabits shoreline reefs. A distinct black band across the eyes. Caudal peduncles are black. Body has rows of black spots. Dorsal fin has 13 spines and 20–23 rays, anal fin has 3 spines and 17–20 rays. Aquarium fish. Up to 13cm in total length.

Distribution — Hawaii

Chaetodontidae
Chaetodon auripes Jordan et Snyder ⇧

Oriental Butterflyfish, Whitecollar Coralfish, Golden Butterflyfish

Diagnosis — Found in schools about shoreline reefs and coral reefs. Resists low water temperatures. A black band followed by a white band across the eyes. Body has brown stripes. Dorsal fin has 12 spines and 22–27 rays, anal fin has 3 spines and 17–22 rays, fin margins are black. Feeds on coral polyp and benthos. Aquarium fish. Up to 20cm in total length.

Distribution — Japan, Ryukyu, Formosa, Korea, Philippines, Indonesia

Chaetodontidae
Chaetodon kleinii Bloch ⇩

Klein's Coralfish, Klein's Butterflyfish, White Spotted Butterfly Fish

Diagnosis — Inhabits rather deep coral reefs. A distinct black band across the eyes. Snout is pointed. Body has 2 indistinct brown bands. Caudal fin is yellow with transparent margin. Dorsal fin has 13–14 spines and 20–23 rays, anal fin has 3 spines and 17–20 rays. Feeds on coral polyp and seaweed. Aquarium fish. Up to 18cm in total length.

Distribution — Japan, Ryukyu, Formosa, Indo-Pacific, Hawaii, South Africa

Chaetodontidae
Chaetodon citrinellus Cuvier ⇧

Citron Coralfish, Speckled Butterflyfish

Diagnosis — Inhabits shoreline reefs and coral reefs. Has a distinct black band across the eyes. Snout is pointed. Body is profusely dotted with small dark color. Anal fin margin is black. Dorsal fin has 13–15 spines and 20–22 rays, anal fin has 3 spines and 15–17 rays. Feeds on seaweed and ascidian. Aquarium fish. Up to 16cm in total length.

Distribution — Japan, Ryukyu, Formosa, Indo-Pacific, Hawaii, East Africa

Chaetodontidae
Chaetodon lunula (Lacépède) ⇩

Moon Butterflyfish, Halfmoon Butterflyfish, Red Striped Butterflyfish, Raccoon Butterflyfish

Diagnosis — Inhabits coral reefs. A broad black band across the eyes. Opercular margins are white. Caudal peduncles have a black blotch. Dorsal fin has 11–13 spines and 22–25 rays, anal fin has 3 spines and 17–19 rays. Nocturnal. Eats all kinds of food. Aquarium fish. Up to 25cm in total length.

Distribution — Japan, Ryukyu, Formosa, Indo-Pacific, Hawaii, South Africa

Chaetodontidae
Chaetodon ulietensis Cuvier ⇧

Pacific Double-saddle Butterflyfish

Diagnosis — Inhabits shallow coral reefs. Resembles *C. lineolatus*, but distinguished by 2 indistinct black bands on back. Caudal peduncles are black. Dorsal fin has 12–13 spines and 23–25 rays, anal fin has 3 spines and 19–21 rays. Feeds on coral polyp and calcareous algae. Aquarium fish. Up to 20cm in total length.

Distribution — Japan, Ryukyu, Formosa, Pacific, Hawaii

Chaetodontidae
Chaetodon sp. ⇩

Half-blooded Butterflyfish

(New Common Name)

Diagnosis — Inhabits coral reefs. Body has irregular dark bands, profusely dotted with black. A short orange band across the eyes. Posterior head has a black blotch. Caudal peduncles are orange. Considered to be natural hybrid between *C. punctatofasciatus* and *C. pelewensis*. Aquarium fish. Up to 13cm in total length.

Distribution — Philippines, New Britain, Australia

Chaetodontidae
⇩ *Chaetodon unimaculatus* Bloch

[*Lepidochaetodon unimaculatus* (Bloch)]

Limespot Butterflyfish, One Spot Coralfish, Teardrop Butterflyfish

Diagnosis — Inhabits shallow coral reefs. A broad black band across the eyes. Snout is pointed. Back has a large black blotch. A black band runs from dorsal end to anal fin base. Dorsal fin has 13 spines and 21–23 rays, anal fin has 3 spines and 16–20 rays. Feeds on coral polyp. Aquarium fish. Up to 20cm in total length.

Distribution — Japan, Ryukyu, Formosa, Indo-Pacific, Hawaii, Red Sea, South Africa

Chaetodontidae

⇦ *Chaetodon trifasciatus* Mungo Park ⇩

[*Rhabdophorus trifasciatus* (Mungo Park)]

Lineated Butterfly Fish, Purple Butterflyfish, Three-banded Coralfish, Three Striated Butterfly Fish, Red-fin Butterflyfish

Diagnosis — Inhabits shoreline reefs and coral reefs. Body is oval, has dark longitudinal stripes. Snout is slightly pointed. A narrow black band across the eyes. Caudal margin is transparent with a black band. Dorsal fin has 13–14 spines and 20–22 rays, anal fin has 3 spines and 18–21 rays. Feeds on coral polyp and seaweed. Aquarium fish. Up to 15cm in total length.

Distribution — Japan, Ryukyu, Formosa, Indo-Pacific, Hawaii, South Afric

Chaetodontidae

⇦ *Chaetodon bennetti* Cuvier

Archer Butterflyfish, Bennett's Butterflyfish, Bennett's Coralfish

Diagnosis — Inhabits shallow coral reefs. Body is rounded, back has a white-margined black blotch. 2 white stripes crossing pectoral fins. A brown band across the eyes is margined with white. Dorsal fin has 13–14 spines and 16–20 rays, anal fin has 3 spines and 15–17 rays. Feeds on coral polyp and seaweed. Aquarium fish. Up to 18cm in total length.

Distribution — Japan, Ryukyu, Indo-Pacific, South Africa

Chaetodontidae

Chaetodon plebeius Cuvier ⇧

Blue-blotched Butterflyfish, Two-spot Coralfish, Blue-spot Butterflyfish

Diagnosis — Inhabits shallow coral reefs. Body is oval, back has an indistinct bluish-black blotch. A black band across the eyes is distinct. Upper part of caudal peduncles have a black blotch, caudal margin is transparent. Dorsal fin has 13–15 spines and 16–19 rays, anal fin has 4–5 spines and 14–17 rays. Feeds on coral polyp. Aquarium fish. Up to 15cm in total length.

Distribution — Japan, Ryukyu, Formosa, Indo-Pacific

Chaetodontidae

⇦ *Chaetodon speculum* Cuvier ⇧

Mirror Butterflyfish, One Spotted Butterflyfish, Oval-spot Coralfish, Oval-spot Butterflyfish

Diagnosis — Inhabits singly shoreline reefs and coral reefs. A black band across the eyes is somewhat narrow. Snout is pointed. Back has a large black blotch. Caudal margin is transparent. Dorsal fin has 14 spines and 17–18 rays, anal fin has 3 spines and 15–16 rays. Feeds on coral polyp. Aquarium fish. Up to 18cm in total length.

Distribution — Japan, Ryukyu, Formosa, Indo-Pacific

Chaetodontidae

Chaetodon auriga Forskål ⇨

[*C. setifer* Bloch]

[*Linophora auriga* (Forskål)]

Filamented Coralfish, Spined Butterfly Fish, Threadfin Butterflyfish, Thread Fin Coralfish

Diagnosis — Inhabits shallow shoreline reefs and coral reefs. A broad black band across the eyes is indistinct on upper side. Snout is projected. Body has dark stripes meeting at right angles. Dorsal fin ray has a black blotch, anterior rays extended like thread. Dorsal fin has 12–13 spines and 22–25 rays, anal fin has 3 spines and 18–21 rays. Feeds on coral polyp and sea anemone. Has records of ciguatera poisoning. Aquarium fish. Up to 23cm in total length.

Distribution — Japan, Ryukyu, Formosa, China, Indo-Pacific, Hawaii, Red Sea, South Africa

Chaetodontidae

Chaetodon trifascialis Quoy et Gaimard ⇧

[*Megaprotodon trifascialis* (Quoy et Gaimard)]

Rightangle Butterflyfish, Right-angled Butterflyifish, V-lined Butterflyfish, Chevroned Butterflyfish

Diagnosis — Inhabits shoreline reefs and coral reefs. Body is oval, has V-shaped dark stripes. Snout is pointed. A broad black band across the eyes. Caudal fin is dark with yellow margin. Dorsal fin has 14 spines and 14–17 rays, anal fin has 4–5 spines and 15–17 rays. Lateral line is not complete. Feeds on coral polyp. Up to 18cm in total length.

Distribution — Japan, Ryukyu, Formosa, Indo-Pacific, Hawaii, Red Sea, South Africa

Chaetodontidae

⇩ *Chaetodon ornatissimus* Cuvier

Clown Butterfly Fish, Ornate Coralfish,Ornate Butterflyfish

Diagnosis — Inhabits shallow coral reefs. A narrow black band across the eyes, similar black band is present around mouth. Body has 6–7 yellowish-brown oblique bands. Dorsal, anal, caudal fins have 2 black stripes. Dorsal fin has 12–13 spines and 24–28 rays, anal fin has 3 spines and 20–23 rays. Feeds on coral polyp and calcareous algae. Aquarium fish. Up to 20cm in total length.

Distribution — Japan, Ryukyu, Bonin Islands, Formosa, Indo-Pacific, Hawaii

Chaetodontidae

Chaetodon meyeri Bloch et Schneider ⇧

Maypole Butterflyfish, Meyer's Coralfish

Diagnosis — Inhabits shallow coral reefs. Resembles *C. ornatissimus*, but distinguished by black curved oblique bands on body. Dorsal fin has 12–13 spines and 23–25 rays, anal fin has 3 spines and 18–20 rays. Feeds on coral polyp. Aquarium fish. Up to 20cm in total length.

Distribution — Ryukyu, Formosa, Indo-Pacific, Hawaii, South Africa

Chaetodontidae

⇦ *Chaetodon octofasciatus* Bloch

Eight-banded Butterflyfish, Eight-banded Coralfish

Diagnosis — Inhabits bays and brackish areas of river mouths. Body is rounded, has 8 distinct black bands. Snout is pointed. Caudal fin is transparent, pelvic fins are yellow. Dorsal fin has 10–12 spines and 17–20 rays, anal fin has 3 spines and 14–17 rays. Feeds on coral polyp. Aquarium fish. Up to 12cm in total length.

Distribution — Ryukyu, Formosa, Indo-West Pacific

Chaetodontidae

Chaetodon adiergastos Seale ⇨

Philippine Butterflyfish, Bantayan Butterflyfish

Diagnosis — Inhabits shallow coral reefs. Body is rounded with oblique dark stripes. Snout is pointed. A broad black band across the eyes. In young, dorsal fin has a black blotch. Dorsal fin has 12 spines and 23−26 rays, anal fin has 3 spines and 18−21 rays. Feeds on coral polyp and benthos. Aquarium fish. Up to 16cm in total length.

Distribution — Ryukyu, Formosa, West Pacific

Chaetodontidae

Chaetodon reticulatus Cuvier ⇧

Reticulated Butterflyfish

Diagnosis — Inhabits shallow coral reefs. Body is rounded and blackish. Has a broad black band across the eyes. Snout is also black. Caudal margin is yellow. Dorsal fin has 12−13 spines and 23−29 rays, anal fin has 3 spines and 18−22 rays, ray part of fins are well-developed. Feeds on coral polyp. Has records of ciguatera poisoning. Aquarium fish. Up to 16cm in total length.

Distribution — Ryukyu, Bonin Islands, Formosa, Pacific, Hawaii

Chaetodontidae

⇦ *Chaetodon quadrimaculatus* Gray

Fourspot Butterflyfish

Diagnosis — Inhabits coral reefs. Has a yellow band across the eyes. Back is dark with 2 white blotches. Pectoral, pelvic, caudal fins are yellow. Dorsal fin has 13−15 spines and 20−23 rays, anal fin has 3 spines and 16−18 rays. Feeds on coral polyp. Aquarium fish. Up to 16cm in total length.

Distribution — Ryukyu, Bonin Islands, Pacific, Hawaii

Chaetodontidae

⇦ *Chaetodon xanthurus* Bleeker

Yellow-tail Butterflyfish

Diagnosis — Inhabits shallow coral reefs. Body is white with dark reticulated patterns. A black band across the eyes reaches the throat. Posterior head has a black blotch. Posterior part of body is orange. Caudal fin also has an orange band. Dorsal fin has 13−14 spines and 20−23 rays, anal fin has 3 spines and 16−17 rays. Omnivorous. Aquarium fish. Up to 14cm in total length.

Distribution — Japan, Ryukyu, Formosa,Philippines, Indonesia

Chaetodontidae

Chaetodon baronessa Cuvier ⇧

Herring-bone Butterflyfish, Triangular Coralfish

Diagnosis — Inhabits coral reefs. Body is rounded. Dorsal and anal fin rays are developed. A black band across the eyes is followed by another black band. Snout is pointed and reddish. Dorsal fin has 11−12 spines and 22−26 rays, anal fin has 3 spines and 20−22 ryas. Feeds on coral polyp. Aquarium fish. Up to 15cm in total length.

Distribution — Japan, Ryukyu, Formosa, Pacific

Chaetodontidae

Chaetodon fremblii Bennett ⇨

Blue-lined Butterfly Fish, Bluestripe Butterflyfish

Diagnosis — Inhabits shoreline reefs. Body is yellow with blue longitudinal stripes. Back of head and posterior end of body have large black blotches. Snout is pointed. Dorsal fin has 13 spines and 20−21 rays, anal fin has 3 spines and 17−18 rays. Feeds on coral polyp and benthos. Aquarium fish. Up to 13cm in total length.

Distribution — Hawaii

Chaetodontidae

Chaetodon nippon Döderlein ⇩

Blackloin Butterflyfish

Diagnosis — Temperate species found in schools about shoreline reefs. Has no black band across the eyes. Body is light-brown, posterior part is dark. Pelvic and caudal fins are yellow with transparent margins. Dorsal fin has 12−13 spines and 18−21 rays, anal fin has 3 spines and 15−17 rays. Spawns between spring and fall. Aquarium fish. Up to 15cm in total length.

Distribution — Japan, Izu Islands, Formosa, Korea, Philippines

Chaetodontidae

⇦ ***Chaetodon tinkeri*** Schultz

Tinker's Butterfly Fish

Diagnosis — Inhabits singly rather deep rocky reefs. Anterior half of body is white, has rows of black spots, posterior part is black. A band across the eyes is dark-yellow. Snout and caudal fin are yellow. Dorsal fin has 13−14 spines and 19−21 rays, anal fin has 3 spines and 16−17 rays. Feeds on benthos. Aquarium fish. Up to 15cm in total length.

Distribution — Hawaii

Chaetodontidae

Chaetodon modestus

Temminck et Schlegel ⇨

Brownbanded Butterflyfish

Diagnosis — Inhabits rather deep rocky reefs. Body is rounded with 3 broad brown bands. Dorsal fin ray has a black blotch. Snout is projected. Caudal fin is transparent. Dorsal fin has 11 spines and 21−25 rays, anal fin has 18 spines and 21 rays. Aquarium fish. Up to 17cm in total length.

Distribution — Japan, Ryukyu, Formosa, Indo-Pacific, Hawaii, Persian Gulf

Chaetodontidae

⇦ ***Chaetodon melannotus*** Schneider

[*C. melanotus* Bloch et Schneider]

[*Chaetodontops melanotus* (Bloch)]

Blackback Butterflyfish, Black-backed Coralfish, Striped Butterfly Fish, Two-spot Coralfish

Diagnosis — Inhabits shoreline reefs and coral reefs. A narrow black band across the eyes. Upper part of caudal peduncles and anterior part of anal fin base have a black blotch. Body has oblique rows of black spots. Dorsal fin has 12−13 spines and 17−20 rays, anal fin has 3 spines and 17−18 yellow rays. Feeds on seaweed and coral polyp. Aquarium fish. Up to 18cm in total length.

Distribution — Japan, Ryukyu, Formosa, Indo-Pacific, Red Sea, South Africa

Chaetodontidae

Coradion altivelis McCulloch ⇨

[*C. fulvocinctus* Tanaka]

Tawnygirdled Butterflyfish, Highfin Coralfish

Diagnosis — Inhabits shoreline reefs. Body is deep, snout is pointed. A dark brown band crossing pectoral fins is divided in two. Dorsal and anal fins are developed with vertical hind margins. In young, dorsal fin has a white-margined black blotch. Dorsal fin has 8 spines and 30−33 rays, anal fin has 3 spines and 18−22 rays. Aquarium fish. Up to 17cm in total length.

Distribution — Japan, Izu Islands, South China Sea, West Pacific

Chaetodontidae

⇦ ***Chaetodon argentatus*** Smith et Radcliffe

Threeband Butterflyfish, Asian Butterflyfish

Diagnosis — Inhabits shoreline reefs and coral reefs. Body is covered with black reticulated patterns, has 3 dark brown bands. A black band across the eyes. Dorsal fin has 13−14 spines and 21−24 rays, anal fin has 3 spines and 16−17 rays. Feeds on seaweed and coral polyp. Aquarium fish. Up to 20cm in total length.

Distribution — Japan, Izu Islands, Ryukyu, Formosa, Hong Kong, Philippines

Chaetodontidae

⇦ ***Hemitaurichthys polylepis*** (Bleeker)

[*H. zoster* (Bennett)]

Brushytooth Butterflyfish, Pyramid Butterflyfish

Diagnosis — Found in schools about outer slope of coral reefs. Body is compressed and deep. Anterior part of body, dorsal and anal fins are dark-yellow. Sickle-shaped pectoral fins are transparent. Scales are small. Dorsal fin has 11–12 spines and 23–26 rays, anal fin has 3 spines and 19–22 rays. Feeds on zooplankton. Aquarium fish. Up to 18cm in total length.

Distribution — Japan, Ryukyu, Formosa, Pacific, Hawaii

Chaetodontidae

Johnrandallia nigrirostris (Gill) ⇨

Barberfish, Blacknose Butterflyfish

Diagnosis — Found in schools about shoreline reefs. Body is rounded; snout is pointed. Head is white. Surface around eyes, mouth and posterior head is black. Dorsal fin base has a broad black band. Dorsal fin has 12 spines and 24–25 rays, anal fin has 3 spines and 18–20 rays. Feeds on benthos. Aquarium fish. Up to 17cm in total length.

Distribution — Gulf of California, Mexico, Panama, Columbia, East Pacific, Galapagos Islands

Chaetodontidae

⇦ ***Parachaetodon ocellatus*** (Cuvier)

Ocellated Coralfish, Six-spined Butterflyfish

Diagnosis — Inhabits shoreline reefs and rather deep muddy-sand bottoms. Body is rounded. Dorsal and anal fins are developed. A brown band across the eyes is margined with black, other bands are indistinct. Dorsal fin has a black blotch on the middle. Caudal peduncles also have a black blotch. Caudal fin is transparent. Dorsal fin has 6 spines and 28–30 rays, spines are not stiff. Anal fin has 3 spines and 18–20 rays. Scales are large. Omnivorous. Aquarium fish. Up to 17cm in total length.

Distribution — Ryukyu, Bonin Islands, Formosa, China, Indo-Pacific

Chaetodontidae

Chelmon rostratus (Linnaeus) ⇩

[*Chelmo rostratus* (Linnaeus)]

Beaked Coral Fish, Black-fin Coralfish, Longnose Butterflyfish, Copper-banded Butterflyfish

Diagnosis — Inhabits muddy-sand bottoms and seaweed beds of bays. Snout is projected, tube-shaped. Feeds on benthos. Dorsal and anal fin rays are developed, have a white-margined black blotch. Dorsal fin has 9 spines and 28–30 rays, anal fin has 3 spines and 19–21 rays. Scales are somewhat large. Aquarium fish. Up to 20cm in total length.

Distribution — Ryukyu, Formosa, Indo-Pacific, East Africa

Chaetodontidae

Forcipiger flavissimus Jordan et McGregor ⇧

[*F. longirostris* (Broussonet)]

Long-beaked Butterflyfish, Long-nosed Butterflyfish

Diagnosis — Found in schools about shoreline reefs and coral reefs. Snout is remarkably long like fountain pen ink-filler. Feeds on small benthos. Caudal fin is transparent. Dorsal fin has 12 spines and 22–24 rays, spine part is long. Anal fin has 3 spines and 17–18 rays, has a black blotch. Aquarium fish. Up to 18cm in total length.

Distribution — Japan, Ryukyu, Formosa, Indo-Pacific, Hawaii, Mexico, Red Sea, South Africa

Chaetodontidae
Heniochus monoceros Cuvier ⇧

Horned Coachman, Unicorn Pennant Coralfish, Masked Bannerfish

Diagnosis — Inhabits shoreline reefs and coral reefs. Body is large and deep. 3 black bands extend down from back. A black band across the eyes reaches mouth. Dorsal and caudal fins are yellow. Dorsal fin has 13 spines and 23–27 rays, anal fin has 3 spines and 18–19 rays. Aquarium fish. Up to 25cm in total length.

Distribution — Japan, Izu Islands, Ryukyu, Formosa, Indo-Pacific, South Africa

Chaetodontidae
Heniochus singularis ⇧ Smith et Radcliffe

Philippine Pennant Coralfish, Singular Bannerfish

Diagnosis — Inhabits shoreline reefs and coral reefs. Resembles *H. monoceros*, but distinguished by indistinct black bands on body, a narrow short black band across the eyes. Dorsal fin has 11–12 spines and 25–28 rays, anal fin has 3 spines and 16–18 rays. Aquarium fish. Up to 25cm in total length.

Distribution — Japan, Ryukyu, Formosa, Pacific

Chaetodontidae
⇧ *Heniochus diphreutes* Jordan

[*H. macrolepidotus* Bleeker]

[*H. bifasciatus* Shaw]

[*H. mycteryzans* Gray]

Pennant Coralfish, Coachman, Featherfin Coralfish, Schooling Bannerfish, Threadbacked Coralfish

Diagnosis — Resembles *H. acuminatus*, but found in schools in rather deep waters. Snout is somewhat short. Upper jaw has 3–4 rows of teeth. Dorsal fin has 12–13 spines and 23–25 rays, anal fin has 3 spines and 17–19 rays. Feeds on zooplankton. Aquarium fish. Up to 16cm in total length.

Distribution — Japan, Ryukyu, Indo-Pacific, Hawaii, Red Sea, South Africa

Chaetodontidae
Heniochus acuminatus (Linnaeus) ⇧ ⇨

[*H. diphreutes* Jordan]

Angel Fish, Butterfly Fish, Coachman, Pennant Coral Fish, Thread-backed Butterflyfish, Feather-fin Bullfish

Diagnosis — Inhabits shoreline reefs and coral reefs. Body is deep with 2 broad black bands. Snout is pointed. A black band across the eyes. Fourth spine of dorsal fin is extended remarkably long. Dorsal fin has 11 spines and 23–27 rays, anal fin has 3 spines and 16–19 rays. Omnivorous. Has records of ciguatera poisoning. Aquarium fish. Up to 25cm in total length.

Distribution — Japan, Ryukyu, Korea, Formosa, Indo-Pacific, Hawaii, Persian Gulf, Red Sea, South Africa

Chaetodontidae

⇦ ***Heniochus chrysostomus*** Cuvier

Threeband Pennant Coralfish, Pennant Bannerfish

Diagnosis — Inhabits singly shallow coral reefs. Resembles *H. acuminatus*, but snout is yellow and short. A broad black band across the eyes extends to pelvic fins. Middle black band extends to the longest dorsal spine. Dorsal fin has 12 spines and 21–22 rays, anal fin has 3 spines and 17–19 rays. Feeds on coral polyp. Aquarium fish. Up to 16cm in total length.

Distribution — Japan, Izu Islands, Ryukyu, Formosa, Indo-Pacific

Chaetodontidae

Heniochus varius (Cuvier) ⇩

Horned Bull-fish, Horned Pennant Coralfish, Hunchback Coralfish

Diagnosis — Inhabits shallow coral reefs. Body is deep, has 2 white bands extending down from back. Snout and posterior head are projected. A spine is present above eyes. A black blotch on abdominal side is indistinct. Dorsal fin has 11–12 spines and 21–24 rays, anal fin has 3 spines and 17–18 rays. Aquarium fish. Up to 20cm in total length.

Distribution — Japan, Ryukyu, Formosa, Pacific

Chaetodontidae

⇦ ***Heniochus pleurotaenia*** Ahl

[*H. varius* Kner]

Indian Bannerfish,
Indian Ocean Bannerfish

Diagnosis — Found in schools about coral reefs. Resembles *H. varius*, but distinguished by a distinct black blotch on abdominal side, posterior part is white. 2 broad white bands extending down from back. Dorsal fin has 11 spines and 23–25 rays, anal fin has 3 spines and 17–18 rays. Aquarium fish. Up to 18cm in total length.

Distribution — Indian Ocean

Scatophagidae

Scatophagus argus (Linnaeus)

[*S. argus* var. *rubrifrons*]

[*S. rubrifrons*]

Butterfish, Spadefish, Spotted Butter Fish, Spotted Scat, Green Scat, Tiger Scat, Red Scat

Diagnosis — Inhabits shoreline waters and bays, also found in brackish and freshwater areas. Body is rounded, profuesely blotched with black. Snout is slightly pointed. Dorsal fin has 10–11 spines and 16–18 rays, anal fin has 4 spines and 14–15 rays, spines are venomous. Omnivorous. Aquarium fish. Up to 35cm in total length.

Distribution — Japan, Ryukyu, Indo-Pacific, Persian Gulf

Scatophagidae

Scatophagus tetracanthus

(Lacépède) ⇨

[*S. multifasciatus* Richardson]

[*Selenotoca multifasciatus* (Richardson)]

Southern Butterfish, Northern Butterfish, John Dory, Striped Scat, African Scat, Scatty, Banded Scat, Silver Scat

Diagnosis — Resembles *S. argus*, but distinguished by dark stripes on body. Dorsal fin has 11 spines and 15–18 rays, anal fin has 4 spines and 14–15 rays, spines are venomous. Aquarium fish. Up to 40cm in total length.

Distribution — Indo-West Pacific, South Africa

◀Young

Zanclidae

Zanclus cornutus (Linnaeus)

[*Z. canescens* (Linnaeus)]

Moorish Idol, Toby, Spined Moorish Idol

Diagnosis — Found in schools about shallow shoreline reefs and coral reefs. Body is disk-shaped. Snout is projected. Third spine of dorsal fin extends remarkably long. Eyes have a pair of spines. Dorsal fin has 12 spines and 40−43 rays, anal fin has 3 spines and 33−36 rays. Omnivorous. Edible. Has records of ciguatera poisoning. Aquarium fish. Up to 30cm in total length.

Distribution — Japan, Ryukyu, Indo-Pacific, Hawaii, Mexico, South Africa

▼ Dorsal fin forms a sucking disk.

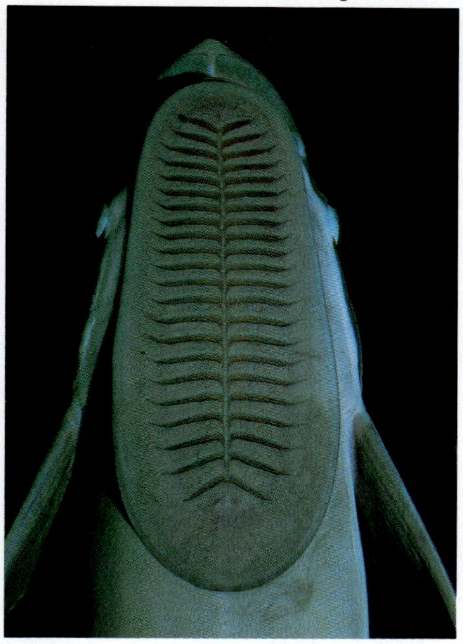

Echeneididae

Echeneis naucrates Linnaeus

Indian Remora, Long Sucking-fish, Pilot Fish, Pilot Sucker, Remora, Shark Remora, Shark's Pilot, Shark Sucker, Slender Sucker, White-tailed Remora

Diagnosis — Move by adhering to large migratory fishes, whales and turtles by sucking disk on head. Body is elongated with a dark brown band. Dorsal fin has 32−42 rays and anal fin has 32−38 rays, correspond both fins. Lower jaw and pectoral fins are pointed. Up to 1m in total length.

Distribution — Japan, Ryukyu, Formosa, Indo-Pacific-Atlantic, Persian Gulf, Red Sea, South Africa

▲ Migrates with shark.

▲ Migrates with Manta.

Ophidiidae
⇦ ***Sirembo imberbis*** (Temminck et Schlegel)
Loach Brotula
Diagnosis — Inhabits waters more than 100m deep. Body is elongated with rows of yellowish-brown blotches. A yellowish-brown band runs from eyes. Dorsal fin has 92−102 rays, anal fin has 68−72 rays, joins caudal fin. Pelvic fins have 1 ray. Edible. Up to 25cm in total length.
Distribution — Japan, Korea, Formosa

Ophidiidae
Brutula multibarbata Temminck et Schlegel ⇨
Barbed Brotula, Barbelled Eel-pout, Weasel Fish
Diagnosis — Inhabits shoreline reefs. Body is somewhat deep and extended. Head is small; snout and lower jaw have 6 barbels. Pelvic fin rays branched. Dorsal fin has 109−139 rays, anal fin has 80−106 rays, margined with white. Spawns between July and August. Edible. Up to 70cm in total length.
Distribution — Japan, Formosa, Indo-Pacific, Hawaii, Red Sea, South Africa

Ophidiidae
⇦ ***Monomitopum pallidus*** Smith et Radcliffe
Double-toothed Brotula (New Common Name)
Diagnosis — Inhabits waters more than 200m deep. Body is elongated and compressed. Tail is pointed. Tip of snout is truncated. Eyes are large. Preoperculum have 2 small, strong spines. Jaws have rows of enlarged outer teeth. Edges of dorsal and anal fins, caudal margin are black. Dorsal fin has 91−95 rays, anal fin has 71−76 rays. Up to 20cm in total length.
Distribution — Ryukyu, East China Sea, Philippines

Istiophoridae
Tetrapturus audax (Philippi) ⇨
[*Makaira audax* (Philippi)]
Barred Marlin, Marlinspike Fish, Mitsukuri's Marlin, New Zealand Marlin, Pacific Marlin, Pacific Striped Marlin, Spear Fish, Striped Marlin
Diagnosis — Swims in upper and middle layers of oceanic waters. Body is elongated, head is developed, snout is extended, lower jaw is also long. Caudal fin is falcated, caudal peduncles have 2 keels. First dorsal fin is elevated, short pelvic fins are string-shaped. First dorsal fin has 37−42 spines, second has 5−6 rays. First anal fin has 13−18 spines, second has 5−6 rays. Feeds on saury and sardine. Edible. Up to 3.8m in total length.
Distribution — Japan, Ryukyu,Formosa, Korea, Indo-Pacific, East Africa

Istiophoridae
Makaira mazara
(Jordan et Snyder) ⇨
Black Marlin, Black Spearfish, Blue Marlin, Marlin, Real Swordfish, Pacific Blue Marlin, Spearfish
Diagnosis — Swims in upper and middle layers of oceanic waters. Resembles *Tetrapturus audax*, but distinguished by deeper body, more shorter string-shaped pelvic fins, short lower jaw. Caudal peduncles have 2 keels. First dorsal fin has 40−44 spines, second has 6−7 rays. First anal fin has 12−15 spines, second has 6−7 rays. Edible. Up to 4.5m in total length.
Distribution — Japan, Formosa, Indo-Pacific, West America

Istiophoridae

Istiophorus platypterus (Shaw et Nodder)

Banana Sailfish, Bayonet Fish, Billfish, Indian Sailfish, Japanes Sailfish, Pacific Sailfish, Peacock-fish, Spike Fish, Swordfish

Diagnosis — Migrates in oceanic waters, also found in shoreline waters. Body is compressed and extended. Snout i projected remarkably long. Caudal fin is falcated. Well-developed dorsal fin is sail-like. Pelvic fins are string-like, long enough to reach anus. First dorsal fin has 42–48 spines, second has 6–7 rays. First anal fin has 12–15 spines, second has 6–7 rays. Edible. Up to 3.3m in total length.

Distribution — Southern Japan, Ryukyu, Indo-Pacific

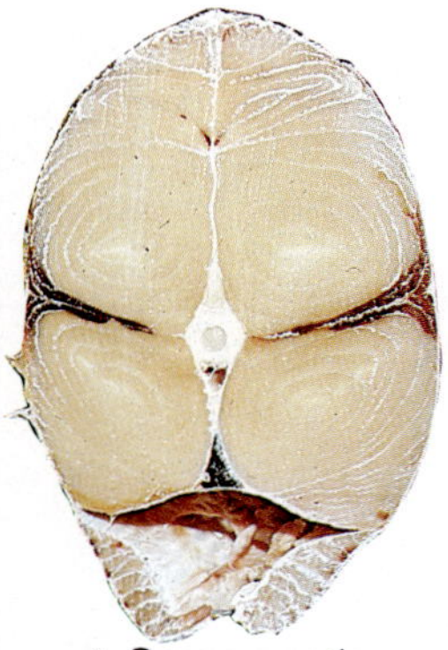

▲ Cross section of Marlin.

Scombridae

Thunnus obesus (Lowe)

[*T. sibi* (Temminck et Schlegel)]

[*T. mebachi* Kishinouye]

[*Parathunnus sibi* (Temminck et Schlegel)]

Albacore, Big Eye, Bigeyed Tuna, Coffrey, False Albacore, Japanese Tuna, Southern Bluefin Tuna, Tuna

Diagnosis — Pelagic fish migrating in lower layers of oceanic waters. Resembles *T. thynnus*, but distinguished by blunt-body and larger eyes. Pectoral fins are long enough to reach anal fin. In young, body has 5–6 cross stripes. First dorsal fin has 14–15 spines, second has 13–15 rays. Anal fin has 13–15 rays with 8–9 finlets on each side. Spawns in waters near equator. Edible. Up to 2m in total length.

Distribution — Japan, Ryukyu, Indo-Pacific-Atlantic, Hawaii, South Africa

Scombridae

Thunnus albacarres (Bonnaterre)

[*Neothunnus albacora* (Lowe)]

[*N. macropterus* (Temminck et Schlegel)]

[*Germo albacore* (Lowe)]

Yellowfin Tuna

Diagnosis — Swims in schools in middle layers more than 40m deep. Most common tropical bluefin tunas. Spawns in waters near the equator. Grows in 2 years. Resembles *T. obesus*, but distinguished by somewhat smaller eyes, yellowish long-extended second dorsal and anal fins. Pectoral fins are long. In young, yellowish body has more than 10 vertical stripes. First dorsal fin has 12–14 spines, second has 14–15 rays, anal fin has 14–15 rays, with 8–9 finlets on each side. Edible. Up to 3m in total length.

Distribution — Japan, Ryukyu, Indo-Pacific-Atlantic, South Africa

▼ Testis of Tuna.

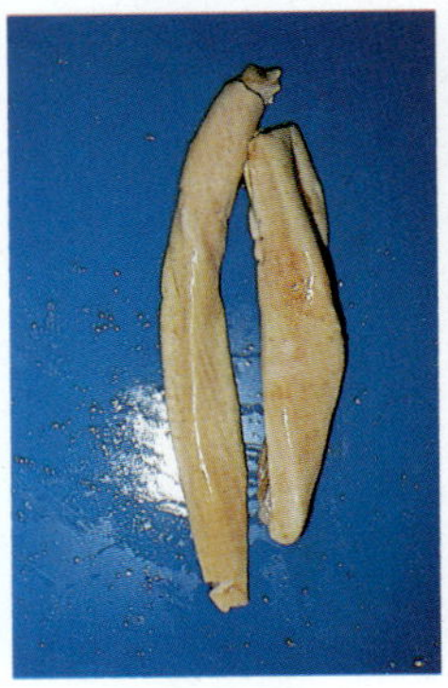

▼ Ovary of Tuna.

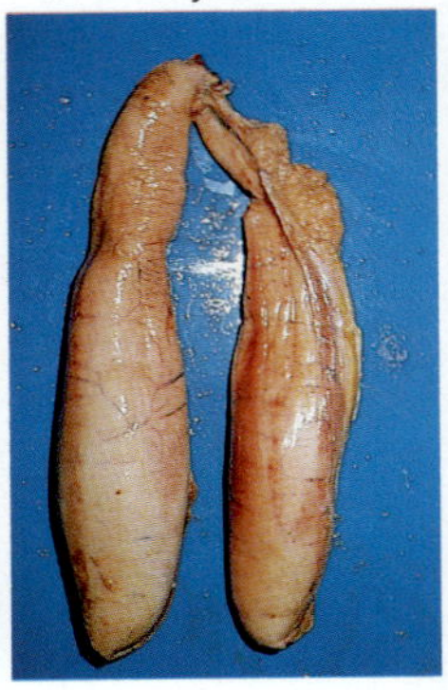

▲ Mouth has small conical teeth.

Scombridae

Thunnus alalunga (Bonnaterre)

[*Germo alalunga* (Bonnaterre)]

Albacore, Alilonghi, Atlantic Albacore, Avoador, Bastard, Longfin, Long-fin Albacore, Long Finned Tuna, Pacific Albacore, White Tuna

Diagnosis — Swims in schools in upper layers of oceanic waters. Small species of bluefin tuna. Snout is pointed. Have longest pectoral fins within tuna groupe, extending beyond anal fin base. First dorsal fin has 13–14 spines, second has 14–16 rays, anal fin has 14–15 rays, with 7–8 finlets on each side. Feeds on small fish, squid, crustacea. Spawns in summer. Edible. Up to 1.2m in total length.

Distribution — Japan, Ryukyu, Indo-Pacific-Atlantic, South Africa

Scombridae

Thunnus tonggol (Bleeker)

Long Tail Tuna

Diagnosis — Small littoral species inhabiting continental shelves. Body is fusiform, snout is pointed. Pectoral fins are short. Tail is elongated. First dorsal fin has 12–14 spines, second has 14–15 rays, anal fin has 13–14 rays, with 8–9 finlets on each side. Edible. Up to 1m in total length.

Distribution — Southern Japan, Ryukyu, Indo-West Pacific, Red Sea

Scombridae

Gymnosarda unicolor (Rüppell)

[*Pelamys nuda* (Günther)]

Scaleless Tuna, Dog-toothed Tuna, Dogtooth Tuna, Pig-tooth Tuna, Rüppell's Bonito, Whiteflesh Tuna

Diagnosis — Migratory fish found in small schools in middle layers of coral reefs and rocky reefs. Body is fusiform; head is large. Around pectoral fins and on lateral lines have scales. Posterior part of lateral lines are curved. First dorsal fin has 13–15 spines, second has 12–14 rays. Anal fin has 12–13 rays with 6–7 finlets on each side. Spawns between May and June. Feeds on small fish and squid. Edible but may cause ciguatera poisoning. Up to 2.5m in total length.

Distribution — Japan, Ryukyu, Bonin Islands, Indo-West Pacific, Red Sea, East Africa

▲ Cross section of Tuna.

Scombridae

Thunnus thynnus (Linnaeus)

Atlantic Tuna, Black Tuna, Bluefin Tunny, Common Tunny, Giant Bluefin Tuna, Great Albacore, Leaping Tuna, Horse Mackerel, School Tuna, Pacific Bluefin Tuna

Diagnosis — Swims in schools in middle layers of oceanic waters. Body is fusiform; snout is pointed. Eyes are small. Pectoral fins are short. First dorsal fin has 13–15 spines. second has 13–15 rays. Anal fin has 13–15 rays with 7–9 finlets on each side. Spawns between April and July in waters of the Philippines, Formosa and Ryukyus. Grows during 3–5-year migration in Pacific Ocean. Edible. Up to 3m in total length.

Distribution — Japan, Ryukyu, Formosa, Korea, Indo-Pacific-Atlantic, South Africa

Scombridae

Euthynnus pelamis (Linnaeus) ⇨

[*Katsuwonus pelamis* (Linnaeus)]

Ocean Bonito, Oceanic Skipjack, Arctic Bonito, Mush Mouth, Skipjack, Striped-bellied Tunny, Lesser Tunny, Skipjack Tuna, Striped Bonito

Diagnosis — Swims in schools in upper layers. Body is fusiform, back is blackish, abdominal side has several indistinct black stripes. First dorsal fin has 15–18 spines, second has 2 spines and 12–14 rays. Anal fin has 2 spines and 13–15 rays with 6–8 finlets on each side. Feeds on sardine and crustacea. Edible but may cause histamine poisoning. Up to 1.1m in total length.

Distribution — Japan, Ryukyu, Formosa, Korea, Indo-Pacific-Atlantic, South Africa

Scombridae

⇦ ***Euthynnus affinis*** (Cantor)

[*E. affinis yaito* (Kishinouye)]

[*E. yaito* Kishinouye]

[*E. affinis affinis* (Cantor)]

Wahoo, Black Skipjack, Little Tuna, Eastern Little Tuna, Little Tunny, Mackerel Tuna, Oceanic Bonito, Yaito Tuna

Diagnosis — Swims in small schools in upper and middle layers of offshore waters. Body is fusiform, back has oblique dark stripes. Lower part of pectoral fins have several black spots. Scales are present around pectoral fins and on lateral lines. First dorsal fin has 15–17 spines, second has 12–13 rays, anal fin has 12–14 rays, with 7–8 finlets on each side. Spawns between May and June. Edible. Up to 1m in total length.

Distribution — Japan, Ryukyu, Formosa, Indo-Pacific, Hawaii, Persian Gulf, South Africa

Scombridae

⇦ ***Sarda orientalis*** (Temminck et Schlegel)

[*S. chilensis* (Cuvier et Valenciennes)]

[*Pelamys orientalis* Temminck et Schlegel]

Belted Bonito, Mexican Bonito, Oriental Bonito, Oriental Tuna, Skipjack, Striped Bonito

Diagnosis — Swims in schools in upper layers of shoreline waters. Body is fusiform; snout is long. Back has some 10 dark stripes. Jaws have large, strong, inwardly curved teeth. Body is covered with scales. First dorsal fin has 17–19 spines, second has 14–17 rays. Anal fin has 14–16 rays, with 6–9 finlets on each side. Feeds on small fish. Edible. Up to 1m in total length.

Distribution — Japan, Ryukyu, Korea, Indo-Pacific-Atlantic, Hawaii, South Africa, North-South America

Scombridae

Auxis rochei (Risso) ⇨

[*A. tapeinosoma* Bleeker]

Round Frigate Mackerel, Frigate Mackerel, Bullet Mackerel, Two-finned Bonito

Diagnosis — Swims in schools in upper layers of offshore waters. Feeds on sardine and horse mackerel. Body is rounded and elongated. Scales are absent on back and abdomen. A black spot on upper surface of opercles is connected. First dorsal fin has 10–11 spines, second has 10–12 rays. Anal fin has 12–13 rays with 7–9 finlets on each side. Edible but often causes histamine poisoning. Up to 55cm in total length.

Distribution — Japan, Ryukyu, Korea, Indo-Pacific-Atlantic

Scombridae

⇦ ***Auxis thazard*** (Lacépède)

[*A. hira* Kishinouye]

Frigate Mackerel, Mexican Skipjack, Leadenall, Bullet Mackerel, Plain Bonito

Diagnosis — Swims in schools in shoreline waters. Body is fusiform, somewhat deep. No scales except on posterior part of eyes and around pectoral fins. Opercles have a black spot on upper surface. First dorsal fin has 11–12 spines, second has 10–12 rays. Anal fin has 12–14 rays with 7–8 finlets on each side. Edible but may cause histamine poisoning. Up to 60cm in total length.

Distribution — Japan, Ryukyu, Korea, Indo-Pacific-Atlantic, Hawaii, Mediterranean, South Africa

Scombridae

⇦ *Scomber japonicus* Houttuyn ⇩

[*Pneumatophorus japonicus* (Houttuyn)]

Spanish Mackerel, Southern Mackerel, Japanese Mackerel, Tinker Mackerel Fish, Chub Mackerel, Common Mackerel, Pacific Mackerel, Slimy Mackerel, Mackreel, Japanese Chubmackerel, Thimble-eye Mackerel

Diagnosis — Swims in schools in middle layers of shoreline waters. Surfaces at night, feeds on small fish and crustacea. Body is fusiform, back has irregular dark stripes. First dorsal fin has 9–10 spines, second has 11–12 rays. Anal fin has 1 spine and 11–12 rays with 5 finlets on each side. Spawns between spring and summer. Edible but has records of histamine poisoning. Up to 50cm in total length.

Distribution — Japan, Ryukyu, Formosa, China, Korea, Indo-Pacific-Atlantic, Hawaii, California, South Africa

Scombridae

Grammatorcynus bicarinatus (Quoy et Gaimard) ⇩

[*G. bilineatus* (Rüppell)]

Couta, Double-lined Mackerel, Large-scaled Tuna, Large-scaled Tunny, Salmon Mackerel, Scaly Kingfish, Shark Mackerel

Diagnosis — Found in schools in upper layers of shoreline waters. Body is fusiform. Caudal fin is deeply falcated. Lateral lines branched, lower branching crossing abdomen. Caudal peduncles have 3 keels. First dorsal fin has 12 spines, second has 10–14 rays. Anal fin has 10–14 rays with 6–7 finlets on each side. Edible but may cause histamine poisoning. Up to 70cm in total length.

Distribution — Southern Japan, Ryukyu, Indo-West Pacific, Red Sea

Scombridae

⇦ *Rastrelliger kanagurta* (Cuvier) ⇧

[*R. canagurta* (Cuvier)]

[*Scomber kanagurta* Rüppell]

Long-jawed Mackerel, Rake-gilled Mackerel, Striped Mackerel

Diagnosis — Swims in schools in upper layers of shoreline waters. Body is fusiform, somewhat deep. Back has several longitudinal stripes. First dorsal fin has 9–11 spines, second has 11–13 rays. Anal fin has 11–12 rays with 5 finlets on each side. Edible but may cause histamine poisoning. Up to 45cm in total length.

Distribution — Southern Japan, Ryukyu, Formosa, China, Indo-Pacific, Persian Gulf, South Africa

Scombridae

Scomber australasicus Cuvier ⇨

[*S. tapeinocephalus* Bleeker]

[*Pneumatophorus japonicus tapeinocephalus* (Bleeker)]

Slimy Mackerel, Spotted Mackerel, Spotted Chub Mackerel

Diagnosis — Swims in schools in upper layers of shoreline waters. Resembles *S. japonicus*, but distinguished by somewhat slender and rounded body. Abdominal side has many dark spots. First dorsal fin has 11–12 spines, second has 1 spine and 11–12 rays. Anal fin has 1 spine and 11–13 rays with 5 finlets on each side. Spawns between winter and spring. Edible but has records of histamine poisoning. Up to 50cm in total length.

Distribution — Japan, Ryukyu, Formosa, Korea, China, Pacific, Hawaii

Scombridae

⇦ ***Scomberomorus commerson*** (Lacépède)

[*Cybium commersoni* (Lacépède)]

Barred Spanish Mackerel, Banded Tuna, Commerson's Mackerel, Seerfish, Giant Mackerel, Indo-Pacific King Mackerel, Leaping Tuna, Snook, Doggie

Diagnosis — Swims in schools in upper layers of shoreline waters. Body is shallow and extended with many vertical stripes. Snout is pointed. Caudal fin is deeply falcated, caudal peduncles have a keel. Lateral lines are curved in posterior part. First dorsal fin has 15−18 spines, second has 15−20 rays. Anal fin has 16−21 rays with 8−9 finlets on each side. Edible but may cause ciguatera poisoning. Up to 2.3m in total length.

Distribution — Southern Japan, Ryukyu, Formosa, Indo-Pacific, Persian Gulf, Red Sea, South Africa

Scombridae

⇦ ***Scomberomorus sinensis*** (Lacépède)

Chinese Mackerel

Diagnosis — Inhabits upper layers of shoreline waters. Also found in brackish areas. Body is elongated, has 2 rows of indistinct spots. Snout is pointed. Pectoral fins are broad, round-margined. Caudal fin is large, deeply falcated. First dorsal fin has 15−17 spines, second has 15−17 rays. Anal fin has 16−19 rays with 6−8 finlets on each side. Edible. Up to 2m in total length.

Distribution — Japan, Formosa, Korea, China, South East Asia

Scombridae

Scomberomorus niphonius (Cuvier) ⇨

[*Sawara niphonia* (Cuvier et Valenciennes)]

Japanese Spanish Mackerel, Spanish Mackerel, Spotted Mackerel, Spotted Spanish Mackerel

Diagnosis — Swims in schools in upper layers of shoreline waters. Body is shallow and elongated with rows of indistinct dark blotches. Scales are remarkably small. Lateral lines have many fine branches. First dorsal fin has 19−21 spines, second has 15−19 rays. Anal fin has 16−20 rays with 6−9 finlets on each side. Feeds on small fish. Spawns in inland seas between May and June. Edible. Up to 1m in total length.

Distribution — Japan, Korea, China, East China Sea, Australia

Scombridae

⇦ ***Scomberomorus guttatus*** (Bloch et Schneider)

[*S. kuhlii* (Cuvier et Valanciennes)]

[*Cybium guttatum* Kishinouye]

[*Indocybium guttatum* (Bloch et Schneider)]

Spotted Spanish Mackerel, Indo-Pacific Spanish Mackerel

Diagnosis — Swims in schools in upper layers of shoreline waters. Resembles *S. niphonius*, but distinguished by somewhat deeper body and large head. Lateral lines have many fine branches. First dorsal fin has 15−18 spines, second has 18−24 rays. Anal fin has 19−23 rays with 7−10 finlets on each side. Edible. Up to 80cm in total length.

Distribution — Southern Japan, Formosa, Indo-West Pacific, Persian Gulf

Scombridae

Acanthocybium solandri (Cuvier) ⇨

[*Cybium sara* Lay et Bennett]

Wahoo, Pacific Kingfish, Queenfish, Bastard Mackerel, Guachu, Jack Mackerel, Kingfish, Mongrel Mackerel, Peto, Pike Mackerel

Diagnosis — Swims in upper layers of oceanic waters. Body is elongated, has many cross bands. Snout is pointed. Caudal fin is developed, caudal peduncles have a keel. Lateral lines are curved in middle. First dorsal fin has 23−28 spines, second has 12−16 rays. Anal fin has 12−14 rays with 8−9 finlets on each side. Edible but may cause ciguatera poisoning. Up to 2.2m in total length.

Distribution — Japan, Ryukyu, Formosa, Korea, Indo-Pacific-Atlantic, South Africa

Gempylidae

⇦ ***Lepidocybium flavobrunneum*** (Smith)

[*Xenogramma carinatum* Waite]

Escolar, Smith's Escolar Fish

Diagnosis — Inhabits deep waters. Surfaces at night for feeding. Body is fusiform; snout is pointed. Caudal peduncles have a keel. Lateral lines are curved like wave, covered with scales. Dorsal fin has 8–10 spines and 16–19 rays, anal fin has 12–15 rays with 4–5 finlets on each side. Flesh contains enough wax to cause wax poisoning. Up to 1.5m in total length.

Distribution — Japan, Ryukyu, Indo-Pacific-Atlantic, Hawaii, California, South Africa

Gempylidae

⇩ ***Promethichthys prometheus*** (Cuvier)

Bermuda Catfish, Black Tuna, Rabbit Fish, Black Tuna Kamasu, Snake Mackerel

Diagnosis — Inhabits waters more than 200m deep. Surfaces at night for feeding. Body is elongated; snout is pointed. Mouth is large, lower jaw is long. Pelvic fins are degenerated, forming 1 small spine. Dorsal and anal fins have 2 finlets on posterior parts. No branched lateral lines. Dorsal fin has 17–19 spines and 18–20 rays, is compressed between spine and ray part. Anal fin has 2 spines and 15–18 rays. Spawns in April. Edible. Up to 60cm in total length.

Distribution — Japan, Indo-Pacific-Atlantic

▼ Upper jaw has sharp canines.

Gempylidae

⇧⇦ ***Thyrsitoides marleyi*** Fowler

[*Mimasea taeniosoma* Kamohara]

Blacksail Snake Mackerel

Diagnosis — Inhabits waters more than 300m deep. Body is compressed and extended. Lower jaw is projected, upper jaw has 3 pairs of canines. Eyes are somewhat small. Pectoral fins are short. Lateral lines are branched. Scales are small, easily removable. Dorsal fin has 17–19 spines and 16–18 rays, anal fin has 2 spines and 15–17 rays. Edible. Up to 1.5m in total length.

Distribution — Japan, Ryukyu, Indo-West Pacific

Trichiuridae

Trichurus lepturus Linnaeus ⇨

Ribbonfish, Bandfish, Cutlassfish, Snakefish, Scabbardfish, Japanese Hairtail, Silverfish, Sable, Silver Eel, Silvery Hairtail, Bladefish, Machete

Diagnosis — Inhabits muddy-sand bottoms about 100m deep. Surfaces at night, feeds on fish, crustacea, squid. Body is long and narrow. Mouth is large, has sharp-pointed canines, lower jaw is projected. Dorsal fin extends from head to tip of whip-like tail. Pelvic and caudal fins are absent. Scales are also absent. Epidermis contains Guanine and is brilliant as silver. Spawns in shallow waters between June and September. Edible. Up to 1.5m in total length.

Distribution — Japan, Ryukyu, Formosa, Korea, Indo-Pacific-Atlantic, Red Sea, South Africa

▲ A white spine on caudal peduncles.

Acanthuridae
Acanthurus dussumieri Valenciennes ⇩

Dussumier's Surgeonfish, Pencilled Surgeon, Bronze Tang, Tang, Lancetfish, Surgeonfish

Diagnosis — Inhabits shallow coral reefs. Resembles *A. bariene*, but distinguished by absence of a black spot on upper surface of opercles, a yellow band across the eyes, caudal fin profusely dotted with black, a white spine on caudal peduncles. Margins of opercles are black. Dorsal fin has 9 spines and 25−27 rays, anal fin has 3 spines and 24−28 rays with blue stripes. Edible but may cause ciguatera poisoning. Up to 40cm in total length.

Distribution — Japan, Ryukyu, Indo-Pacific, Hawaii, South Africa

Acanthuridae
Acanthurus mata (Cuvier) ⇩

[*Hepatus mata* (Valenciennes)]

Tailring Surgeon

Diagnosis — Inhabits coral reefs. Posterior to eyes have a yellow blotch. Caudal peduncles have a black spine. Caudal fin base has a white band. Resembles *A. xanthopterus*, but distinguished by more black stripes on dorsal and anal fins. Dorsal fin has 9 spines and 25−27 rays, anal fin has 3 spines and 23−25 rays. Edible. Up to 40cm in total length.

Distribution — Ryukyu, Indo-Pacific, Hawaii, East Africa

Acanthuridae
Acanthurus bariene Lesson ⇩

[*A. dussumieri* Cuvier et Valenciennes]

[*Hepatus bariene* (Lesson)]

Dussumier's Surgeonfish, Pencilled Surgeonfish, Black-spot Surgeonfish

Diagnosis — Inhabits shoreline reefs and coral reefs. Body is compressed and oval. In adult, forehead is protruded. Opercles have a black spot on upper surface, an orange band on behind. Caudal fin base is pale, lobes are extended. Caudal peduncles have a spine. Dorsal fin has 9 spines and 26−28 rays, has no stripes. Anal fin has 3 spines and 25−26 rays. Feeds on calcareous algae. Edible. Up to 45cm in total length.

Distribution — Japan, Ryukyu, Formosa, Indo-Pacific, Hawaii, South Africa

Acanthuridae
⇩ *Acanthurus xanthopterus* Valenciennes

Yellowfin Surgeonfish, Ring-tailed Surgeonfish, Ring-tail Surgeonfish

Diagnosis — Inhabits coral reefs. Resembles *A. mata*, but distinguished by smaller eyes, absence of a yellow blotch behind eyes. Black stripes on dorsal and anal fins are 4 in number. Caudal peduncles have a short and black spine. Pectoral fins are yellow. Dorsal fin has 9 spines and 25−27 rays, anal fin has 3 spines and 23−25 rays. Edible but may cause ciguatera poisoning. Up to 55cm in total length.

Distribution — Japan, Ryukyu, Indo-Pacific, South Africa

Acanthuridae
Acanthurus maculiceps (Ahl) ⇨

Pale-lined Surgeonfish

Diagnosis — Inhabits shallow coral reefs. Body is oval, has many blue wavy-lines. Upper part of opercles has a black blotch. Pectoral fin tip is yellow. Pectoral fin lobes are yellow. Caudal lobes are extended. Dorsal fin has 4 spines and 24−26 rays, anal fin has 4 spines and 22−24 rays. Edible. Up to 25cm in total length.

Distribution — Ryukyu, Pacific

Acanthuridae

⇦ *Acanthurus nigrofuscus* (Forskål)

Blackish Brown Surgeon Fish, Spot-cheeked Surgeonfish, Yellowfin Surgeonfish

Diagnosis — Inhabits coral reefs. Body is oval, head and pectoral side are profusely dotted with orange. A black spot is present on posterior ends of dorsal and anal fins. Caudal fin is lunated. Caudal peduncles have a spine. Dorsal fin has 9 spines and 24−27 rays, anal fin has 3 spines and 22−24 rays. Edible but may cause ciguatera poisoning. Up to 20cm in total length.

Distribution — Southern Japan, Ryukyu, Formosa, Indo-Pacific, Hawaii, Red Sea, East Africa

Acanthuridae

Acanthurus bleekeri Günther ⇨⇩

Bleeker's Surgeonfish, Bluelined Surgeonfish, Striped Surgeonfish, Bleeker's Lined Surgeon-fish

Diagnosis — Found in schools about coral reefs. Body is oblong with many wavy stripes. A yellow band across the eyes. Caudal peduncles have a spine. Caudal fin base is pale, lobes extended. Dorsal fin has 9 spines and 24−26 rays, anal fin has 3 spines and 23−24 rays. Edible but may cause ciguatera poisoning. Up to 40cm in total length.

Distribution — Southern Japan, Ryukyu, Indo-Pacific, Red Sea, East Africa

Acanthuridae

Acathurus thompsoni (Fowler) ⇨

Yellowtail Surgeonfish, Thompson's Surgeon Fish

Diagnosis — Inhabits outer slope of coral reefs. Body is oblong. Head is rounded; snout is short. Caudal fin is white or yellow, lobes are extended. Pectoral fins are bluish. Dorsal fin has 9 spines and 23−26 rays, anal fin has 3 spines and 23−26 rays. Edible. Up to 20cm in total length.

Distribution — Southern Japan, Ryukyu, Indo-Pacific, Hawaii, East Africa

Acanthuridae

Acanthurus gahhm (Forskål) ⇨⇩

[*A. nigricans* Jordan et Seale]
[*A. nigricans* (Linnaeus)]
[*A. nigricauda* Dunker et Mohr.]

Dusky Surgeonfish, Black-barred Surgeonfish, Brown-eared Surgeonfish

Diagnosis — Inhabits singly coral reefs. Body is oval. Caudal base is white, lobes are extended. A short black band is present on surface posterior to eyes and on tail spines. Pectoral fins are margined with dark yellow. Dorsal fin has 9 spines and 24−28 rays, anal fin has 3 spines and 23−26 rays. Edible. Up to 40cm in total length.

Distribution — Ryukyu, Indo-West Pacific, Red Sea, South Africa

Acanthuridae

Acanthurus pyroferus Kittlitz ⇦⇧

Orange-gilled Surgeonfish, Yellow Spot Surgeonfish

Diagnosis — Inhabits singly coral reefs. Body is oval. Snout is slightly pointed. Caudal lobes are extended, margined with yellow. Caudal peduncles have a spine. Margins of opercles are black. In young, body is yellow, resembles that of *Centropyge flavissimus*. Dorsal fin has 8 spines and 27−30 rays, anal fin has 3 spines and 24−28 rays. Edible. Up to 25cm in total length.

Distribution — Japan, Ryukyu, Indo-Pacific

Acanthuridae
Acanthurus guttatus
Schneider ⇧

Spotted Surgeon Fish
Diagnosis — Inhabits outer slopes of coral reefs. Body is deep with posterior part profusely dotted with white. Snout is projected. Pelvic fins are yellow. Dorsal fin has 9 spines and 27–30 rays, anal fin has 3 spines and 23–26 rays. Edible but may cuase ciguatera poisoning. Up to 30cm in total length.
Distribution — Ryukyu, Indo-Pacific, Hawaii

Acanthuridae
Acanthurus achilles Shaw ⇧
Achilles Surgeon Fish, Achilles Tang
Diagnosis — Inhabits singly coral reefs. Body is oval with a large orange blotch on posterior part. Caudal fin is orange with white margin, lobes are extended. Opercles are margined with white. Dorsal fin has 9 spines and 29–33 rays. Caudal peduncles have a spine. Edible. Up to 25cm in total length.
Distribution — China, Pacific, Hawaii

Acanthuridae
Acanthurus leucosternon
(Bennett) ⇧

Bluetang, Painted Surgeon, Blue Surgeonfish, Powder-blue Surgeonfish, White-breasted Surgeonfish
Diagnosis — Inhabits singly shallow coral reefs. Body is oval. Opercles are black. Caudal margin is lunated. Black and yellow dorsal fin has 9 spines and 28–29 rays. Anal fin has 3 spines and 26–27 rays. Caudal peduncles have a spine. Edible. Up to 23cm in total length.
Distribution — Indo-West Pacific, Red Sea, South Africa

Acanthuridae
Acanthurus glaucopareius
(Cuvier) ⇧

Lesson's Surgeonfish, Philosophe, Whitecheek Surgeonfish, White-cheeked Surgeonfish
Diagnosis — Inhabits coral reefs. Body is oval, somewhat deep. Caudal fin is white with yellow stripes. Bases of dorsal and anal fins are yellow. Caudal peduncles have a spine. Lower part of eyes and margins of mouth are white. Dorsal fin has 9 spines and 28–32 rays, anal fin has 3 spines and 26–29 rays. Edible but may cause ciguatera poisoning. Up to 25cm in total length.
Distribution — Ryukyu, Indo-Pacific, Hawaii, Mexico

Acanthuridae
⇧ *Acanthurus leucopareius* (Jenkins)
Head-band Surgeonfish
Diagnosis — Inhabits shallow rocky reefs and coral reefs. Body is somewhat deep. Head and caudal peduncles have a white band. Posterior end of dorsal fin base has a black blotch. Dorsal fin has 4 spines and 25–27 rays, anal fin has 3 spines and 23–25 rays. Edible. Up to 25cm in total length.
Distribution — Japan, Ryukyu, Pacific, Hawaii, Easter Island

⇧ Acanthuridae
⇦ *Acanthurus triostegus* (Linnaeus)
[*A. sandviscens* Jenkins]
[*Hepatus triostegus* (Linnaeus)]
Fiveband Surgeonfish, Barred Surgeonfish, Lancetfish, Banded Surgeonfish, Convict Surgeonfish, Five-banded Surgeonfish
Diagnosis — Found in schools about reef fronts of coral reefs. Body is oval with 5 black bands. Caudal margin is truncated, caudal peduncles have a short spine. Dorsal fin has 9 spines and 22–24 rays, anal fin has 3 spines and 19–22 rays. Edible but may cause ciguatera poisoning. Up to 25cm in total length.
Distribution — Japan, Ryukyu, Formosa, Indo-Pacific, Hawaii, South Africa

Acanthuridae
⇦ ***Acanthurus lineatus*** (Linnaeus) ⇩
[*Hepatus lineatus* (Linnaeus)]

Bluelined Surgeonfish

Diagnosis — Inhabits singly passage or edges of coral reefs. Body is oblong. Caudal lobes are extended. Caudal peduncles have a long spine. Pelvic fin margins are black. Dorsal fin has 9 spines and 27−30 rays, anal fin has 3 spines and 25−28 rays. Feeds on seaweed. Edible but may cause ciguatera poisoning. Aquarium fish. Up to 35cm in total length.

Distribution — Japan, Ryukyu, China, Indo-Pacific, South Africa

Acanthuridae
Paracanthurus hepatus (Linnaeus) ⇩
[*P. theuthis* (Lacépède)]

[*Acanthurus theuthis* Lacépède]

Flagtail Surgeonfish, Tang, Tangbarbero, Doctor Fish, Lancetfish, Blue Surgeon, Wedge-tailed Blue Tang, Common Surgeon

Diagnosis — Inhabits passage or front of coral reefs. Body is oblong. Snout is roundish. Upper and lower edges of cuadal fin are black, lobes slightly extended. Caudal peduncles have a spine. Dorsal fin has 9 spines and 19−20 rays, anal fin has 3 spines and 18−19 rays. Edible. Aquarium fish. Up to 30cm in total length.

Distribution — Japan, Ryukyu, Indo-Pacific, South Africa

Acanthuridae
Acanthurus olivaceus Schneider ⇩
[*Hepatus olivaceus* (Schneider)]

Orangespot Surgeonfish, Orangespot Tang, Lemon Surgeonfish, Orange-epaulette Surgeonfish

Diagnosis — Found in schools about coral reefs. Body is oblong. Caudal lobes are extended long. Caudal peduncles have a sharp-pointed spine. Posterior part of opercles have an orange blotch with black margin. In young, body is yellow. Dorsal fin has 9 spines and 23−25 rays, anal fin has 3 spines and 22−24 rays. Edible but may cause ciguatera poisoning. Aquarium fish. Up to 38cm in total length.

Distribution — Japan, Ryukyu, Formosa, Indo-Pacific, Hawaii

▼ Mouth has incisor teeth.

▲ Caudal peduncles have bony plates.

⇦⇧ Acanthuridae
Prionurus scalprum Valenciennes ⇧⇨
[*P. microlepidotus* Lacépède]

Sawtail, Surgeon-fish

Diagnosis — Found in schools about shoreline reefs. Body is oval, snout is projected, skin is thick, scales are small. Caudal peduncles have 4−5 bony plates. Caudal margin is truncated. Dorsal fin has 9 spines and 22−24 rays, anal fin has 3−4 spines and 21−23 rays. Feeds on calcareous algae. Spawns in spring. Edible. Up to 50cm in total length.

Distribution — Japan, Ryukyu, Formosa, Korea, China

Acanthuridae
Zebrasoma xanthurum (Blyth) ⇨

Yellowtail Surgeonfish, Dusky Sail-finned Surgeon-fish

Diagnosis — Inhabits shallow coral reefs. Body is deep, anterior half is profusely dotted with dark color. Snout is pointed. Pectoral fin margins and caudal fin are yellow. Caudal peduncles have a spine. Dorsal fin has 5 spines and 24–55 rays, anal fin has 3 spines and 19–20 rays. Edible. Up to 22cm in total length.

Distribution — Indian Ocean, Persian Gulf, Red Sea, East Africa

Acanthuridae
Zebrasoma desjardinii (Bennett) ⇩

Sailfin Surgeonfish

Diagnosis — Resembles *Z. veliferum*, but distinguished by pale spots on head and abdomen, also narrower cross bands on body, fewer rays of dorsal and anal fins. Dorsal fin has 4 spines and 27–31 rays, anal fin has 3 spines and 22–24 rays. Up to 40cm in total length.

Distribution — Indian Ocean, Red Sea, East Africa

Acanthuridae
⇦ *Zebrasoma veliferum* (Bloch)

Sail-finned Surgeon-fish, Purple-finned Sailfin-tang, Purple Sailfinned Tang, Sailfin Tang

Diagnosis — Inhabits singly coral reefs. Body is oval with 6 black bands. Snout is projected. Dorsal fin has 4 spines and 27–33 rays, anal fin has 3 spines and 22–26 rays, fins are well developed. In young, body is yellowish. Caudal peduncles have a spine. Up to 45cm in total length.

Distribution — Japan, Ryukyu, Indo-Pacific, Hawaii, Red Sea, South Africa

Acanthuridae
Zebrasoma gemmatus (Cuvier et Valenciennes) ⇨

Spotted Tang

Diagnosis — Inhabits shoreline reefs. Body is oval, blackish-brown, profusely dotted with white. Snout is pointed. Caudal peduncles have a spine. Up to 20cm in total length.

Distribution — South Africa

Acanthuridae
⇦ *Zebrasoma scopas* (Cuvier) ⇧

Bicolorsail Surgeonfish, Blue-lined Sailfin Tang, Brown Sailfin-tang

Diagnosis — Inhabits coral reefs. Body is deep, has a pale stripe. Snout is pointed. Caudal peduncles have a white spine. Dorsal fin has 5 spines and 23–26 rays, anal fin has 3 spines and 19–21 rays, fins are well developed. In young, body has vertical stripes. Edible. Up to 20cm in total length.

Distribution — Japan, Ryukyu, Indo-Pacific, East Africa

Acanthuridae
Zebrasoma flavescens (Bennett) ⇧ ⇨

Yellow Surgeon Fish, Yellow Tang, Sombre Surgeon-fish, Yellow Surgeon, Yellow Sailfinned Surgeon-fish

Diagnosis — Inhabits coral reefs. Closely resembles *Z. scopas*, but distinguished by yellow body. Caudal peduncles have a white spine. Dorsal fin has 5 spines and 23–26 rays, anal fin has 3 spines and 19–22 rays. Feeds on seaweed. Edible but may cause ciguatera poisoning. Up to 20cm in total length.

Distribution — Japan, Ryukyu, Indo-Pacific, Hawaii, South Africa

Acanthuridae

Ctenochaetus binotatus Randall ⇩

Black-spotted Bristletooth (New Common Name).

Diagnosis — Resembles *C. striatus*, but distinguished by eyes margined with blue, caudal lobes are extended, a black spot present on posterior ends of dorsal and anal fins. In young, caudal fin is yellow. Dorsal fin has 8 spines and 24–27 rays, anal fin has 3 spines and 22–25 rays. Edible. Up to 18cm in total length.

Distribution — Ryukyu, Formosa, Philippines, Indonesia

Acanthuridae

Ctenochaetus strigosus (Bennett) ⇧

Goldring Surgeonfish, Bristletooth Surgeonfish, Spotted Bristletooth, Slendertoothed Surgeonfish, Yellow-eyed Surgeon Fish

Diagnosis — Inhabits shallow coral reefs. Resembles *C. hawaiiensis*, but distinguished by yellow-margined eyes, head profusely dotted with blue. Lower jaw is blue. Pectoral fins are transparent. Up to 20cm in total length.

Distribution — Philippines, Indo-Pacific, Hawaii

Acanthuridae

⇧ *Ctenochaetus hawaiiensis* Randall

Hawaiian Surgeon Fish, Black Surgeonfish

Diagnosis — Inhabits outer slope of rather deep coral reefs. Body is oval. Snout is slightly pointed. Caudal margin is emarginated and dark, pectoral fin margins also black. Dorsal and anal fins have oblique stripes. Caudal peduncles have a spine. Up to 25cm in total length.

Distribution — Hawaii, Micronesia, Polynesia

Acanthuridae

Ctenochaetus striatus (Quoy et Gaimard) ⇧ ⇨

[*C. strigosus* (Bennett)]

Surgeon Fish, Striped Bristletooth, Orange-dotted Hair-toothed Tang

Diagnosis — Found in schools about shallow coral reefs. Body is oval with blue wavy stripes. Head is profusely dotted with orange. Caudal lobes are extended. Caudal peduncles have a spine. Dorsal fin has 8 spines and 27–31 rays, anal fin has 3 spines and 24–28 rays, fin rays are angular. Feeds on calcareous algae. Edible but has records of ciguatera poisoning. Up to 26cm in total length.

Distribution — Southern Japan, Ryukyu, Formosa, China, Indo-Pacific, Hawaii, Red Sea, South Africa

Acanthuridae

Naso brevirostris (Valenciennes)

Short-nosed Unicornfish, Short-snouted Unicornfish, Unicorn Fish

Diagnosis — Found in schools about shoreline reefs and coral reefs. Resembles *N. unicornis*, but distinguished by extended body, shorter snout, remarkably longer horn-like process on forehead. Caudal margin is rounded with a yellow band. Dorsal fin has 5–6 spines and 27–29 rays, anal fin has 2 spines and 25–30 rays. Edible but may cause ciguatera poisoning. Up to 70cm in total length.

Distribution — Japan, Ryukyu, Indo-Pacific, Red Sea, South Africa

Acanthuridae

Naso annulatus (Quoy et Gaimard)

Ring-tailed Unicorn, Ringed Unicorn Surgeon Fish, Banded Unicornfish

Diagnosis — Resembles *N. unicornis*, but distinguished by long body, shorter horn-like processes on head. Caudal margin is white, caudal bony plates are movable. Dorsal fin has 4 spines and 28–32 rays, anal fin has 3 spines and 22–26 rays. Edible. Up to 50cm in total length.

Distribution — Japan, Ryukyu, Indo-Pacific, Hawaii, Caribbean Sea, South Africa

Acanthuridae

Naso unicornis (Forskål)

Brown Unicornfish, Longnosed Scraper, Longsnouted Unicornfish, Longhorn Unicorn, Nosefish, Bluespine Unicornfish

Diagnosis — Found in schools about shoreline reefs and coral reefs. Body is compressed and oblong. Snout is pointed. Head has horn-like processes. Caudal peduncles have 2 bony plates, caudal lobes extended like thread. Dorsal fin has 5–6 spines and 27–31 rays, anal fin has 2 spines and 26–30 rays. Edible. Up to 70cm in total length.

Distribution — Japan, Ryukyu, Korea, China, Indo-Pacific, Hawaii, Red Sea, South Africa

▲ Bony plates of Brown Unicornfish

▲Young Adult

▼Adult

Acanthuridae

Naso lituratus (Schneider)

[*Callicanthus lituratus* (Bloch et Schneider)]

Smooth-head Unicornfish, Barcheek Surgeonfish, Poll Unicornfish, Stripe-faced Unicornfish, Green Unicorn

Diagnosis — Inhabits singly coral reefs. Body is oval, snout is pointed. Caudal margin is emarginated, lobes extended like thread. Caudal peduncles have 2 orange bony plates. Dorsal fin has 6–7 spines and 27–30 rays, anal fin has 2 spines and 28–30 rays. Edible but may cause ciguatera poisoning. Up to 60cm in total length.

Distribution — Japan, Ryukyu, Indo-Pacific, Hawaii, Red Sea, South Africa

Acanthuridae
Naso vlamingii (Valenciennes) ⇨
[*N. vlamingi* (Cuvier et Valenciennes)]

Zebra Unicorn Fish, Bignose Unicorn, Vlaming's Unicornfish

Diagnosis — Found in schools about outer slope of coral reefs. Body is oblong, snout is projected and roundish. Caudal lobes extended like thread. A blue band on surface anterior to eyes and on caudal base. Caudal peduncles have 2 bony plates. Dorsal fin has 6 spines and 26−27 rays, anal fin has 2 spines and 26−29 rays. Edible but may cause ciguatera poisoning. Up to 75cm in total length.

Distribution — Japan, Ryukyu, Indo-Pacific, Hawaii, South Africa

Acanthuridae
Naso tuberosus Lacépède ⇩
[*Cyphomycter tuberosus* (Lacépède)]

Hump-headed Unicornfish, Hump-nosed Unicornfish

Diagnosis — Body is somewhat deep, tail is extended, forehead is elevated. Caudal peduncles have 2 bony plates. Caudal margin is emarginated. Dorsal fin has 5 spines and 27−29 rays, anal fin has 2 spines and 26−27 rays. Edible. Up to 90cm in total length.

Distribution — Ryukyu, Indo-West Pacific, South Africa

Young

Acanthuridae
Naso hexacanthus (Bleeker) ⇖⇧
Black Unicorn Fish, Sixspine Surgeonfish

Diagnosis — Found in schools about outer slope of coral reefs. Body is compressed and extended, surface is smooth and velvet-like. Head is not elevated. Caudal peduncles have 2 large bony plates, caudal margin is lunated. Dorsal fin has 6 spines and 25−29 rays, anal fin has 2 spines and 27−31 rays. Edible but may cause ciguatera poisoning. Up to 75cm in total length.

Distribution — Japan, Ryukyu, Indo-Pacific, Red Sea, South Africa

Young Adult

Acanthuridae
Naso thynnoides (Valenciennes) ⇧⇨
[*Axinurus thynnoides* (Valenciennes)]

Oneknife Unicorn

Diagnosis — Resembles *N. hexacanthus*, but distinguished by dark stripes on body, single bony plate on caudal peduncles. Dorsal fin has 4 spines and 28−30 rays, anal fin has 2 spines and 27−29 rays. Edible. Up to 40cm in total length.

Distribution — Japan, Ryukyu, Formosa, Indo-West Pacific, South Africa

▼ Adult

Acanthuridae
⇦ *Naso maculatus* Randall et Struhsaker
Spotted Black Unicornfish (New Common Name)

Diagnosis — Found in schools about outer slope of coral reefs. Resembles *N. hexacanthus*, but distinguished by somewhat deeper body, rounded head. Body is profusely dotted and striped with dark color. Dorsal fin has 6−7 spines and 26−28 rays, anal fin has 2 spines and 26−28 rays. Edible. Up to 50cm in total length.

Distribution — Japan, Ryukyu, Indo-Pacific, Hawaii

Acanthuridae

Naso lopezi Herre ⇨

[*N. vomer* (Klunzinger)]

Long-bodied Black Unicornfish (New Common Name)

Diagnosis — Found in schools among rather deep coral reefs. Resembles *N. maculatus*, but distinguished by extended body. Caudal peduncles have 2 small bony plates, caudal margin is lunated. Dorsal fin has 5 spines and 28–31 rays, anal fin has 2 spines and 27–29 rays. Edible. Up to 50cm in total length.

Distribution — Japan, Ryukyu, Indo-Pacific, Red Sea

Siganidae

Siganus fuscescens (Houttuyn) ⇧⬀⇨

Dusky Spinefoot, Rabbit Fish, Fuscous Spinefoot, Slimy Spinny

Diagnosis — Inhabits shoreline reefs and bays. Body is oblong with irregular blotches and black spots. Posterior part of opercles have a black blotch. Snout is roundish, mouth is small. Scales are remarkably small. Pectoral fins are somewhat small. Caudal margin is emarginated, caudal peduncles are slender. Dorsal fin has 13 spines and 10 rays, anal fin has 7 spines and 9 rays. Spines are venomous. Juvenile feeds on zooplankton, adult feeds on seaweed. Edible. Up to 40cm in total length.

Distribution — Japan, Formosa, China, Indo-Pacific, South Africa

▲ Dorsal spines are venomous.

Siganidae

Siganus oramin (Bloch et Schneider) ⇨

[*S. canaliculatus* (Park)]

Whitespotted Spinefoot, Pearl-spotted Spinefoot, Rabbit Fish, Slimy Spinny

Diagnosis — Found in schools about shallow coral reefs and in bays. Resembles *S. fuscescens*, but distinguished by body profusely dotted with white, opercles margined with black, a black blotch on posterior surface of opercles, slightly pointed snout, larger pectoral fins, more deeply compressed caudal fin. Dorsal fin has 13 spines and 10 rays, anal fin has 7 spines and 9 rays. Spines are venomous. Edible. Up to 28cm in total length.

Distribution — Ryukyu, Formosa, China, Indo-Pacific, Persian Gulf, South Africa

Siganidae

⇦ *Siganus vermiculatus* (Valenciennes)

Reticulated Rabbitfish, Scribbled Spinefoot, Vermiculated Spinefoot

Diagnosis — Inhabits coral reefs, also found in brackish areas. Body is oval. Resembles *S. guttatus*, but distinguished by body covered with wavy stripes, also by absence of a yellow blotch on anterior of caudal peduncles. Dorsal fin has 13 spines and 10 rays, anal fin has 7 spines and 9 rays. Spines are venomous. Edible. Up to 40cm in total length.

Distribution — Ryukyu, Indo-Pacific

Siganidae

⇦ ***Siganus rostratus*** (Valenciennes) ⇩

[*S. argenteus* (Quoy et Gaimard)]

Roman-nose Spinefoot, Rabbit-faced Spinefoot

Diagnosis — Found in schools about coral reefs. Body is oval, snout is pointed. Caudal fin is falcated, caudal peduncles are slender. Opercles have a black band on margin. Dorsal fin has 13 spines and 10 rays, spines are not stiff, compressed between spine and ray parts. Anal fin has 7 spines and 9 rays. Edible but has records of ciguatera poisoning. Up to 40cm in total length.

Distribution — Japan, Ryukyu, Indo-Pacific, Red Sea, East Africa

Siganidae

Siganus spinus (Linnaeus) ⇩

Marbled Spinefoot, Black Spinefoot, Black Trevally, Blue-spotted Trevally, Happy Moments, Blunt-nosed Spinefoot, Stinging Bream

Diagnosis — Inhabits coral reefs. In young, found in large schools, often moves in close to coastal areas. Body is oblong with irregular reticulated patterns. Caudal margin is truncated or emarginated. Dorsal fin has 13 spines and 10 rays, is compressed between spine and ray parts. Anal fin has 7 spines and 9 rays. Spines are venomous. Edible but may cause ciguatera poisoning. Up to 20cm in total length.

Distribution — Japan, Ryukyu, Indo-Pacific, South Africa

Large Adult

Siganidae

Siganus javus (Linnaeus) ⇩

Blue Spotted Spinefoot, Bluespotted Trevally, Rabbit Fish, Streaked Spinefoot, Streaky Spinefoot

Diagnosis — Inhabits coral reefs. Body is oval, back is profusely dotted with pale color. Abdomen has yellowish-brown stripes. Caudal fin is lunated with white margin. Dorsal fin has 13 spines and 10 rays, anal fin has 7 spines and 9 rays. Spines are venomous. Edible. Up to 33cm in total length.

Distribution — Indo-West Pacific, Persian Gulf

Siganidae

Siganus sutor (Valenciennes) ⇧

Brown Spinefoot

Diagnosis — Body is oval and dark-brown. Caudal fin is falcated, lobes are pointed. Ray parts of dorsal and anal fins are angular. Dorsal fin has 13 spines and 10 rays, anal fin has 7 spines and 9 rays. Spines are venomous. Edible. Up to 37cm in total length.

Distribution — Ryukyu, Indo-West Pacific

Siganidae

⇦ ***Siganus chrysospilos*** (Bleeker) ⇧

Goldspotted Spinefoot, Yellowspotted Spinefoot, Rabbitfish

Diagnosis — Found in schools about coral reefs. Body is oval, profusely dotted with orange. Posterior part of opercles have a large black blotch. Caudal fin is falcated, lobes rounded at pointed end. Dorsal fin has 13 spines and 10 rays, anal fin has 7 spines and 9 rays. Spines are venomous. Edible. Up to 38cm in total length.

Distribution — Ryukyu, Indo-Pacific, South Africa

Siganidae

⇧⇩ ***Siganus virgatus*** (Valenciennes)

Bluelined Spinefoot, Rabbit Fish

Diagnosis — Found in schools about shallow coral reefs. Body is oval, somewhat deep, back is yellow. Caudal fin is yellow, emarginated. 2 black bands across the eyes and opercles. Yellow dorsal fin has 13 spines and 10 rays, anal fin has 7 spines and 9 rays. Spines are venomous. Edible. Up to 26cm in total length.

Distribution — Japan, Ryukyu, Formosa, China, Indo-Pacific

Siganidae

⇧⇩ ***Siganus guttatus*** (Bloch)

Deepbody Spinefoot, Golden Spinefoot, Golden-lined Spinefoot, Rabbit Fish

Diagnosis — Found in schools about coral reefs, also in brackish areas. Body is oval, profusely dotted with yellow. Anterior caudal peduncles have a yellow blotch. Ray parts of dorsal and anal fins are angular. Dorsal fin has 13 spines and 10 rays, anal fin has 7 spines and 9 rays. Edible but has records of ciguatera poisoning. Up to 40cm in total length.

Distribution — Ryukyu, Formosa, Indo-Pacific, South Africa

Siganidae

Siganus doliatus (Valenciennes) ⇧

Barred Spinefoot, Pencil-streaked Spinefoot

Diagnosis — Found in schools about shallow coral reefs. Body is oval with many yellow stripes. Dorsal and caudal fins are yellow. A black band is present between eyes and lower jaw. Dorsal fin has 13 spines and 10 rays, anal fin has 7 spines and 9 rays. Spines are venomous. Edible. Up to 25cm in total length.

Distribution — Ryukyu, Pacific

Siganidae

Siganus corallinus (Valenciennes) ⇨⇨

Coral Spinefoot, Ocellated Orange Spinefoot, Orange Spinefoot

Diagnosis — Found in schools about coral reefs. Body is oval, profusely dotted with blue on yellowish-brown skin. An indistinct dark band across the eyes. Snout is somewhat long. Throat is depressed. Dorsal fin has 13 spines and 10 rays, anal fin has 7 spines and 9 rays. Spines are venomous. Feeds on seaweed. Edible. Up to 27cm in total length.

Distribution — Ryukyu, Indo-Pacific, East Africa

Siganidae

⇦ ***Siganus puellus*** (Schlegel) ⇨

Blue-lined Spinefoot, Maiden Spinefoot

Diagnosis — Found in schools about coral reefs. Body is oblong, yellowish with 6 blue stripes on anterior, many blue stripes on posterior surfaces. A black band across the eyes. Caudal fin is falcated. Dorsal fin has 13 spines and 10 rays, anal fin has 7 spines and 9 rays. Spines are venomous. Edible. Up to 30cm in total length.

Distribution — Ryukyu, West Pacific

Jaws have incisor teeth. ▼

Siganidae

⇦ ***Siganus lineatus*** (Valenciennes)

Golden-spot Spinefoot, Gold-lined Spinefoot

Diagnosis — Found in schools about shallow coral reefs. Body is oval. Resembles *S. guttatus*, for a yellow blotch on anterior surface of caudal peduncles, but distinguished by stripes on body. Dorsal fin has 13 spines and 10 rays, anal fin has 7 spines and 9 rays. Spines are venomous. Edible. Up to 38cm in total length.

Distribution — Ryukyu, Bonin Islands, West Pacific

Siganidae

⇧⇦ ***Siganus vulpinus*** (Schlegel et Müller)

Foxface, Foxfish

Diagnosis — Found in schools among coral reefs. Body is yellow, has a large black blotch which is absent in some individuals. Snout is projected. Throat is depressed. Dorsa fin has 13 spines and 10 rays, anal fin has 7 spines and 9 rays Spines are venomous. Edible. Aquarium fish. Up to 23cm i total length.

Distribution — Ryukyu, West Pacific

Centrolophidae

⇩ ***Psenopsis anomala*** (Temminck et Schlegel)

Butter Fish, Japanese Butterfish, Melon Seed, Wart Perch

Diagnosis — Inhabits bottoms more than 100m deep. Surfaces at night for feeding on pelagic copepod. Body is oblong, snout is rounded. Posterior part of opercles have a black blotch. Caudal fin is falcated. Dorsal fin has 6 spines and 27−33 rays, anal fin has 3 spines and 24−28 rays. 55−63 lateral scales. Young live with jelly fish in symbiosis. Edible. Up to 30cm in total length.

Distribution — Japan, Formosa, East China Sea

Centrolophidae

Hyperoglyphe japonica (Döderlein) ⇧

[*Ocycrius japonicus* (Döderlein)]

Japanese Butterfish

Diagnosis — Inhabits waters more than 100m deep. Body is compressed and oblong. Eyes are large. Caudal fin is falcated. Dorsal fin has 7−8 spines and 22−26 rays, anal fin has 3 spines and 17−19 rays. 99−103 lateral scales. Edible. Up to 1m in total length.

Distribution — Japan, Ryukyu

Ariommidae

Ariomma evermanni Jordan et Snyder ⇨

Evermann's Nomeid Fish

Diagnosis — Inhabits rather deep waters. Body is elongated, snout is rounded. Eyes are large. Scales are large cycloid. First dorsal fin has 11 spines, second has 1 spine and 15 rays, anal fin has 3 spines and 14 rays. 44 lateral scales. Edible. Up to 80cm in total length.

Distribution — Southern Japan, Ryukyu, South China Sea, Hawaii

Gobiidae

Eleotris fusca

(Bloch et Schneider) ⇨

[*Poecilia fusca* Bloch et Schneider] ⇲

Brown Gudgeon, Dusky Sleeper, Indian Spined Sleeper

Diagnosis — Inhabits freshwater and brackish areas in river-mouths. Body is elongated and dark-brown, snout is pointed. Caudal margin is rounded. Longitudinal scale rows are 65 in number. First dorsal fin has 6 spines, second has 1 spine and 8 rays, anal fin has 1 spine and 8 rays. Carnivorous. Up to 18cm in total length.

Distribution — Ryukyu, Formosa, Indo-West Pacific, South Africa

Stromateidae

Pampus argenteus (Euphrasen) ⇧

[*Stromateoides argenteus* (Euphrasen)]

Butterfish, Harvest Fish, Pomfret, Silver Pomfret, Silvery Pomfret, White-butter-fish, White Pomfret

Diagnosis — Inhabits middle layers of oceanic waters. Body is compressed and oval. Dorsal fin corresponds to anal fin. Caudal fin is falcated. Scales are remarkably small, easily removable. Pelvic fins are absent in adult. Wavy ridge is present between posterior head and lateral lines. Dorsal fin has 5–10 spines and 38–43 rays, anal fin has 5–7 spines and 34–43 rays. Edible. Up to 60cm in total length.

Distribution — Japan, Ryukyu, Formosa, Korea, East China Sea, Indo-West Pacific, Persian Gulf

Gobiidae

Eleotris oxycephala

Temminck et Schlegel ⇲

[*E. balia* Jordan et Seale]

Spined Sleeper

Diagnosis — Inhabits freshwater and brackish areas in river-mouths. Body is elongated, snout is slightly pointed. Fins are dark. 2 stripes run from eyes. First dorsal fin has 6 spines, second has 1 spine and 8 rays, anal fin has 1 spine and 8 rays. Up to 20cm in total length.

Distribution — Japan, Ryukyu, Formosa, China, Indo-Pacific

Gobiidae

⇦ ***Eleotris acanthopoma*** Bleeker

[*E. oxycephala melanosoma*]

Brown Sleeper (New Common Name)

Diagnosis — Inhabits freshwater and brackish areas in river-mouths. Resembles *E. fusca*, but distinguished by yellow bands on first dorsal fin, fewer longitudinal scale rows on body (51). First dorsal fin has 6 spines, second has 1 spine and 8 rays, anal fin has 1 spine and 8 rays. Up to 15cm in total length.

Distribution — Japan, Ryukyu, Bonin Islands, Formosa, Indonesia, Malaysia

Gobiidae

⇦ ***Ptereleotris zebra*** (Fowler)

[*Pogonoculis zebra* Fowler]

[*Pogonoculius zebra* Fowler]

Zebra Goby (New Common Name)

Diagnosis — Inhabits sandy bottoms about coral reefs, often enters burrows inhabited by shrimps. Body is elongated with many yellow stripes. Cheeks and pectoral fin bases have blue stripes. First dorsal fin has 6 spines, second has 1 spine and 27–28 rays, margin is black. Anal fin has 1 spine and 26–27 rays. Up to 10cm in total length.

Distribution — Ryukyu, Indo-West Pacific

Gobiidae

Favonigobius gymnauchen (Bleeker) ⇨

[*Acentrogobius gymnauchen* (Bleeker)]

Naked-headed Goby, Whitebait Goby

Diagnosis — Inhabits sandy bottoms of coastal waters. Body is elongated, profusely dotted with dark color. In male, dorsal fin has a yellow stripe, second spine is extended like thread. First dorsal fin has 6 spines, second has 1 spine and 9 rays, anal fin has 1 spine and 9 rays. Scales are absent on cheeks and opercles. Up to 10cm in total length.

Distribution — Japan, Ryukyu, Korea, China

Gobiidae

Chasmichthys gulosus (Guichenot) ⇨

[*C. dolichognathus gulosus* (Guichenot)]

Gluttonous Goby

Diagnosis — Inhabits shoreline reefs and tidepools. Body is elongated, profusely dotted with white. Caudal margin is white. First dorsal fin has 6 spines, second has 1 spine and 11 rays, anal fin has 1 spine and 9 rays. Carnivorous. Up to 10cm in total length.

Distribution — Japan, Ryukyu, Korea

Gobiidae
⇦ *Amblyeleotris fasciata* (Herre)
Red-necked Goby (New Common Name)
Diagnosis — Symbiosis with *Alpheus*, found in holes dug out in shallow sandy bottoms. Body has 7 reddish-brown bands, snout is short. First dorsal fin has 6 spines, second has 1 spine and 12 rays, sparsely dotted with red. Anal fin has 1 spine and 12 rays. Up to 12cm in total length.
Distribution — Ryukyu, Bonin Islands, West Pacific

Gobiidae
Bathygobius fuscus (Rüppell) ⇩
[*Gobius fuscus* Rüppell]
Brown Goby, Frill Goby
Diagnosis — Inhabits shallow shoreline reefs and tidepools. Body is elongated with irregular dark blotches, head is angular. First dorsal fin has 6 spines, margined with white, second has 1 spine and 9 rays, anal fin has 1 spine and 8 rays. Spawns in summer. Up to 12cm in total length.
Distribution — Japan, Ryukyu, Korea, Indo-Pacific, Hawaii, Red Sea, South Africa

Gobiidae
Pterogobius zacalles Jordan et Snyder ⇩
Beauty Goby
Diagnosis — Inhabits rocky reefs and muddy-sand bottoms between 5−60m deep. Body is elongated with 6 dark bands. Caudal margin is black and rounded. Dorsal and anal fin margins are yellow. First dorsal fin has 8 spines, second has 1 spine and 25 rays, anal fin has 1 spine and 26 rays. Up to 15cm in total length.
Distribution — Japan

Gobiidae
Pterogobius zonoleucus Jordan et Snyder ⇧
Whitegirdled Goby
Diagnosis — Found in schools in coastal seaweed beds, swims in middle layers. Body is brown with 7 orange bands. Dorsal and anal fins also have orange bands. First dorsal fin has 8 spines, second has 1 spine and 20 rays, anal fin has 1 spine and 20 rays. Spawns in spring. Up to 9cm in total length.
Distribution — Japan, Korea

Gobiidae
Cryptocentroides insignis (Seale) ⇩
Diagonal Striped Goby (New Common Name)
Diagnosis — Inhabits brackish areas in river-mouths. Body is elongated with oblique dark bands, profusely dotted with blue in male, with dark-red in female. Scales are absent on cheeks and opercles. First dorsal fin has 6 spines, is extended like thread, second has 1 spine and 12 rays, anal fin has 1 spine and 12 rays. Up to 9cm in total length.
Distribution — Ryukyu, West Pacific

Gobiidae
Rhinogobius giurinus (Rutter) ⇩
Paradise Goby
Diagnosis — Found in schools in freshwater and brackish areas of river-mouths. Body is elongated with black splotches. Head has wavy stripes. First dorsal fin has 6 spines, second has 1 spine and 8 rays, anal fin has 1 spine and 10 rays. Spawns in river mouths between July and October. Carnivorous. Edible. Up to 10cm in total length.
Distribution — Japan, Ryukyu, Formosa, Korea, China

▲ Male

Gobiidae
Cryptocentrus filifer (Valenciennes) ⇧
Gafftopsail Goby
Diagnosis — Inhabits sandy bottoms of bays. Body is elongated with brown bands. Head is profusely dotted with blue. First dorsal fin has 6 spines, is extended like thread with black spots on lower surface, second has 1 spine and 10 rays, anal fin has 1 spine and 9 rays. Up to 15cm in total length.
Distribution — Japan, Formosa, Korea, China, Indo-West Pacific

Gobiidae
Nemateleotris magnificus Fowler ⇧
Firefish

Diagnosis — Inhabits sandy bottoms about shallow coral reefs. Body is elongated, posterior part is red. Caudal margin is rounded with 2 black stripes. First dorsal fin has 6 spines, is extended like thread, second has 1 spine and 28 rays, anal fin has 1 spine and 26 rays. Up to 6cm in total length.

Distribution — Ryukyu, West Pacific

Gobiidae
Ophieleotris sp. ⇧
Mud Cod, Mud Gudgeon, Snakeheaded Gudgeon

Diagnosis — Inhabits brackish and freshwater areas. Body is large, head is especially large. Radial stripes run from eyes. In male, fins have yellow bands. First dorsal fin has 6 spines, second has 1 spine and 8 rays, anal fin has 1 spine and 9 rays. Up to 20cm in total length.

Distribution — Ryukyu

Gobiidae
Ptereleotris hanae (Jordan et Starks) ⇧
[*Vireosa hanae* Jordan et Snyder]

Flower Goby, Threadtail Gudgeon

Diagnosis — Inhabits sandy bottoms of shallow waters, often enters burrows inhabited by shrimps. Body is elongated, snout is short, lower jaw has flaps. Caudal lobes and second spine of dorsal fin are extended like thread. First dorsal fin has 6 spines, second has 1 spine and 23–26 rays, anal fin has 1 spine and 23–26 rays. Up to 12cm in total length.

Distribution — Japan, Ryukyu, Korea, Indonesia

Gobiidae
Ptereleotris evides (Jordan et Hubbs) ⇩
[*P. tricolor* Smith]

Scissor-tailed Goby, Blackfin Gudgeon

Diagnosis — Inhabits in pairs sandy bottoms among coral reefs, often enters burrows inhabited by shrimps. Body is elongated. Dorsal and anal fins are developed, margins are black. Upper and lower ends of caudal fin are also black. First dorsal fin has 6 spines, second has 1 spine and 24–26 rays, anal fin has 1 spine and 24–26 rays. In young, caudal peduncles have a black blotch on lower surface. Up to 16cm in total length.

Distribution — Japan, Ryukyu, Indo-West Pacific

Gobiidae
Acentrogobius caninus (Valenciennes) ⇧
[*Gobius caninus* Valenciennes]

Dog-toothed Goby

Diagnosis — Inhabits muddy-sand bottoms of bays. Body is profusely blotched with dark color and dotted with blue. Snout is roundish. Upper surface of opercles have a blue blotch. First dorsal fin has 6 spines, second has 1 spine and 9 rays, anal fin has 1 spine and 9 rays. Up to 13cm in total length.

Distribution — Formosa, Korea, China, Indo-West Pacific

Gobiidae
Priolepis semidoliatus (Valenciennes) ⇧
[*Zonogobius semidoliatus* (Cuvier et Valenciennes)]

Girdled Goby

Diagnosis — Inhabits shallow coral reefs. Head has orange stripes. Scales are absent on posterior head. First dorsal fin has 6 spines, second has 1 spine and 8 rays, is extended. Anal fin has 1 spine and 7 rays. Up to 4cm in total length.

Distribution — Southern Japan, Ryukyu, Indo-Pacific, Red Sea

Gobiidae
⇧ *Ptereleotris heteroptera* (Bleeker)
Blue Gobies, Indigo Hover Goby

Diagnosis — Resembles *P. evides*, but distinguished by a large black blotch on caudal fin. Dorsal and anal fin margins are not black. First dorsal fin has 6 spines, second has 1 spine and 30 rays, anal fin has 1 spine and 29–30 rays. Up to 13cm in total length.

Distribution — Ryukyu, Pacific, Hawaii

Gobiidae
Nemateleotris decora Allen et Randall ⇨
Flame Goby

Diagnosis — Inhabits shallow coral reefs. Body is elongated, purplish-rosy on posterior part. Dorsal, anal, caudal fins are red. First dorsal fin has 6 spines, is extended, second has 1 spine and 30 rays, anal fin has 1 spine and 29 rays. Up to 10cm in total length.

Distribution — Ryukyu, West Pacific

Gobiidae
Butis amboinensis (Bleeker) ⇧

Amboina Goby (New Common Name)

Diagnosis — Inhabits brackish areas. Head is flat, snout is pointed. Pectoral fin bases have 2 yellowish-brown spots. Body is similarly dotted with yellowish-brown. Lower jaw is projected. First dorsal fin has 6 spines, second has 1 spine and 8 rays. Anal fin has 1 spine and 9 rays. Up to 8cm in total length.

Distribution — Ryukyu, Formosa, Fiji

Gobiidae
Rhinogobius brunneus (Temminck et Schlegel) ⇧

Lake Goby, Common Freshwater Goby

Diagnosis — Inhabits rivers and lakes. Juvenile descend to sea, after living there return to river. Pelvic fins form a sucker to adhere to rocks, etc. In male, first dorsal fin is long. Varies widely in color according to habit. Photo shows male of black type with yellow-margined fins. First dorsal fin has 6 spines, second has 1 spine and 8−9 rays, anal fin has 1 spine and 8 rays. Up to 10cm in total length.

Distribution — Japan, Ryukyu, Formosa, Korea, China

Gobiidae
Yongeichthys criniger (Valenciennes) ⇦⇧

[*Gobius criniger* Cuvier et Valenciennes]
[*Acentrogobius criniger* (Cuvier et Valenciennes)]
[*Rhinogobius nebulosus* (Forskål)]

Fringyfin Goby, Horny Goby, Shadow Goby, Hair-finned Goby

Diagnosis — Inhabits shallow bays and brackish areas in river-mouths. Body is elongated with dark blotches. Snout is rounded. Pelvic fins form a sucker. Scales are absent on cheeks and opercles. Margins of pelvic and anal fins are black. First dorsal fin has 6 spines and second spine is extended, second has 1 spine and 9 rays, anal fin has 1 spine and 9 rays. Omnivorous. Edible but skin and internals contain virulent poison, has records of death from poisoning. Up to 15cm in total length.

Distribution — Ryukyu, Bonin Islands, Formosa, Indo-Pacific, Red Sea, South Africa

Gobiidae
Gobiodon quinquestrigatus (Valenciennes) ⇧

Bicolored Coral-goby (New Common Name)

Diagnosis — Inhabits coral reefs. Body is dark-brown, head is reddish with 5 blue lines. First dorsal fin has 6 spines, second has 1 spine and 10−11 rays, anal fin has 1 spine and 8−9 rays. Up to 4cm in total length.

Distribution — Japan, Ryukyu, Indo-West Pacific

Gobiidae
Acanthogobius flavimanus (Temminck et Schlegel) ⇧⇩⇨

Yellowfin Goby, Genuine Goby, Common Brackish Goby

Diagnosis — Inhabits muddy-sand bottoms of bays and river-mouth. Body is elongated; snout is somewhat long. Cheeks and opercles have scales. Dorsal and caudal fins have dark spots. Pelvic fins form a sucker. First dorsal fin has 8 spines, second has 1 spine and 13 rays, anal fin has 1 spine and 11 rays. Spawns in holes dug out in muddy bottoms between December and May. Omnivorous. Edible. Up to 30cm in total length.

Distribution — Japan, Korea, China, Australia, California

▲Pelvic fins form a sucker.

Gobiidae
Gobiodon okinawae
Sawada, Arai et Abe ⇨

Yellow Coral-goby (New Common Name)

Diagnosis — Inhabits branching corals. Body is yellow, cheeks are pale. First dorsal fin has 6 spines, second has 1 spine and 10 rays, anal fin has 1 spine and 9 rays. Skin secretes ichthyocrinotoxin. Up to 3.5cm in total length.

Distribution — Japan, Ryukyu, China

Gobiidae
⇩ *Periophthalmus vulgaris* Eggert

Barred Mud-skipper

Diagnosis — Resembles *P. cantonensis*, but distinguished by cheeks profusely dotted with white. First dorsal fin has 14 spines and white spots, margin is depressed, second has 1 spine and 11 rays, has a black band. Anal fin has 1 spine and 11 rays. Edible. Up to 15cm in total length.

Distribution — Ryukyu, West Pacific

Gobiidae
⇩ *Boleophthalmus pectinirostris* (Linnaeus)
[*B. chinensis* (Osbeck)]

[*Gobius pectinirostris* Günther]

Bluespotted Mud Hopper, Pond-skipper

Diagnosis — Inhabits burrows dug out in muddy bottoms of shallow waters. Has habit of creeping about in mud at low tide. Feeds on diatom. First dorsal fin has 5 spines, spread out like flag, second has 26 rays. Anal fin has 25 rays. Spawns in July. Edible. Up to 20cm in total length.

Distribution — Southern Japan, Formosa, Korea, China, Burma

Gobiidae
Periophthalmus barbarus
(Linnaeus) ⇗

[*Periophthalmodon barbarus* (Linnaeus)]

Tropical Mud-skipper
(New Common Name)

Diagnosis — Resembles *P. cantonensis*, but scales are large. Longitudinal scale row is 50 in number. Spines on first dorsal fin are as few as 4–7, second has 1 spine and 12 rays, white margined. Anal fin has 1 spine and 11–12 rays. Edible. Up to 27cm in total length.

Distribution — China, Indo-Pacific

Blenniidae
⇩ *Salarias sinuosus* Snyder

Crinkle-lipped Blenny, Fringe-lip Blenny

Diagnosis — Inhabits tidepools in coral reefs. Body has rows of small black and blue spots. Abdomen has rows of pale blotches. Cirrus on upper surface of eyes are slender. Dorsal fin has 12 spines and 16–17 rays, slightly compressed between spine and ray parts. Anal fin has 2 spines and 17–18 rays. In male, anterior rays are extended long. Up to 7cm in total length.

Distribution — Ryukyu, Indo-West Pacific

Gobiidae
⇧ *Periophthalmus cantonensis* (Osbeck)

Mud Skipper, Beach Skipper, Dusky Mud Hopper, Mudhopper

Diagnosis — Inhabits muddy bottoms of bays, also found in tidepools. Skin secrets slime, adapted for living in air. Pelvic fins form a sucker. Can walk on land by movement of pelvic fins which are muscled. First dorsal fin has 14 spines and round-margined, second has 1 spine and 12 rays, anal fin has 1 spine and 11 rays. Spawns between June and July. Carnivorous. Edible. Up to 15cm in total length.

Distribution — Japan, Ryukyu, Formosa, Korea, China, Indo-Pacific, Red Sea, South Africa

▲ Female

Blenniidae

⇦ ***Istiblennius lineatus*** (Valenciennes)

[*Halmablennis lineatus* (Cuvier et Valenciennes)]

Black-lined Blenny, Lined Blenny, Lumphead Blenny, Narrowlined Blenny

Diagnosis — Inhabits tidepools in coral reefs. Body is elongated with dark longitudinal lines. Cirrus on upper surface of eyes are short and broad. Dorsal fin has 12–13 spines and 21–24 rays, compressed between spine and ray parts. Anal fin has 2 spines and 22–25 rays. Up to 16cm in total length.

Distribution — Japan, Ryukyu, Indo-Pacific, Red Sea

Blenniidae

Salarias fasciatus (Bloch) ⇨

Banded Blenny, Barred Blenny, Jewelled Rockskipper, Lineated Blenny, Painted Blenny, Yellow-banded Blenny

Diagnosis — Inhabits tidepools in coral reefs. Body is somewhat blunt with dark-brown stripes and pale blotches. Cirrus on upper surface of eyes are branched out. Uncompressed dorsal fin has 12 spines and 18–20 rays, anal fin has 2 spines and 19–21 rays. In male, anterior rays are extended long. Up to 15cm in total length.

Distribution — Ryukyu, Formosa, Indo-Pacific, Red Sea, East Africa

Blenniidae

Cirripectes castaneus (Valenciennes) ⇩

[*C. variolosus* (Valenciennes)]

Collared Blenny, Muzzled Rockskipper, Red-tipped Fringed-blenny

Diagnosis — Inhabits shallow coral reefs. Body is blackish-brown, has no blotches. Eyes are large and protruded. Dorsal fin has 12 spines and 14 rays, anterior spines are extended like thread. Anal fin has 2 spines and 15 rays. Caudal fin is pale. Up to 9cm in total length.

Distribution — Japan, Ryukyu, Indo-Pacific, Hawaii, Red Sea, East Africa

Blenniidae

⇧ ***Istiblennius bilitonensis*** (Bleeker)

Blackcheek Blenny, Brevoort's Blenny

Diagnosis — Inhabits tidepools in coral reefs. Body has dark blotches and cross bands. In male, occipital crest are present on posterior head, cheeks are black. Slender cirrus are present on upper end of eyes. Dorsal fin has 12–13 spines and 19–22 rays, anal fin has 2 spines and 20–22 rays. Up to 12cm in total length.

Distribution — Ryukyu, China, West Pacific

Blenniidae

Istiblennius zebra (Vaillant et Sauvage) ⇧

Zebra Blenny Fish, Zebra Rockskipper, Rock Skipper

Diagnosis — Inhabits shallow rocky reefs. Body is elongated, bluish-brown with dark cross bands. Occipital crest is present on posterior head. Cirrus is present on upper surface of eyes. Compressed between spine and ray parts of dorsal fin. Up to 15cm in total length.

Distribution — Hawaii

Blenniidae

Lophalticus kirki (Günther) ⇩

Kirki's Blenny (New Common Name)

Diagnosis — Inhabits shallow rocky reefs and tidepools. Body is elongated, blackish-brown with many white stripes. Occipital crest is present on posterior head. Deeply compressed between spine and ray parts of dorsal fin. Up to 9cm in total length.

Distribution — Indian Ocean, East Africa

Blenniidae

⇧ ***Praealticus margaritarius*** (Snyder)

[*Istiblennis margaritarius* (Snyder)]

Pearlyspot Blenny

Diagnosis — Inhabits shallow shoreline reefs and coral reefs. Body is elongated, sparsely dotted with white. Cirrus on upper surface of eyes is branched out like wing. Dorsal fin has 12–13 spines and 18–19 rays, compressed between spine and ray parts. Anal fin has 2 spines and 20–21 rays. Up to 10cm in total length.

Distribution — Japan, Ryukyu

Tripterygiidae

Tripterygion bapturum Jordan et Snyder ⇧

Blue-faced Blenny

Diagnosis — Inhabits shallow rocky reefs. Body is elongated, has 5 brown bands, profusely dotted with yellowish-brown. Head is large. Caudal fin is black. First dorsal fin has 3 spines, second has 17 spines, third has 12 rays. Anal fin has 2 spines and 26 rays. Body has 2 lateral lines in one side. Up to 7cm in total length.

Distribution — Japan

Blenniidae

⇩ ***Meiacanthus atrodorsalis atrodorsalis*** (Günther)

[*M. atrodorsalis* (Günther)]

[*M. herlihyi* (Fowler)]

[*Petroscirtes atrodorsalis* Günther]

Goldfin Blenny

Diagnosis — Inhabits shallow coral reefs. Posterior part of body is yellow. A black band runs from eyes to dorsal fin. Upper and lower ends of caudal fin are yellow, lobes are extended. Mouth has large canines connected with venom gland. Dorsal fin has 4 spines and 25–28 rays, anal fin has 2 spines and 15–18 rays. Up to 7cm in total length.

Distribution — Ryukyu, Indo-Pacific

Blenniidae

Plagiotremus tapeinosoma (Bleeker) ⇧

[*Aspidontus maroubrae* (Ogilby)]

[*Runula tapeinosoma* (Bleeker)]

Calico Blenny, Piano Blenny, Violetbanded Blenny, Mimic Blenny

Diagnosis — Inhabits holes in shallow coral reefs. Body is elongated. A black band is present between snout tip and caudal fin. Snout is pointed, mouth is situated on ventral side. Lateral lines and scale are absent. Black-margined dorsal fin has 7–9 spines and 34–39 rays, anal fin has 2 spines and 28–33 rays. Up to 13cm in total length.

Distribution — Japan, Ryukyu, Indo-Pacific, South Africa

Blenniidae

Petroscirtes breviceps (Valenciennes) ⇗

[*Dasson trossulus* (Jordan et Snyder)]

Black-banded Blenny, Dandy Blenny, Sabretooth Blenny, White-banded Blenny

Diagnosis — Adult inhabit gravel bottoms of shoreline waters, young are found in seaweed bed. Body is elongated with 2 brown bands. Head is profusely dotted with red. Caudal margin is serrated, lobes are extended. Dorsal fin has 10–12 spines and 17–21 rays, anal fin has 2 spines and 17–21 rays. Spawns in early summer. Up to 12cm in total length.

Distribution — Japan, Korea, West Pacific

Mugiloididae

Parapercis xanthozona (Bleeker) ⇧

[*Neopercis xanthozona* (Bleeker)]

Spottyfin Weever, Yellow-spot Grubfish

Diagnosis — Body is cylindrical with 8–9 blackish-brown blotches. Opercles have 2 spines. Upper caudal lobe is extended. Black spots are present on dorsal, anal, caudal fins. Dorsal fin has 5 short spines and 23–24 rays, anal fin has 20–21 rays. 62 pored lateral scales. Edible. Up to 26cm in total length.

Distribution — Southern Japan, Ryukyu, Indo-Pacific, East Africa

Mugiloididae

Parapercis sexfasciata (Temminck et Schlegel) ⇧

Saddled Weever

Diagnosis — Inhabits rather deep muddy-sand bottoms. Body is cylindrical with 4–5 dark blotches. Eyes are crossed by a dark stripe. Opercles have 2 spines. Dorsal fin has 5 short spines and 23 rays, anal fin has 1 spine and 19 rays. 60–64 pored lateral scales. Edible. Up to 25cm in total length.

Distribution — Japan, Ryukyu, Formosa, Korea, China, Hong Kong

Mugiloididae

Parapercis multifasciata Döderlein ⇨

Bicolorbarred Weever, Redbarred Weever

Diagnosis — Inhabits rather deep muddy-sand bottoms. Body is cylindrical with 10 brown bands. Upper jaw is red. 2 yellow stripes are present on surface posterior to eyes. Caudal peduncles have a black blotch. Dorsal fin has 5 spines and 22–24 rays, anal fin has 1 spine and 19–20 rays. 60 pored lateral scales. Edible. Up to 20cm in total length.

Distribution — Japan, Formosa, Korea

Mugiloididae

Parapercis polyphthalma (Cuvier) ⇩⇨

Black-tailed Weever (New Common Name)

Diagnosis — Inhabits sandy bottoms of shallow coral reefs. Body is cylindrical, profusely dotted with blackish-brown. Caudal fin has a large black blotch. Dorsal fin has 5 short spines and 21–22 rays, anal fin has 1 spine and 17–18 rays. 58–60 pored lateral scales. Edible. Up to 25cm in total length.

Distribution — Ryukyu, Indo-West Pacific, East Africa

Mugiloididae

Parapercis pulchella (Temminck et Schlegel) ⇩

[*Cilias pulchella* (Temminck et Schlegel)]

Rosy Grub Fish

Diagnosis — Inhabits muddy-sand bottoms of shallow waters. Body is cylindrical with 6 brown cross bands. Lower jaw is long. Upper caudal lobe is extended. Dorsal fin has 5 short spines and 21–22 rays, anal fin has 1 spine and 17–18 rays. 56–64 pored lateral scales. Carnivorous. Edible. Up to 22cm in total length.

Distribution — Japan, Korea, Indo-West Pacific, South Africa

Stichaeidae

Chirolophis japonicus Herzenstein ⇨

[*Azuma emmnion* Jordan et Snyder]

Fringed Blenny

Diagnosis — Inhabits seaweed bed of bays. Body is elongated. Head is small. Many flaps are present on crown of head and lower jaw. Jaws have incisor teeth. Dorsal fin is extended from back of head as far as caudal peduncles, has 60–63 rays. Anal fin has 1 spine and 45–47 rays. Edible. Up to 55cm in total length.

Distribution — Northern Japan, Korea, North China

Anarhichadidae

Anarhichas sp. ⇩

Black-spotted Wolffish (New Common Name)

Diagnosis — Resembles *A. orientalis*, but distinguished by 1 row of black circular blotches on back. 2 black spots are present on upper surface of eyes. Edible. Up to 1.5m in total length.

Distribution — Canada, North America

Pholididae

Enedrias nebulosa (Temminck et Schlegel) ⇧

[*Pholis nebulosus* (Temminck et Schlegel)]

Blenny, Clouded Blenny, Silver Tail, Tidepool Gunnel

Diagnosis — Inhabits shallow rocky reefs, also in seaweed bed of bays. Body is elongated with indistinct dark-brown blotches. Dorsal fin base has rows of black triangular blotches. Caudal margin is white. Dorsal fin has 76–83 rays, anal fin has 2 spines and 35–42 rays. Edible but ovary may cause poisoning. Up to 35cm in total length.

Distribution — Japan, Sakhalin

Anarhichadidae

⇦ ***Anarhichas orientalis*** Pallas

Bering Wolffish, North Pacific Wolffish, Okhotsk Wolffish, Wolf Fish

Diagnosis — Inhabits rather deep rocky reefs. Body is elongated. Head is rounded. Mouth is large, jaws have 4–6 canines. Pelvic fins are absent. Dorsal fin has 83–88 spines, extended from back of head as far as caudal peduncles. Anal fin has 53–54 rays. Feeds on shellfish and crustacea. Spawns between October and November. Edible. Up to 1.5m in total length.

Distribution — Northern Japan, Okhotsk Sea, Bering Sea, North America

Uranoscopidae

⇩ ***Uranoscopus oligolepis*** Bleeker

Blue Stargazer, Yellowfinned Stargazer

Diagnosis — Inhabits rather deep sandy bottoms. Resembles *U. japonicus*, but distinguished by absence of blotches on back. Also distinguished from *U. tosae* found in waters of mainland Japan by absence of scales on back. First dorsal fin has 4–5 spines, second has 13 rays. Anal fin has 12–13 rays. Dorsal spines are venomous. Edible. Up to 30cm in total length.

Distribution — Ryukyu, Korea, China, West Pacific

Uranoscopidae

⇧ ***Uranoscopus japonicus*** Houttuyn

Star-gazer, Japanese Stargazer, Mishima Pufferfish

Diagnosis — Inhabits rather deep muddy-sand bottoms. Hides in sand. Body is somewhat flat, head is large. Mouth is large, opened upward. Eyes are small, situated on back. Upper end of pectoral fins have a sharp-pointed venomous spine. Black first dorsal fin has 4–5 spines, second has 13–15 rays. Anal fin has 14–15 rays. Edible. Up to 30cm in total length.

Distribution — Japan, Formosa, Korea, China, Philippines, Singapore

Uranoscopidae
Uranoscopus bicinctus
Temminck et Schlegel ⇧

Blackbanded Stargazer, Spectacles Fish

Diagnosis — Inhabits rather deep sand-gravel bottoms. Body has 2 blackish-brown blotches. Head has a blackish-brown blotch. Caudal margin is black. First dorsal fin has 4–5 spines, second has 12–14 rays. Anal fin has 13 rays. Carnivorous. Dorsal spines are venomous. Up to 30cm in total length.

Distribution — Japan, Ryukyu, Korea, Indonesia

Agonidae
⇩ *Podothecus sachi* (Jordan et Snyder)

Sailfin Poacher

Diagnosis — Inhabits rather deep muddy bottoms. Body is elongated, covered with developed bony plates. Snout is pointed with many barbels on underside and around mouth. First dorsal fin has 8–10 spines, second has 12–14 rays. Anal fin has 13–17 rays. In male, dorsal and anal fins are remarkably developed. Edible. Up to 50cm in total length.

Distribution — Northern Japan, Korea

Uranoscopidae
Gnathagnus elongatus
(Temminck et Schlegel) ⇧

Bluespotted Stargazer, Blue Mishima Pufferfish

Diagnosis — Inhabits muddy-sand bottoms more than 200m deep. Body is slightly elongated, profusely dotted with brown. Head is flat, spines on back of head are small. Dorsal fin is single with 13–14 rays, anal fin has 16–18 rays. Feeds on small fish. Edible. Up to 35cm in total length.

Distribution — Japan, Formosa, Korea, East China Sea

Agonidae
Agonomalus proboscidalis
(Valenciennes) ⇩

Barbed Hunchback Poacher

Diagnosis — Inhabits sandy bottoms of shallow waters. Body is covered with developed bony plates. First dorsal fin is well-developed, has 9 spines, second has 5–7 rays. Anal fin has 11–13 rays. Snout has a barbel. Air bladder is absent. Up to 20cm in total length.

Distribution — Northern Japan, Okhotsk Sea

Congiopodidae
Hypodytes rubripinnis
(Temminck et Schlegel) ⇩

Redfin Velvetfish, Tiny Stinger

Diagnosis — Inhabits shoreline reefs and seaweed bed. Body is somewhat flat with fine buried scales on posterior part. Well-developed dorsal fin has 14–15 spines and 6–7 rays, spines are venomous. Anal fin has 3 spines and 3–4 rays. Spawns in seaweed bed in early summer. Up to 11cm in total length.

Distribution — Japan, Korea

Dactylopteridae
Dactyloptena orientalis (Cuvier) ⇧⇨
[*Dactylopterus orientalis* Cuvier]

Flying Gurnard, Purple Flying Gurnard

Diagnosis — Inhabits shallow waters and also deep waters. Body is slightly extended, head is angular, covered with hard carapace. Interorbital space is deeply depressed. Pectoral fins are well-developed, long enough to reach caudal fin. Caudal margin is truncated. First dorsal fin has 5 spines with 2 separated spines on anterior surface, second has 1 spine and 8 rays. Anal fin has 7 rays. Edible. Up to 40cm in total length.

Distribution — Japan, Ryukyu, Indo-Pacific, Hawaii, South Africa

Dactylopteridae
Dactyloptena gilberti Snyder ⇧
Gilbert's Flying Gunard

Diagnosis — Resembles *D. orientalis*, but distinguished by short snout, slightly depressed interorbital space. Head is flat, covered with hard carapace. First dorsal fin has 5 spines with 2 separated spines on anterior surface, second has 1 spine and 8 rays. Anal fin has 6 rays. Edible. Up to 30cm in total length.

Distribution — Southern Japan, Ryukyu

Anoplopomatidae
Erilepis zonifer (Lockington) ⇩
Skilfish, Giant Skilfish, Fat Priest

Diagnosis — Inhabits rocky reefs in deep-sea. Body is blunt, dark-brown with irregular pale blotches.Caudal margin is truncated. First dorsal fin has 11–12 spines, second has 13–16 rays. Anal fin has 3 spines and 10–12 rays. 120 lateral scales. Flesh contains so much fat and oil that wax poisoning may be caused. Up to 3m in total length.

Distribution — Northern Japan, Bering Sea, Alaska, California

Hexagrammidae
Pleurogrammus monopterygius (Pallas) ⇨
Atka Fish, Yellowfish, Atka Mackerel

Diagnosis — Resembles *P. azonus*, but body has 5 broad cross bands. 5 lateral lines in one side, covered with fine scales. Dorsal fin has 21 spines and 29 rays, anal fin has 1 spine and 24–26 rays. Edible. Up to 70cm in total length.

Distribution — Northern Japan, Sakhalin, Okhotsk Sea, Bering Sea

Hexagrammidae
⇦ *Pleurogrammus azonus* Jordan et Metz
Arabesque Greenling, Atka Mackerel

Diagnosis — Inhabits shallow waters and also deep waters. Body is flat and extended. Caudal fin is falcated. 5 lateral lines in one side. Uncompressed dorsal fin has 21–23 spines and 28–29 rays, anal fin has 27–32 rays. Spawns between September and February. Grows in 3 years. Edible. Up to 70cm in total length.

Distribution — Japan, Yellow Sea, Okhotsk Sea

▼Female

Hexagrammidae
Hexagrammos agrammus (Temminck et Schlegel) ⇧⇨

[*Agrammus agrammus* (Temminck et Schlegel)]

Spottybelly Greenling

Diagnosis — Inhabits seaweed bed of shoreline waters. Resembles *H. otakii*, but distinguished by single lateral line, irregular pale blotches on body. Caudal margin is rounded. Dorsal fin has 17–19 spines and 19–23 rays, anal fin has 18–21 rays. Edible. Up to 35cm in total length.

Distribution — Japan, Korea, Yellow Sea

▼Male

⇧ Hexagrammidae
⇧ *Hexagrammos otakii* Jordan et Starks

Fat Cod, Fat Greenling, Greenling

Diagnosis — Inhabits shoreline reefs. Body is compressed, slightly elongated, color varies widely. In male, body is yellowish. Head has 2 pairs of flaps. 5 lateral lines in one side, covered with fine scales. Dorsal fin has 19–21 spines and 21–23 rays, anal fin has 21–23 rays. Spawns between October and December. Edible. Up to 40cm in total length.

Distribution — Japan, Korea, Yellow Sea

Hexagrammidae
Hexagrammos lagocephalus (Pallas) ⇧

Rock Greenling

Diagnosis — Body is compressed, head is roundish. Caudal margin is rounded. Scales are present on cheeks and opercles. 5 lateral lines in one side. In male, body is reddish-brown with a black blotch on upper pectoral fins. In female, body is light-brown. Dorsal fin has 20–23 spines and 22–25 rays, anal fin has 20–24 rays. Spawns between November and December. Edible. Up to 80cm in total length.

Distribution — Northern Japan, Okhotsk Sea, Bering Sea

Hoplichthyidae
Hoplichthys gilberti Jordan et Richardson ⇧

Gilbert's Spiny Flathead, Spiny Flathead

Diagnosis — Inhabits rather deep sandy bottoms. Head is flat. Tail is extended. Snout length and eye diameter are approximately equal. Interorbital space is narrow. A row of spines are present on bony plates. Pectoral fins are short. First dorsal fin has 6 spines, second has 15 rays. Anal fin has 16–18 rays. Up to 25cm in total length.

Distribution — Japan, Formosa, Indo-West Pacific

Platycephalidae
Parabembras curtus (Temminck et Schlegel) ⇨

Matron Flathead

Diagnosis — Inhabits rather deep muddy-sand bottoms. Body is somewhat flat. Snout is projected. Head is large. Lower jaw is long. Scales are large. Second dorsal and anal fins are short. First dorsal fin has 9–10 spines, second has 1 spine and 7–9 rays. Anal fin has 3 spines and 5 rays. Spawns in winter. Edible. Up to 30cm in total length.

Distribution — Southern Japan, Formosa, Korea, China, Indian Ocean

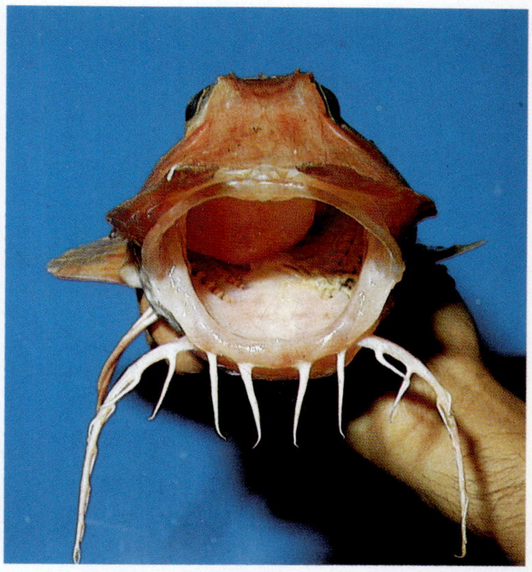

Peristediidae
Satyrichthys sp.

Sea Robin, Offshore Crocodilefish

Diagnosis — Inhabits waters more than 200m deep. Body is elongated, covered with hard bony plates. Head is large, snout end has a pair of rostral projections. Lower jaw has 3 pairs of branching barbels. First dorsal fin has 7 spines, second has 17 rays. Anal fin has 17 rays. 27 lateral scales. Up to 50cm in total length.

Distribution — Southern Japan, Ryukyu

Platycephalidae
Inegocia guttata (Cuvier)

Crocodile Flathead (New Common Name)

Diagnosis — Inhabits muddy-sand bottoms of shoreline waters. Body is flat with 5−6 dark bands. Snout is long, mouth is large. Pelvic fins have dark blotches. First dorsal fin has 7−8 spines with 1 separated spine on anterior surface, second has 11 rays. Anal fin has 11 rays. Edible. Up to 80cm in total length.

Distribution — Japan, Ryukyu, Formosa, China, South China Sea

▼Hide in sand for hunting food.

Platycephalidae

Thysanophrys arenicola Schultz ⇨⇩

Sand Flathead (New Common Name)

Diagnosis — Inhabits sandy bottoms of shoreline waters. Body is flat, covered closely with small light-brown spots. Interorbital space is broad. Preoperculum have 3 short spines. First dorsal fin has 8 spines with 1 separated spine on anterior surface, second has 12 rays. Anal fin has 12−13 rays. Edible. Up to 30cm in total length.

Distribution — Ryukyu, Micronesia

Platycephalidae

Cociella crocodila (Tilesius) ⇩

[*Thysanophrys crocodilus* (Tilesius)]

Spotted Flathead

Diagnosis — Body is elongated and flat. Snout is pointed, lower jaw is long. Scales are small, easily removable. First dorsal fin has 8 spines and blackish, second has 11 rays. Anal fin has 11 rays. Black spots are present on upper surface of body, second dorsal, anal and caudal fins. Edible. Up to 40cm in total length.

Distribution — Japan, Ryukyu, Korea, Formosa, China, Indian Ocean

Platycephalidae

Inegocia japonica (Tilesius) ⇧

Lizard Flathead

Diagnosis — Inhabits muddy-sand bottoms of shoreline waters. Body is flat, light-brown, profusely dotted with black. Snout is somewhat short. 2 spines are present on margin of opercles. First dorsal fin has 8 spines with 1 separated spine on anterior surface, second has 12 rays. Anal fin has 12 rays. Edible. Up to 25cm in total length.

Distribution — Japan, Yellow Sea, East China Sea

Platycephalidae

Suggrundus meerdervoorti (Bleeker) ⇩⇩

Big-eyed Flat-head

Diagnosis — Inhabits rather deep muddy-sand bottoms. Body is flat, profusely dotted with black. Head has many spines. First dorsal fin has 7−8 spines with 1 separated spine on anterior surface, second has 11 rays. Anal fin has 11 rays. Transsexual. Spawns in spring. Edible. Up to 25cm in total length.

Distribution — Japan, Formosa, China, Yellow Sea, East China Sea

Platycephalidae

Platycephalus indicus (Linnaeus) ⇧⇨

Bar Tailed Flathead, Flat Head, Flathead Fish, Indian Flathead, River Gurnard, Sandgurnard, Sand Scraper

Diagnosis — Inhabits muddy-sand bottoms of shoreline waters. Body is flat. Tail is extended. Scales are small. Head is large, has no spines. First dorsal fin has 7 spines with 2 separated spines on anterior surface, second has 1 spine and 13 rays. Anal fin has 13 rays. Spawns between spring and summer. Carnivorous. Up to 1m in total length.

Distribution — Japan, Ryukyu, Formosa, Korea, China, Indo-Pacific, Persian Gulf, Red Sea, South Africa

Cottidae

⇧ ***Alcichthys alcicornis*** (Herzenstein)

Elkhorn Sculpin

Diagnosis — Inhabits rather deep waters. Body is covered with slime. Head has flaps. Opercles have 4 spines. First dorsal fin has 9–10 spines, second has 14–17 rays. Anal fin has 13–16 rays. Edible. Up to 40cm in total length.

Distribution — Northern Japan, Okhotsk Sea

Cottidae

⇦ ***Pseudoblennius percoides*** Günther

Perch Sculpin, Thick-lipped Blenny

Diagnosis — Inhabits inshore seaweed bed and tidepools. Body is somewhat flat, color and blotch vary widely. Snout is pointed, mouth is large. First dorsal fin has 9–10 spines, second has 17–20 rays. Anal fin has 15–17 rays. Spawns in winter. Up to 20cm in total length.

Distribution — Japan

Cottidae

Hemitripterus villosus (Pallas) ⇨

[*H. americanus villosus* (Pallas)]

Sea Raven, Shaggy Sea Raven

Diagnosis — Inhabits rather deep rocky reefs. Body is covered with small processes with irregular dark bands. Head has many fleshy flaps. First dorsal fin has 17–19 spines with deeply falcated membranes, second has 11–13 rays. Anal fin has 12–15 rays. Spawns in winter. Edible. Up to 40cm in total length.

Distribution — Northern Japan, Bering Sea

Cottidae

⇦ ***Hemilepidotus jordani*** Bean

Irish Lord, Yellow Irish Lord, Yellow Sculpin

Diagnosis — Head is large. Body has irregular dark blotches. Fins are dark. Dorsal fin has 11–12 spines and 18–23 rays, first spine is short. Anal fin has 16–18 rays. 59–68 lateral scales. Edible. Up to 50cm in total length.

Distribution — Northern Japan, Okhotsk Sea, Bering Sea, Alaska

Scorpaenidae

Sebastes joyneri Günther

Joyner Stingfish, Saddled Brown Rockfish

Diagnosis — Inhabits rather deep rocky reefs. Resembles *S. inermis*, but body is pale, back has 5 black bands. Caudal margin is emarginated. Dorsal fin has 13 spines and 14–15 rays, anal fin has 3 spines and 7 rays. 47–53 lateral scales. Ovoviviparous. Produces youngs between winter and spring. Edible. Up to 20cm in total length.

Distribution — Japan, Korea, Formosa

Scorpaenidae

Sebastes inermis Cuvier

[*S. güntheri* Jordan et Starks]

[*Sebastoides inermis* (Cuvier et Valenciennes)]

Black Rockfish, Darkbanded Rockfish, Japanese Stingfish

Diagnosis — Inhabits shoreline reefs. Body is compressed, brown with 5–6 cross bands, color varies according to habitat. Eyes and mouth are large. Caudal margin is truncated. Small scales are present on lower jaw. Dorsal fin has 13 spines and 13–14 rays, anal fin has 3 spines and 7–8 rays. 42–45 lateral scales. Ovoviviparous. Produces youngs between January and February. Edible. Up to 35cm in total length.

Distribution — Japan, Korea

Scorpaenidae

Sebastes taczanowskii (Steindachner)

White-edged Rockfish

Diagnosis — Inhabits shoreline reefs, also found brackish areas in river-mouths. Body is dark-brown, has small white blotches in young. Caudal margin is rounded with white margin. Dorsal fin has 13 spines and 13–15 rays, anal fin has 3 spines and 6–8 rays. 40–49 lateral scales. Edible. Up to 30cm in total length.

Distribution — Japan, Sakhalin

Scorpaenidae

Sebastes oblongus Günther

Oblong Rockfish

Diagnosis — Inhabits shoreline reefs. Body is somewhat shallow with irregular dark blotches. Interorbital space is flat. No spines anterior to eyes. Dorsal fin has 13 spines and 10–13 rays, anal fin has 3 spines and 5–7 rays. 39–49 lateral scales. Edible. Up to 40cm in total length.

Distribution — Japan, Korea

Scorpaenidae

Sebastes thompsoni (Jordan et Hubbs)

Goldeye Rockfish

Diagnosis — Inhabits rather deep offshore reefs. Body is somewhat compressed, back has 5 dark bands. Pectoral fins are reddish. Caudal margin is truncated. Dorsal fin has 13 spines and 14–15 rays, anal fin has 3 spines and 7 rays. 52–56 lateral scales. Edible. Up to 40cm in total length.

Distribution — Japan, Korea

Scorpaenidae

⇦ *Sebastes nivosus* Hilgendorf

Snowy Rockfish, Spotted Rockfish

Diagnosis — Inhabits shoreline reefs. Body is somewhat deep, covered closely with white spots. Head has sharp-pointed spines. Dorsal fin has 13 spines and 11−12 rays, anal fin has 3 spines and 6 rays. 35−38 lateral scales. Edible. Up to 40cm in total length.

Distribution — Japan

Scorpaenidae

Sebastes schlegeli Hilgendorf ⇩

Jacopever, Rockfish, Sting Fish, Schlegel's Black Rockfish

Diagnosis — Inhabits shallow reefs. Body is somewhat shallow. Spines on head are developed. 2 black bands extend from eyes. Caudal margin is rounded, upper and lower ends white. Dorsal fin has 13 spines and 11−13 rays, anal fin has 3 spines and 6−8 rays. 39−52 lateral scales. Ovoviviparous. Produces youngs between May and June. Edible. Up to 50cm in total length.

Distribution — Japan, Korea, China

Scorpaenidae

⇦ ⇙ *Sebastes pachycephalus* Temminck et Schlegel ⇧

[*S. pachycephalus pachycephalus* (Temminck et Schlegel)]

[*S. pachycephalus nudus* Matsubara]

Spotbelly Rockfish, Mottled Yellow Rockfish

Diagnosis — Inhabits shoreline reefs. Body is somewhat deep, color and blotch vary according to habitat. Head has large spines. Interorbital space is remarkably depressed. Caudal margin is rounded. Dorsal fin has 13 spines and 11−13 rays, anal fin has 3 spines and 5−7 rays. 30−35 lateral scales. Ovoviviparous. Produces youngs between April and June. Edible. Up to 40cm in total length.

Distribution — Japan, Korea, China

Scorpaenidae

Sebastes owstoni (Jordan et Thompson) ⇩

Owston Sting Fish, Owston's Rockfish

Diagnosis — Inhabits waters more than 100m deep. Body is elongated with 4 indistinct brown bands. Caudal margin is lunated. Dorsal fin has 14 spines and 12−15 rays, anal fin has 3 spines and 7−11 rays. 31−35 lateral scales. Edible. Up to 30cm in total length.

Distribution — Japan, Korea, Okhotsk Sea

Scorpaenidae

⇖ *Sebastes vulpes* Döderlein

Fox Jacopever

Diagnosis — Inhabits shoreline reefs. Resembles *S. schlegeli*, but distinguished by grayish-brown body profusely dotted with light color. Dorsal fin has 13 spines and 12−13 rays, anal fin has 3 spines and 5−6 rays. 30−37 lateral scales. Ovoviviparous. Produces youngs in spring. Edible. Up to 45cm in total length.

Distribution — Japan, Korea

Scorpaenidae

Sebastes trivittatus Hilgendorf ⇨

Threestripe Rockfish

Diagnosis — Inhabits shoreline reefs. Body is somewhat deep with 2 dark longitudinal bands. Dorsal fin has 13−14 spines and 12−14 rays, anal fin has 3 spines and 6−7 rays. 31−40 lateral scales. Edible. Up to 40cm in total length.

Distribution — Japan, Korea

Scorpaenidae

⇧ *Sebastes zonatus* Chen et Barsukov

Rockfish

Diagnosis — Body is purplish-brown with dark blotches, and 3 dark bands. Dorsal fin has 13 spines and 12−13 rays, anal fin has 3 spines and 6−7 rays. 31−37 lateral scales. Edible. Up to 40cm in total length.

Distribution — Japan

Scorpaenidae

⇦ ***Sebastes matsubarae*** Hilgendorf

[*S. matsubarai* Hilgendorf]

Matsubara's Red Rockfish, Matsubara Stingfish

Diagnosis — Inhabits deep-water reefs. Body is compressed and somewhat deep, head has many spines. Interorbital space is almost undepressed. Dorsal fin has 13 spines and 11–13 rays, anal fin has 3 spines and 6–7 rays. 28–31 lateral scales. Ovoviviparous. Produces youngs between February and April. Edible. Up to 70cm in total length.

Distribution — Japan

Scorpaenidae

⇦ ***Sebastolobus macrochir*** (Günther)

Bighand Thornyhead, Kichiji Rockfish

Diagnosis — Inhabits waters more than 300m deep. Body is somewhat deep, head has sharp-pointed spines. Mouth is large. Lower part of pectoral fins are swollen. Dorsal fin has 15–16 spines and 8–10 rays, has a large black blotch. Anal fin has 3 spines and 5 rays. 30–34 lateral scales. Spawns between March and June. Carnivorous. Edible. Up to 40cm in total length.

Distribution — Japan, Sakhalin, Okhotsk Sea

Scorpaenidae

⇩ ***Sebastiscus marmoratus*** (Cuvier) ⇘⇗

Scorpionfish, Marbled Rockfish, Rockfish, Japanese Stingfish

Diagnosis — Inhabits shoreline reefs. Body is somewhat deep with irregular blotches, blackish-brown in shallow-rock inhabitants, in deep sea inhabitants, very red. Dorsal fin has 12 spines and 10–12 rays, anal fin has 3 spines and 5 rays. 49–54 lateral scales. Ovoviviparous. Produces youngs between December and April. Carnivorous. Edible. Up to 34cm in total length.

Distribution — Japan, Ryukyu, Formosa, Korea, China, South China Sea, Philippines

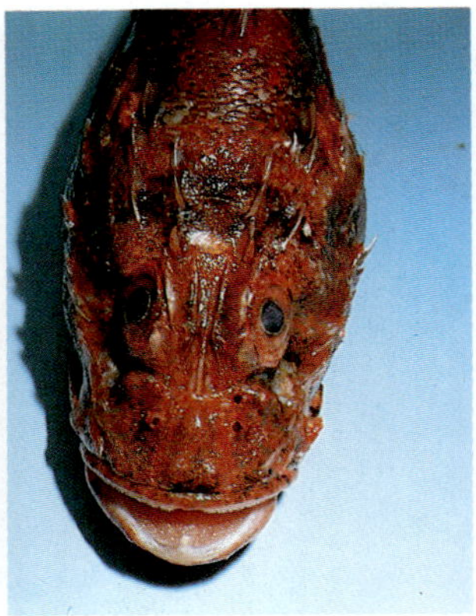

Scorpaenidae

⇦ ***Scorpaenopsis diabolus*** (Cuvier)

[*S. gibbosa* (Bloch et Schneider)]

Humpbacked Scorpionfish, Stingfish, False Stonefish, Humped Scorpionfish, Satsuma Scorpionfish, Devil Scorpionfish

Diagnosis — Inhabits shoreline reefs. Body is somewhat deep, has small fleshy flaps. Spines are present on head. Pectoral fins reach anal fin. Caudal fin has a broad dark band. Dorsal fin has 12 spines and 9 rays, anal fin has 3 spines and 5 rays. 22–25 pored lateral scales. Spines are venomous, has records of having caused death. Edible. Up to 30cm in total length.

Distribution — Japan, Ryukyu, Formosa, Indo-Pacific, Hawaii, Red Sea, South Africa

Scorpaenidae

Scorpaena neglecta fimbriata (Steindachner et Döderlein) ⇨

Fringyhead Scorpionfish

Diagnosis — Resembles *S. onaria*, but distinguished by deeply falcated dorsal spines' membranes, absence of black blotches on body. Pectoral fins reach anus. Dorsal fin has 12 spines and 9 rays, anal fin has 3 spines and 5 rays. Edible. Up to 45cm in total length.

Distribution — Japan, Formosa

Scorpaenidae
⇩ *Scorpaenopsis cacopsis* Jenkins
Jenkins' Scorpion Fish, Titan Scorpionfish
Diagnosis — Inhabits shallow reefs, also rather deep waters. Head is large with sharp-pointed spines. Eyes are small, mouth is large. Body has small flaps. Dorsal fin has 12 spines and 10 rays, anal fin has 3 spines and 5 rays. 54 lateral scales. Spines are venomous. Edible. Up to 50cm in total length.
Distribution — Hawaii

Scorpaenidae
Scorpaena miostoma Günther ⇩
[*S. neglecta miostoma* Günther]
Smallmouth Scorpionfish
Diagnosis — Resembles *S. onaria*, but mouth is slightly small. Posterior tip of upper jaw does not reach below eye centers. Spines, membranes on dorsal fin are less falcated. In male, dorsal fin has a black blotch. Dorsal fin has 12 spines and 9 rays, anal fin has 3 spines and 5 rays. Edible. Up to 20cm in total length.
Distribution — Japan, Korea

Scorpaenidae
Scorpaena kowiensis Smith ⇧
[*S. albo-brunnea* Günther]
[*S. tinkhami* Fowler]
Bigeye Scorpionfish
Diagnosis — Body is small, light brown with irregular brown blotches. Fins are profusely dotted with brown. Dorsal fin has 12 long spines and 9 rays, anal fin has 3 spines and 5 rays. 24–25 pored lateral scales. Edible. Up to 10cm in total length.
Distribution — Japan, Ryukyu, Indo-Pacific, Hawaii

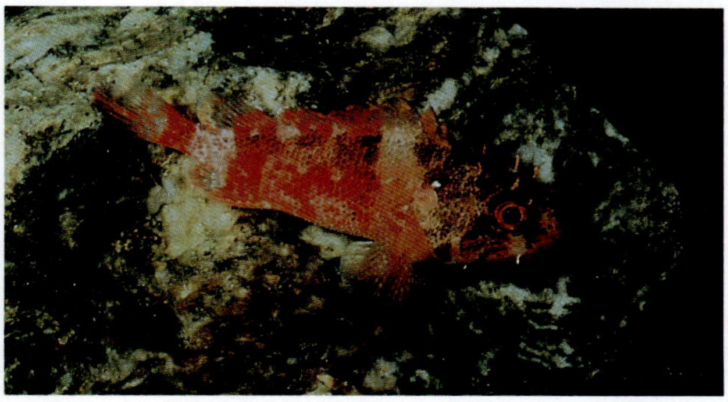

Scorpaenidae
Sebastiscus albofasciatus (Lacépède) ⇩
Yellowbarred Red Rockfish
Diagnosis — Inhabits rather deep shoreline reefs. Resembles *S. marmoratus*, but distinguished by yellowish-red body, 17 rays on pectoral fins. A spine is present under eyes. Dorsal fin has 12 spines and 12–13 rays, anal fin has 3 spines and 5 rays, spines are venomous. Edible. Up to 35cm in total length.
Distribution — Japan, Formosa, Korea, China, Hong Kong, South China Sea

Scorpaenidae
Scorpaenodes littoralis (Tanaka) ⇧
Shore Rockfish
Diagnosis — Inhabits shallow reefs and coral reefs. Dorsal fin has 13 spines and 9 rays, long spines are venomous. Anal fin has 3 spines and 5 rays. Lower ends of opercles have a dark blotch. Edible. Up to 15cm in total length.
Distribution — Japan, West Pacific

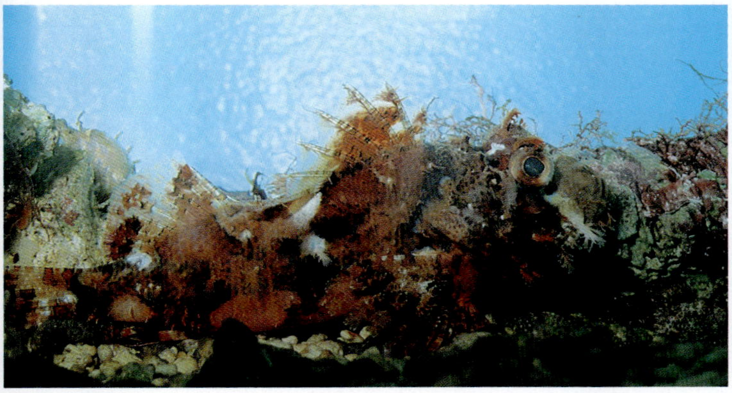

Scorpaenidae
Scorpaenopsis cirrhosa (Thunberg) ⇧ ⇨
[*S. cirrhosus* (Thunberg)]
[*Dendroscorpaena cirrhosa* (Thunberg)]
Hairy Stingfish, Scorpionfish, Poison Arm, Raggy Scorpionfish, Stingfish, Weedy Stingfish
Diagnosis — Inhabits shoreline reefs. Body is shallow with many small flaps, color varies according to habitat. Snout is slightly pointed, interorbital space is deeply depressed. Pectoral fins are not long enough to reach anal fin. Dorsal fin has 12 spines and 8–10 rays, anal fin has 3 spines and 5 rays. Spines are venomous. 23–25 pored lateral scales. Edible. Up to 30cm in total length.
Distribution — Japan, Ryukyu, Formosa, Korea, Indo-Pacific, Hawaii, Red Sea, South Africa

Scorpaenidae
Scorpaena onaria Jordan et Snyder ⇧ ⇦
[*S. neglecta neglecta* Temminck et Schlegel]
[*S. neglecta* Temminck et Schlegel]
[*S. fimbriata* Steindachner et Döderlein]

Spotsail Scorpionfish, Andaman Rockfish

Diagnosis — Inhabits rather deep shoreline reefs. Body is somewhat deep with flaps. Mouth is large, posterior end of upper jaw is just below posterior margin of eyes. Preoperculum have 6 spines. Dorsal fin has 12 spines and 8−10 rays, has a black blotch in male, spines are venomous. Anal fin has 3 spines and 5 rays. 23−24 pored lateral scales. Spawns between summer and fall. Edible. Up to 30cm in total length.
Distribution — Japan, Ryukyu, Formosa, Korea, Indo-West Pacific

Scorpaenidae
Rhinopias frondosa (Günther) ⇩

Goose Scorpionfish

Diagnosis — Inhabits rather deep shoreline reefs. Body is deep, varies widely in color. Body and fins have many flaps. Snout is projected, mouth is directed upward. Dorsal fin has 12 spines and 9 rays, anal fin has 3 spines and 5 rays. Spines are venomous. Up to 25cm in total length.
Distribution — Japan, Indo-West Pacific

Scorpaenidae
Apistus carinatus (Bloch et Schneider) ⇧
[*A. evolans* Jordan et Starks]
[*Hypodytes carinatus* (Bloch et Schneider)]

Ocellated Waspfish, Beared Waspfish, Scorpionfish

Diagnosis — Inhabits muddy-sand bottoms of shoreline waters. Pectoral fins are remarkably long with 1 separated ray on lower end. Lower jaw has 3 barbels. Dorsal fin has 14−15 spines and 8−10 rays, spines are venomous and have a large black blotch. Anal fin has 3 spines and 7−8 rays. Up to 15cm in total length.
Distribution — Japan, Ryukyu, Formosa, Korea, China, Indo-Pacific, Persian Gulf, Red Sea, South Africa

Scorpaenidae
⇦ *Helicolenus hilgendorfi* (Steindachner et Döderlein) ⇧

Hilgendorf Saucord

Diagnosis — Inhabits muddy-sand bottoms more than 100m deep. Resembles *Sebastiscus marmoratus*, but pectoral fin bases have flaps. Opercles are blackish. Caudal margin is truncated. White-margined dorsal fin has 12 spines and 11−13 rays, anal fin has 3 spines and 4−6 rays. Spines are venomous. 25−30 lateral scales. Edible. Up to 20cm in total length.
Distribution — Japan, Ryukyu, Korea

Scorpaenidae
⇦ ***Taenianotus triacanthus***
Lacépède ⇨

[*Taenionotus triacanthus* Lacépède]

Paperfish, Leaf Scorpionfish

Diagnosis — Inhabits outer slopes of coral reefs. Body is compressed, sparsely covered with spines, color varies widely. A pair of flaps are present on upper end of eyes. Well-developed dorsal fin has 12 spines and 10 rays, spines are venomous. Anal fin has 3 spines and 6 rays. Casts off the skin. Up to 10cm in total length.

Distribution — Japan, Ryukyu, Indo-Pacific, Hawaii, South Africa

Scorpaenidae
⇦ ***Parapterois heterurus*** (Bleeker)

Blackfoot Firefish, Threadtail Firefish

Diagnosis — Body is small. No flaps on upper part of eyes. Blackish pectoral fins are not long enough to reach caudal fin. Caudal fin is slender, lobes extended. Dorsal fin has 13 spines and 9 rays, anal fin has 3 spines and 7–8 rays. Spines are venomous. Aquarium fish. Up to 15cm in total length.

Distribution — Japan, Formosa, Indo-West Pacific

Scorpaenidae
⇧ ***Dendrochirus brachypterus*** (Cuvier)

[*Brachirus brachypterus* (Cuvier et Valenciennes)]

Shortfin Lionfish, Scorpionfish, Butterflyfish, Hawaiian Turkeyfish

Diagnosis — Inhabits shallow coral reefs. Resembles *D. zebra*, but distinguished by shorter flaps on upper surface of eyes. Pectoral rays are not extended. Dorsal fin has 13 spines and 9–10 rays, anal fin has 3 spines and 6 rays. Spines are venomous. Aquarium fish. Up to 18cm in total length.

Distribution — Indo-Pacific, Hawaii, Red Sea, South Africa

Scorpaenidae
⇦ ***Dendrochirus zebra*** (Quoy et Gaimard) ⇧

[*Brachirus zebra* (Cuvier et Valenciennes)]

Zebra Firefish, Scorpionfish, Dwarf Lionfish, Zebra Turkeyfish, Butterflyfish, Zebra Butterfly-cod

Diagnosis — Inhabits singly shallow coral reefs. Body has many dark bands. Dorsal, anal, caudal fins have black spots. Pectoral fins are developed with continuous fin membranes. Long flaps are present on upper surface of eyes. Dorsal fin has 13 spines and 10–11 rays, anal fin has 3 spines and 6–7 rays. Spines are venomous. 45–54 lateral scales. Aquarium fish. Up to 30cm in total length.

Distribution — Japan, Ryukyu, Bonin Islands, Indo-Pacific, South Africa

Scorpaenidae
Pterois radiata Cuvier ⇨

[*Pteropterus radiatus* (Cuvier et Valenciennes)]

Longhorn Firefish, Radial Firefish, Clearfin Turkeyfish

Diagnosis — Inhabits shallow coral reefs. Body is reddish-brown with white stripes. Extended dorsal spines are venomous. Pectoral rays are remarkably long. Fine flaps are present on upper surface of eyes. Fins have no blotches. Dorsal fin has 12 spines and 11 rays, anal fin has 3 spines and 6 rays. Aquarium fish. Up to 25cm in total length.

Distribution — Southern Japan, Ryukyu, Indo-Pacific, Hawaii, Red Sea, South Africa

Scorpaenidae
⇦ *Pterois lunulata*
Temminck et Schlegel

Lion Fish, Pink Firefish, Butterflyfish

Diagnosis — Inhabits shoreline reefs and coral reefs. Body has brown cross bands. Venomous dorsal spines are well-developed with falcated fin membranes. Flaps on upper surface of eyes are short. Pectoral fins are long, spreaded like wing. Dorsal fin has 13 spines and 11 rays, anal fin has 3 spines and 7–8 rays. Carnivorous. Spawns in summer. Aquarium fish. Up to 30cm in total length.

Distribution — Japan, Ryukyu, Korea, Indo-Pacific, Red Sea, East Africa

▼Young Adult

Scorpaenidae
⇦ *Pterois volitans* (Linnaeus) ⇨

Red Firefish, Zebrafish, Butterfly Cod, Winged Firefish, Poison Volant, Loffe Volant, Featherfish, Fireworkfish, Stingfish, Devil Firefish, Lionfish, Scorpion-cod

Diagnosis — Resembles *P. lunulata*, but distinguished by longer flaps on upper surface of eyes. Snout is somewhat slender, has flaps. Pectoral fins are long with falcated fin membranes. Venomous dorsal fin has 13 spines and 10–11 rays, anal fin has 3 spines and 6–7 rays. Aquarium fish. Up to 35cm in total length.

Distribution — Japan, Ryukyu, Bonin Islands, Formosa, Indo-Pacific, Persian Gulf, Red Sea, South Africa

▲Adult

Scorpaenidae
⇙ *Pterois antennata* (Bloch) ⇩

[*Pteropterus antennatus* (Bloch)]

Fireworkfish, Devilfish, Sailyard Firefish, Broadbarred Firefish, Ragged-finned Butterfly-cod, Rough-scaled Turkeyfish

Diagnosis — Inhabits shallow coral reefs. Resembles *Dendrochirus zebra*, but distinguished by long-extended rays on pectoral fins, dark spots on fin membranes. Flaps on upper surface of eyes are long. Venomous dorsal fin has 12 spines and 11 rays, anal fin has 3 spines and 6 rays. Aquarium fish. Up to 30cm in total length.

Distribution — Japan, Ryukyu, Bonin Islands, China, Indo-Pacific, South Africa

Synanceiidae

Inimicus japonicus (Cuvier)

Devil Stinger, Lumpfish, Poison Fish

Diagnosis — Inhabits shoreline reefs and approximately 200m deep waters. Body color varies according to depth. Mouth is directed upward. Anterior and posterior ends of eyes are depressed. Lower part of pectoral fins are separated. Venomous dorsal fin has 16–18 spines and 5–8 rays, anal fin has 2 spines and 8–10 rays. Spawns between June and July. Edible. Up to 25cm in total length.

Distribution — Japan, Ryukyu, Formosa, Korea, China, Hong Kong, South China Sea

Synanceiidae

Erosa erosa (Langsdorf)

Monkeyfish, Daruma Stinger

Diagnosis — Inhabits shoreline reefs and sandy bottoms. Head is developed with many warts. Body has no scales. Venomous dorsal fin has 14 spines and 5–6 rays, anal fin has 3 spines and 5–6 rays. Up to 15cm in total length.

Distribution — Japan, Ryukyu, Korea, West Pacific

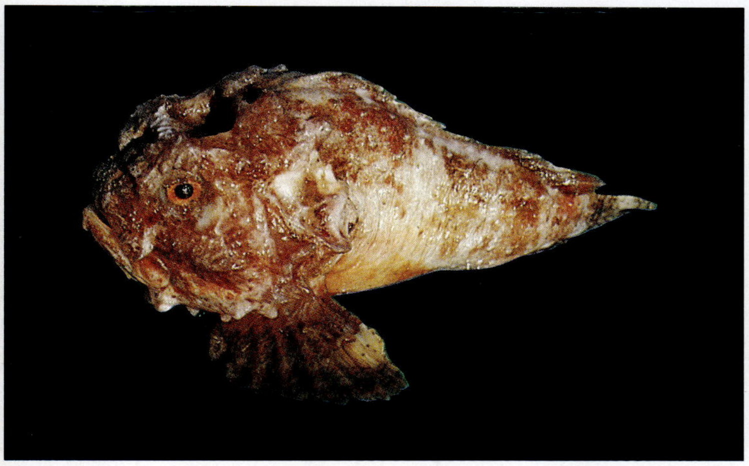

▼ Dorsal spines are virulently venomous.

Synanceiidae

Synanceia verrucosa Bloch et Schneider

[*S. thersites* Seale]

[*S. horrida* Meyer]

[*Synanceichthys verrucosa* (Bloch et Schneider)]

Poison Toadfish, Stonefish, Poison Scorpionfish, Devilfish, Goblinfish, Poison Fish, Reef Stonefish

Diagnosis — Inhabits shallow shoreline reefs and coral reefs. Head and pectoral fins are developed. Body is covered with warts. Mouth is large, directed upward. Dorsal fin has 12–14 spines and 5–7 rays, anal fin has 3 spines and 5–6 rays. Spines are virulently venomous. Nocturnal. Feeds on small fish. Edible. Up to 40cm in total length.

Distribution — Ryukyu, Indo-Pacific, Hawaii, Red Sea, South Africa

Synanceiidae

Minous quincarinatus (Fowler) ⇨

[*M. inermis* Alcock]

Whitetail Goblinfish

Diagnosis — Inhabits rather deep muddy-sand bottoms. Body is somewhat deep, back has dark oblique stripes. Pectoral and anal fins are black, caudal fin is pale. Scales are absent. Venomous dorsal fin has 8–9 spines and 12–14 rays, first and second spines are equal in length. Anal fin has 2 spines and 8–10 rays. Up to 15cm in total length.

Distribution — Southern Japan, Formosa, Indo-West Pacific

Synanceiidae

⇦ ***Minous pusillus*** Temminck et Schlegel

Puny Goblinfish

Diagnosis — Resembles *M. quincarinatus*, but distinguished by soft spines on dorsal fin. Body has no scales. Venomous dorsal fin has 9–11 spines and 9–11 rays, anal fin has 2 spines and 8–9 rays. Up to 7cm in total length.

Distribution — Japan, Formosa, Hong Kong, Philippines

Triglidae

⇦ ***Chelidonichthys spinosus*** McClelland ⇧ ⇩

[*C. kumu* (Lesson et Garnot)]

Bluefin Searobin, Gurnard

Diagnosis — Inhabits rather deep muddy-sand bottoms. Body is extended. Caudal margin is emarginated. Rostral projections are rounded. 3 lower rays of pectoral fins are separated and form legs. Pectoral fins are profusely dotted with blue on inside surface. First dorsal fin has 9–10 rays, second has 15–16 rays. Anal fin has 14–16 rays. 67–71 lateral scales. Spawns in summer. Edible. Up to 50cm in total length.

Distribution — Japan, Ryukyu, Formosa, Korea, West Pacific, South Africa

Triglidae

Lepidotrigla microptera Günther ⇧ ⇨

Gurnard, Redwing Searobin

Diagnosis — Inhabits bays and muddy-sand bottoms about 200m deep. Body is slightly extended. A pair of spinous rostral projections are present on tip of snout. In pectoral fins, 3 lower spines are separated, form legs used in walking, dark red on inside surface. First dorsal fin has 8–10 spines, a large red blotch, second has 16–17 rays. Anal fin has 16–17 rays. 64–67 lateral scales. Spawns in spring. Edible. Up to 40cm in total length.

Distribution — Japan, Formosa, Korea, China, Yellow Sea, South China Sea

Triglidae

Lepidotrigla japonica (Bleeker) ⇧

Longwing Searobin

Diagnosis — Resembles *L. microptera*, but distinguished by remarkably longer blackish pectoral fins. Scales on back are ctenoid. First dorsal fin has 8 spines and 9 rays, a red blotch, second has 14–15 rays. Anal fin has 13–14 rays. 56–60 lateral scales. Spawns in spring. Edible. Up to 20cm in total length.

Distribution — Japan, Formosa, China, South China Sea, Indonesia

Triglidae

Lepidotrigla punctipectoralis Fowler ⇧

Spottyfin Searobin

Diagnosis — Resembles *L. microptera*, but body is somewhat dark. Caudal fin has 2 dark bands. Pectoral fins are profusely dotted with blue. First dorsal fin has 8–9 spines, second has 10–16 rays. Anal fin has 14–16 rays. 61–64 lateral scales. Edible. Up to 20cm in total length.

Distribution — Southern Japan, Ryukyu, Formosa, South China Sea, Philippines

Triglidae

Lepidotrigla kishinouyei Snyder ⇩

[*L. smithi* Regan]

Devil Searobin

Diagnosis — Resembles *L. microptera*, but distinguished by a large black blotch, profusely dotted with white inside the pectoral fins. Scales on back are ctenoid. First dorsal fin has 8–9 spines and a red blotch, second has 14–16 rays. Anal fin has 14–16 rays. 58–62 lateral scales. Spawns in winter. Edible. Up to 20cm in total length.

Distribution — Japan, Formosa, China, East China Sea

Triglidae

Lepidotrigla alata (Houttuyn) ⇩

[*Pachytrigla alata* (Houttuyn)]

Forksnout Searobin

Diagnosis — Resembles *L. microptera*, but distinguished by large rostral projections on snout, greenish-brown inside of pectoral fins. First dorsal fin has 8–10 rays, second has 15–16 rays. Anal fin has 15–16 rays. 59–62 lateral scales. Edible. Up to 25cm in total length.

Distribution — Japan, Formosa, Korea, China, South China Sea, Australia

Triglidae

⇦ ***Pterygotrigla hemisticta*** (Temminck et Schlegel)

Spottyback Searobin

Diagnosis — Inhabits muddy-sand bottoms approximately 200m deep. Resembles *Lepidotrigla microptera*, but distinguished by blunt body with pair of rostral projections. Back is profusely dotted with black. Pectoral fins are black inside. First dorsal fin has 7 spines and a black blotch, second has 10–12 rays. Anal fin has 10–12 rays. 62–72 lateral scales. Edible. Up to 30cm in total length.

Distribution — Japan, Ryukyu, Formosa, China, South China Sea, Indian Ocean

▼Female

Callionymidae

Repomucenus richardsonii (Bleeker) ⇨

[*Callionymus punctatus* Langsdorff]

[*C. richardsoni* Bleeker]

[*C. valenciennesi* (Temminck et Schlegel)]

Richardson's Dragonet, Richardson Dragonets

Diagnosis — Inhabits rather deep muddy-sand bottoms. Body is flat and extended, caudal fin also extended. Mouth is small and pointed. Preoperculum have venomous spines. Body is scaleless. First dorsal fin has 4 spines, black margined in male, has a black blotch in female. Second has 9 rays, anal fin has 9 rays. Spawns between August and October. Feeds on bivalve and crab. Edible. Up to 20cm in total length.

Distribution — Japan, Formoa, Korea, South China Sea

▼Male▶

Callionymidae

⇦⇧ ***Repomucenus huguenini*** (Bleeker)

[*Callionymus doryssus* (Jordan et Fowler)]

[*Calliurichthys doryssus* (Jordan et Fowler)]

Spear Dragonet

Diagnosis — Inhabits rather deep sandy bottoms. Body is flat and extended. Snout is pointed. First dorsal fin has 4 spines, extended like thread in male, second has 9 rays, profusely dotted with black. Anal fin has 9 rays, margin is blackish. Up to 23cm in total length.

Distribution — Japan, Ryukyu, Formosa, East China Sea

Callionymidae

Calliurichthys japonicus (Houttuyn) ⇨

[*Callionymus japonicus* Houttuyn]

Japanese Dragonet

Diagnosis — Inhabits shoreline waters and sandy bottoms some 200m deep. Body is flat. Caudal fin is remarkably long, black on lower end. Snout is pointed, eyes are protruded. First dorsal fin has 4 spines and a black blotch, second has 9 rays. Anal fin has 8 black rays. Up to 30cm in total length.

Distribution — Japan, Formosa, China, Indo-West Pacific

Callionymidae

Pterosynchiropus splendidus (Herre) ⇩

[*Synchiropus splendidus* (Herre)]

Mandarin-fish

Diagnosis — Inhabits shoreline reefs and coral reefs. Body is cylindrical. Head and body have distinctive blotches. Snout is short, eyes are protruded. Pectoral rays are many (30). Caudal fin is large with blue margin. First dorsal fin has 4 spines, first spine is extended like thread in male, second has 8–9 rays. Anal fin has 7–8 rays. Skin secretes ichthyocrinotoxin. Aquarium fish. Up to 10cm in total length.

Distribution — Ryukyu, Formosa, West Pacific

Gobiesocidae

⇩ *Diademichthys lineatus* (Sauvage)

Long-snouted Clingfish

(New Common Name)

Diagnosis — Lives in symbiosis with *Diadema setosum* inhabiting shallow shoreline reefs and coral reefs. Body is elongated, has a yellow line. Snout is long. Dorsal and anal fins are situated close to caudal fin. Caudal fin is rounded, black, middle part yellow. Up to 6cm in total length.

Distribution — Japan, Ryukyu, Indo-West Pacific

Callionymidae

⇧ *Synchiropus picturatus* (Peters)

Spotted Mandarin Fish, Psychedelic Fish

Diagnosis — Inhabits shallow shoreline reefs and coral reefs. Resembles *Pterosynchiropus splendidus*, but distinguished by ocellated blotches on body, also by dorsal and anal fin margins edged with yellow and blue. First dorsal fin has 4 spines, second has 9 rays. Anal fin has 8 rays. Aquarium fish. Up to 10cm in total length.

Distribution — West Pacific

▲ Male

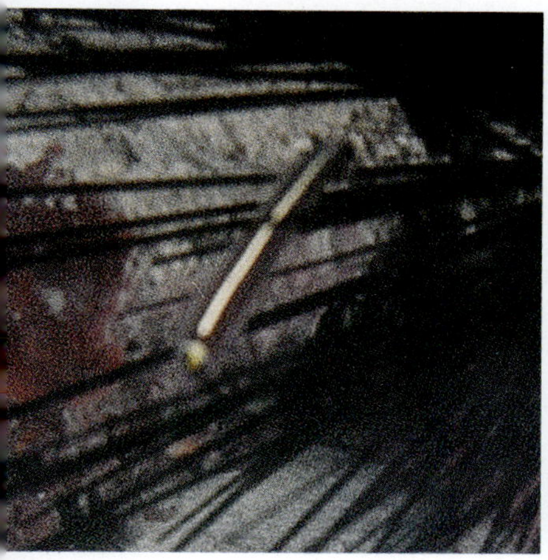

Bothidae

⇦ *Bothus pantherinus* (Rüppell)

Leopard Flounder, Mottled Flounder, Panther Flounder, Spotted Flounder

Diagnosis — Inhabits sandy bottoms about shallow coral reefs. Body is oval, covered closely with indistinct circular blotches. Eyes are large. Dorsal fin has 88–93 rays, anal fin has 65–72 rays. Edible. Up to 45cm in total length.

Distribution — Japan, Ryukyu, Indo-Pacific, Hawaii, Persian Gulf, Red Sea, South Africa

Bothidae

Bothus myriaster (Temminck et Schlegel) ⇨

Disc Flounder, Discoid Flounder, Oval Flounder, Starry Oval Flounder

Diagnosis — Inhabits shallow muddy-sand bottoms. Body is oval, rounded in young. Eyes are separated on left side. Scales are remarkably small. Pectoral fins are filiform. Dorsal fin has 87–97 rays, anal fin has 61–73 rays. Edible. Up to 25cm in total length.

Distribution — Southern Japan, Formosa, Indo-West Pacific, South Africa

Bothidae

⇦ *Bothus mancus* (Broussonet)

Maimed Flounder, Tropical Flounder

Diagnosis — Inhabits sandy bottoms about coral reefs. Body is oblong with scattered pale ocellated blotches. Eyes are small. Scales are ctenoid. Dorsal fin has 96–102 rays, anal fin has 74–81 rays. Edible. Up to 45cm in total length.

Distribution — Japan, Ryukyu, Indo-Pacific, Hawaii, Mexico, South Africa

Paralichthyidae
⇧ ***Pseudorhombus dupliciocellatus*** Regan

Doublespot Flounder, Four-spot Flounder
Diagnosis — Body is oblong, has several ocellated blotches on right side. Scales are ctenoid. Dorsal fin has 72–78 rays, anal fin has 51–63 rays. Edible. Up to 35cm in total length.
Distribution — Southern Japan, Formosa, Indo-West Pacific

Paralichthyidae
Pseudorhombus pentophthalmus Günther ⇧

Fivespot Flounder
Diagnosis — Inhabits rather deep muddy-sand bottoms. Body is oblong with many ocellated blotches. Scales are ctenoid. Dorsal fin has 71–76 rays, anal fin has 53–57 rays. Edible. Up to 20cm in total length.
Distribution — Japan, Ryukyu, Formosa, Korea, China, West Pacific

Paralichthyidae
Paralichthys olivaceus (Temminck et Schlegel) ⇩

Bastard Halibut, False Halibut, Flat Fish, Japanese Flounder, Olive Flounder
Diagnosis — Inhabits sandy bottoms more than 100m deep. Body is oblong. Mouth is large, has canine-like teeth. Scales are small. Dorsal fin has 77–81 rays, anal fin has 59–61 rays. Feeds on fish. Spawns in shallow waters between spring and summer. Edible. Up to 90cm in total length.
Distribution — Japan, Korea, China, Sakhalin, Yellow Sea, South China Sea

Psettodidae
Psettodes erumei (Schneider) ⇩

Indian Halibut, Queensland Halibut
Diagnosis — Body is oval with 4 indistinct dark cross bands. Mouth is large, teeth are sharp-pointed. Eyes are situated on left side or on right side. Scales are ctenoid. Dorsal fin has 51–54 rays, anal fin has 40–42 rays, anterior rays are acanthoid. Edible. Up to 40cm in total length.
Distribution — Formosa, China, Indo-West Pacific, Red Sea, South Africa

Paralichthyidae
⇧ ***Pseudorhombus arsius*** (Hamilton)

False Brill, Large-toothed Flounder, Smooth Scaled Brill
Diagnosis — Inhabits shallow sandy bottoms. Body is oval, profusely dotted with black and white. Scales are ctenoid. Dorsal fin has 74–78 rays, anal fin has 57–60 rays. Edible. Up to 35cm in total length.
Distribution — Japan, Ryukyu, Formosa, Indo-Pacific, Persian Gulf, South Africa

Pleuronectidae
Limanda punctatissima (Steindachner) ⇩

Longsnout Flounder, Sand Flounder
Diagnosis — Inhabits shallow sandy bottoms. Mouth is large, snout is pointed. Bases of fins on left side are yellow. Dorsal fin has 52–67 rays, anal fin has 42–50 rays. Spawns between May and June. Edible. Up to 35cm in total length.
Distribution — Japan, Korea, Sakhalin, Okhotsk Sea

Pleuronectidae

⇦ ***Limanda schrenki*** Schmidt

Cresthead Flounder

Diagnosis — Inhabits rather deep muddy-sand bottoms. Body is dark-brown on right side, white on left side. Dorsal fin has 56–73 rays, anal fin has 42–56 rays, both have black blotches. Spawns in shoreline waters between February and June. Edible. Up to 50cm in total length.

Distribution — Northern Japan, Korea, Yellow Sea, Sakhalin, Okhotsk Sea

Pleuronectidae

Kareius bicoloratus (Basilewsky) ⇨ ⇩

Stone Flounder

Diagnosis — Inhabits rather deep muddy-sand bottoms. Body is scaleless. In adult, body has 2 or 3 rows of rugose bony plates on right side. Lateral lines are straight. Dorsal fin has 62–75 rays, anal fin has 45–55 rays. Feeds on bivalve and crustacea. Spawns between November and February. Edible. Up to 55cm in total length.

Distribution — Japan, Formosa, Korea, China, Sakhalin

Pleuronectidae

⇧ ***Verasper moseri*** Jordan et Gilbert

Barfin Flounder

Diagnosis — Inhabits muddy-sand bottoms about 200m deep. Dorsal, anal, caudal fins have black blotches. In male, body is orangish-yellow on left side; white in female. Dorsal fin has 76–87 rays, anal fin has 53–68 rays. Feeds on fish and crustacea. Spawns between November and February. Edible. Up to 75cm in total length.

Distribution — Japan, Okhotsk Sea, Sakhalin

Pleuronectidae

Pleuronichthys cornutus (Temminck et Schlegel) ⇨

Finespotted Flounder, Frog Flounder

Diagnosis — Inhabits rather deep muddy-sand bottoms of shoreline waters. Body is yellowish-brown on right side with irregular dark blotches, left side is white. Mouth is small. Eyes are protruded. Dorsal fin has 67–83 rays, anal fin has 49–62 rays. Feeds on polychaeta and crustacea. Spawns between September and November. Edible. Up to 35cm in total length.

Distribution — Japan, Formosa, Korea, China, Yellow Sea, South China Sea

Pleuronectidae

⇦ ***Limanda yokohamae*** (Günther)

Marbled Sole

Diagnosis — Inhabits rather deep muddy-sand bottoms. Body is yellowish on right side, profusely dotted with black and white. Lateral line is slightly curved above pectoral fin. Interorbital space is protruded, covered with scales. Dorsal fin has 61–73 rays, anal fin has 48–62 rays. Feeds on crustacea and polychaeta. Spawns in winter. Edible. Up to 50cm in total length.

Distribution — Japan, Korea, Yellow Sea, East China Sea

Pleuronectidae

Eopsetta grigorjewi (Herzenstein) ⇨

Roundnose Flounder, Shothole Halibut, Shotted Halibut

Diagnosis — Inhabits rather deep muddy-sand bottoms. Body is oblong, has many ring-like blotches. Mouth is large. Lateral line is curved above pectoral fins. Dorsal fin has 81–94 rays, anal fin has 61–78 rays. Feeds on fish, squid and crustacea. Spawns between January and June. Edible. Up to 50cm in total length.

Distribution — Japan, Formosa, Korea, Yellow Sea, East China Sea

Pleuronectidae

⇦ ***Platichthys stellatus*** (Pallas) ⇨

California Flounder, Emerywheel, Great Flounder, Grindstone, Long-jaw Flounder, Starry Flounder

Diagnosis — Inhabits shallow waters, brackish areas in river-mouths, freshwater areas in river or lakes. Eyes are situated on left side in Japanese native; on right side in North American native. Dorsal, anal and caudal fins have black bands. Dorsal fin has 52–68 rays, anal fin has 36–51 rays. Feeds on shellfish and crustacea. Spawns between February and March. Edible. Up to 90cm in total length.

Distribution — Japan, Korea, Okhotsk Sea, Bering Sea, North America

Pleuronectidae

⇦ ***Hippoglossus stenolepis*** Schmidt

Butt, Chicken Halibut, Halibut, Common Halibut, Pacific Halibut

Diagnosis — Inhabits deep waters more than 200m deep. Largest of all flatfishes, measuring 2.5m in total length and weighing 250kg. Mouth is large, teeth are sharp-pointed. Lateral lines are curved above pectoral fins. Dorsal fin has 89–109 rays, anal fin has 64–81 rays. Feeds on fish. Spawns between May and June. Edible.

Distribution — Japan, Okhotsk Sea, Bering Sea, North America

Pleuronectidae

Samaris cristatus Gray ⇩

Gray's Crested Flounder, Right-hand Flounder

Diagnosis — Inhabits rather deep muddy-sand bottoms. Body is oblong and dark-brown. 10th to 15th rays of dorsal fin are extended like thread. Anterior part of pelvic fins are also extended. Dorsal fin has 73–86 rays, anal fin has 49–60 rays. Edible. Up to 23cm in total length.

Distribution — Southern Japan, Formosa, China, Indo-West Pacific, South Africa

Cynoglossidae

⇧ ***Arelia bilineata*** (Lacépède)

[*Cynoglossus bilineatus* (Lacépède)]

Four-lined Tongue-sole, Fourline Tonguefish, Two-lined Tongue Sole

Diagnosis — Inhabits rather deep muddy-sand bottoms. Snout hangs as far down as below eyes. Both sides have 2 lateral lines. Scales are ctenoid. Dorsal fin has 110–119 rays, anal fin has 86–95 rays. Feeds on polychaeta and crustacea. Spawns in winter. Edible. Up to 50cm in total length.

Distribution — Japan, Ryukyu, Indo-West Pacific, Red Sea, South Africa

Cynoglossidae

Paraplagusia japonica (Temminck et Schlegel) ⇨

Black Cow Tongue, Black Tonguefish

Diagnosis — Inhabits muddy-sand bottoms of shoreline waters. Body is large, blackish-brown on right side, white on left side. Snout hangs down, barbels are present around mouth. Right side has 3 lateral lines, no lateral line on left side. Dorsal fin has 106–119 rays, anal fin has 85–97 rays. Feeds on bivalve and crustacea. Spawns in summer. Edible. Up to 40cm in total length.

Distribution — Japan, Formosa, South China Sea

Cynoglossidae

⇦ ***Cynoglossus robustus*** Günther

Robust Tonguefish, Tongue Fish

Diagnosis — Inhabits rather deep muddy-sand bottoms. Scales are ctenoid, easily removable. Eyes are small. Mouth is situated remarkably posterior. 2 lateral lines on left side. Dorsal fin has 122–138 rays, anal fin has 93–107 rays. Spawns between July and August. Edible. Up to 45cm in total length.

Distribution — Southern Japan, Formosa, China, Yellow Sea, South China Sea

Soleidae

Aesopia cornuta Kaup ⇨

[*Coryphaesopia cornuta* (Kaup)]

[*Synaptura cornuta* (Kaup)]

Horned Sole

Diagnosis — Inhabits muddy-sand bottoms about 100m deep. Resembles *Zebrias zebra*, but distinguished by indistinct dark bands on body, long-extended first ray on dorsal fin. Orange blotches on caudal fin are indistinct. Dorsal fin has 70–78 rays, anal fin has 57–68 rays. 93–105 lateral scales. Spawns between spring and summer. Edible. Up to 21cm in total length.

Distribution — Japan, Formosa, Indo-West Pacific, South Africa

Soleidae

⇦ ***Zebrias zebra*** (Schneider)

Blend Banded Sole, Striped Sole, Zebra Sole

Diagnosis — Inhabits shoreline waters and muddy-sand bottoms about 100m deep. Body has 12 pairs of dark cross bands. Dorsal and anal fins are connected with caudal fin. Caudal fin is black with orange spots. Dorsal fin has 68–82 rays, anal fin has 56–70 rays. 87–110 lateral scales. Spawns in fall. Feeds on crustacea. Edible. Up to 25cm in total length.

Distribution — Japan, Ryukyu, Formosa, Indo-Pacific, Persian Gulf

▼ Hide in sand for hunting food.

Soleidae

Pardachirus pavoninus (Lacépède) ⇨ ⇩

[*Achirus pavoninus* Lacépède]

Peacock Sole, Broad Sole

Diagnosis — Inhabits sandy bottoms among coral reefs. Body is oblong with many scattered ocellated blotches. Pectoral fin is absent. Dorsal and anal fins are not connected with caudal fin. Eyes are small, close to each other. Bases of dorsal and anal fins have venom glands which secrete ichthyotoxin. Dorsal fin has 66–70 rays, anal fin has 50–53 rays. 75–82 lateral scales. Up to 25cm in total length.

Distribution — Japan, Ryukyu, Formosa, Indo-West Pacific

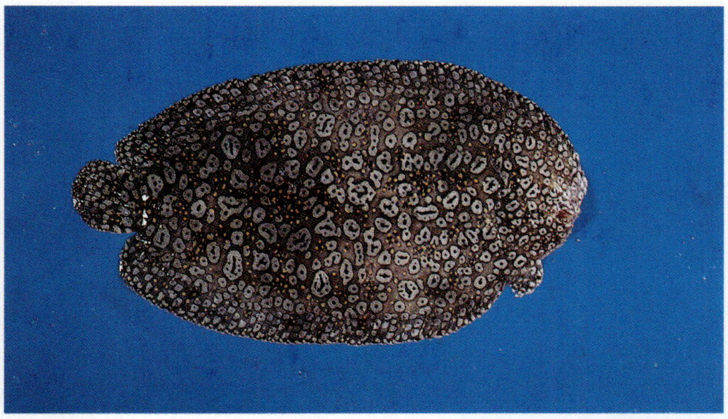

Soleidae

Liachirus melanospilus (Bleeker) ⇨

Blackspotted Sole

Diagnosis — Body is oblong with dark blotches. Dorsal and anal fins are not connected with caudal fin. Pectoral fins are absent. Dorsal fin has 59–62 rays, anal fin has 42–47 rays. 57–65 lateral scales. Edible. Up to 18cm in total length.

Distribution — Southern Japan, Formosa, Indo-Pacific

Soleidae

Pseudaesopia japonica (Bleeker) ⇧

Seto Sole, Wavybanded Sole

Diagnosis — Resembles *Zebrias zebra*, but distinguished by slightly indistinct dark bands on body, dorsal and anal fins not continuous with caudal fin. Caudal margin is black. Dorsal fin has 71–81 rays, anal fin has 59–67 rays. 83–98 lateral scales. Edible. Up to 20cm in total length.

Distribution — Japan, Ryukyu, Formosa, East China Sea

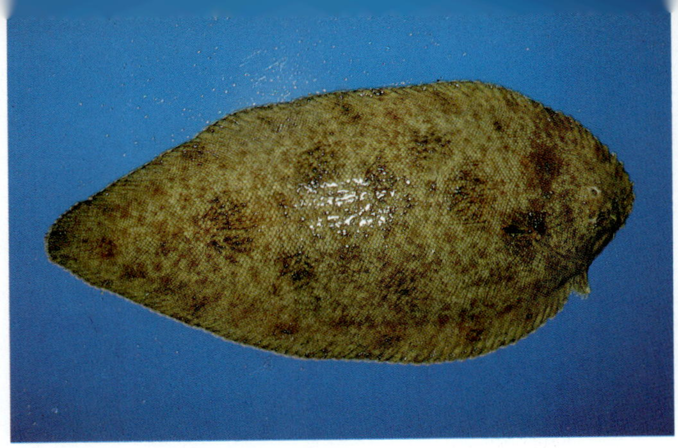

Soleidae

Synaptura marginata Boulenger

Marginated Sole, Shallow Water Sole, Sole, Tong

Diagnosis — Inhabits sandy bottoms about coral reefs. Body is oblong with indistinct brown blotches. Many fleshy flaps are present on body. Scales are ctenoid. Dorsal and anal fins are connected with caudal fin. Dorsal fin has 71–74 rays, anal fin has 56–57 rays. 103–109 lateral scales. Spawns in summer. Edible. Up to 50cm in total length.

Distribution — Ryukyu, South Africa

▲ Hide in sand for hunting food.

Balistidae

Balistoides viridescens (Schneider)

[*Balistes viridescens* Bloch et Schneider]

[*B. nigromarginatus* (Tanaka)]

[*Pseudobalistes viridescens* (Bloch et Schneider)]

Blackedged Triggerfish, Blue-fin Filefish, Blue-finned Triggerfish, Dotty Triggerfish, Leatherjacket, Titan Triggerfish

Diagnosis — Inhabits coral reefs. Resembles *Pseudobalistes flavimarginatus*, but distinguished by black fin margins, a black band on snout. Caudal peduncles are pale with 5–6 rows of small spines. Spawns in summer. First dorsal fin has 3 spines, second has 25–26 rays. Anal fin has 23–25 rays. Edible but may cause ciguatera poisoning. Up to 60cm in total length.

Distribution — Japan, Ryukyu, Indo-Pacific, Red Sea, South Africa

Balistidae

Balistoides conspicillum (Schneider)

[*Balistes conspicillum* Bloch et Schneider]

Bigspotted Triggerfish, Spotted Skinpeeler, Spotted Triggerfish, Clown Triggerfish, Yellow-blotched Triggerfish

Diagnosis — Inhabits singly coral reefs. Snout is pointed. Margin of mouth is orange. Abdominal side has 3–4 rows of white blotches. A yellow band is present under eyes. Caudal peduncles have 3 rows of small spines. First dorsal fin has 3 foldable spines, second has 24–26 rays. Anal fin has 21–23 rays. Edible but may cause ciguatera poisoning. Aquarium fish. Up to 50cm in total length.

Distribution — Japan, Ryukyu, Formosa, Korea, China, Indo-Pacific, Mexico, Red Sea, South Africa

Balistidae

⇦ ***Abalistes stellaris*** (Schneider)

[*A. stellatus* (Lacépède)]

[*A. stellatus* (Bloch et Schneider)]

Flattail Triggerfish, Triggerfish, Leatherjacket, Starry File-fish, Starry Triggerfish

Diagnosis — Inhabits rather deep coral reefs and sandy bottoms. Body is oblong with indistinct pale blotches. Caudal peduncles are slender. Caudal lobes are extended like thread. First dorsal fin has 3 spines, second has 25−27 rays. Anal fin has 24−25 rays. Edible but may cause ciguatera poisoning. Up to 70cm in total length.

Distribution — Japan, Ryukyu, Formosa, China, Indo-Pacific, Persian Gulf, Red Sea, South Africa

Balistidae

Sufflamen bursa (Schneider) ⇨

[*Balistes bursa* Lacépède]

[*Hemibalistes bursa* (Bloch et Schneider)]

Drab Triggerfish, Whiteline Triggerfish, Triggerfish, Green and White Triggerfish

Diagnosis — Inhabits coral reefs. Resembles *S. fraenatus*, but distinguished by white stripes running from mouth to anus. 2 dark bands run from pectoral fins. A spine on pelvic fins are black. First dorsal fin has 3 spines, second has 28−29 rays. Anal fin has 25−26 rays. Edible. Up to 25cm in total length.

Distribution — Japan, Ryukyu, Bonin Islands, Formosa, Indo-Pacific, Hawaii, Red Sea, South Africa

Balistidae

⇦ ***Sufflamen chrysopterus*** (Bloch et Schneider)

[*Balistes chrysopterus* Bloch et Schneider]

[*Hemibalistes chrysoptera* (Bloch)]

Goldenfin Filefish, Black Triggerfish, Golden-finned Filefish, Golden-finned Triggerfish, Halfmoon Triggerfish, Triggerfish

Diagnosis — Inhabits coral reefs. Resembles *S. fraenatus*, but distinguished by absence of pale stripes running from mouth. In some individuals, a white stripe runs from eyes. Caudal margin is white, caudal peduncles have 6−8 rows of small spines. First dorsal fin has 3 spines, second has 26−28 rays. Anal fin has 23−26 rays. Edible but may cause ciguatera poisoning. Up to 30cm in total length.

Distribution — Japan, Ryukyu, Formosa, China, Indo-Pacific, Red Sea, South Africa

Balistidae

⇦ ***Sufflamen fraenatus*** (Latreille) ⇨

[*S. capistratus* (Shaw)]

[*Balistes fraenatus* Latreille]

[*B. mitis* Bennett]

Bridled Triggerfish, Triggerfish, Masked File-fish

Diagnosis — Inhabits coral reefs. Body is dark-brown. Snout is pointed. Pale stripes extended from mouth to posterior. Caudal peduncles have rows of small spines. First dorsal fin has 3 spines, second has 27−30 rays. Anal fin has 24−27 rays. Edible. Up to 50cm in total length.

Distribution — Japan, Ryukyu, Formosa, Indo-Pacific, Hawaii, Galapagos Islands, Red Sea, South Africa

Balistidae

Pseudobalistes fuscus (Bloch et Schneider) ⇖⇨

[*Balistes fuscus* Bloch et Schneider]

Brown File-fish, Triggerfish, Yellowspotted Triggerfish, Rippled Triggerfish, Brown Triggerfish, Blue Triggerfish

Diagnosis — Inhabits coral reefs. Body is brown or dark brown. In some individuals, body has orange spots. Head is rounded. Caudal lobes are extended, have no spines on caudal peduncles. First dorsal fin has 3 spines, second has 25−27 rays. Anal fin has 19−26 rays. Edible but may cause ciguatera poisoning. Up to 53cm in total length.

Distribution — Japan, Ryukyu, Formosa, Indo-Pacific, Hawaii, Red Sea, South Africa

▲ Young Adult

Balistidae
Melichthys buniva (Lacépède) ⇩
[*M. radula* Solander]
[*M. ringens* (Osbeck)]

Black Triggerfish

Diagnosis — Inhabits coral reefs. Body is oval with many yellowish-brown stripes. Stripes of same color extend radially from eyes. Caudal margin is orange. First dorsal fin has 3 spines, second has 31–33 rays. Anal fin has 28–30 rays. Edible. Up to 50cm in total length.

Distribution — China, Indo-Pacific, Hawaii, Mexico, South Africa

Balistidae
Pseudobalistes flavimarginatus (Rüppell) ⇖⇧
[*Balistes flavimarginatus* Rüppell]

Multicoloured Triggerfish, Yellowface Triggerfish, Triggerfish, Green Triggerfish, Yellowmargin Triggerfish

Diagnosis — Inhabits coral reefs. Body is oval. Surface between snout and pectoral side is light-yellow. Margins of dorsal, anal, caudal fins are yellowish. Teeth are white, 2 rows in upper jaw, 1 row in lower. Caudal peduncles have 5–6 rows of small spines. First dorsal fin has 3 spines, second has 24–27 rays. Anal fin has 23–25 rays. Edible but may cause ciguatera poisoning. Up to 60cm in total length.

Distribution — Japan, Ryukyu, Formosa, Indo-Pacific, Red Sea, South Africa

Balistidae
Melichthys vidua (Solander) ⇩⇘
[*Balistes vidua* Richardson]
[*B. vidua* Solander]

Pink Tail Triggerfish, Tattooed Triggerfish, Whitetail Triggerfish, White-tailed Triggerfish, Triggerfish

Diagnosis — Inhabits rather deep coral reefs. Body is dark-brown. Dorsal and anal fins are pale with black margins. Pectoral fins are yellow, caudal fin is reddish. First dorsal fin has 3 spines, second has 31–35 rays. Anal fin has 27–31 rays. Edible. Up to 37cm in total length.

Distribution — Japan, Ryukyu, Formosa, Indo-Pacific, Hawaii, South Africa

Balistidae
Melichthys indicus Randall et Klausewitz ⇨

Blackfinned Triggerfish

Diagnosis — Inhabits coral reefs. Body is oval. Dorsal and anal fins are black with white base. Caudal margin is black. 6 radial stripes extend upward, 2 extend downward from eyes. A dark stripe extends from mouth. Edible. Up to 30cm in total length.

Distribution — Indo-West Pacific, Red Sea

Balistidae
Balistapus undulatus (Mungo Park) ⇧⇨

Undulate Triggerfish, Red-lined Triggerfish, Undulate File-fish, Orangestriped Triggerfish, Rice Pounder, Vermiculated Triggerfish

Diagnosis — Inhabits outer slope of coral reefs. Body has orange oblique stripes. Caudal fin is yellow, caudal peduncles have 2 rows of small black margined spines. First dorsal fin has 3 spines, has a black blotch, second has 24–27 rays. Anal fin has 20–24 rays. Edible but has records of ciguatera poisoning. Aquarium fish. Up to 35cm in total length.

Distribution — Japan, Ryukyu, Formosa, China, Indo-Pacific, Red Sea, South Africa

Balistidae

Rhinecanthus rectangulus (Schneider) ⇨

[*R. echarpe* (Anonymous)]

[*Balistapus rectangulus* (Bloch et Schneider)]

[*B. echarpe* (Anonymous)]

Blackbanded Triggerfish, Wedgetailed Triggerfish

Diagnosis — Inhabits coral reefs. Body is oval. Snout is pointed. A broad black band is present between eyes and anal fin. Caudal peduncles are black with 4 rows of small spines. First dorsal fin has 3 spines, second has 23–26 rays. Anal fin has 20–21 rays. Edible but has records of ciguatera poisoning. Up to 30cm in total length.

Distribution — Japan, Ryukyu, Formosa, Indo-Pacific, Hawaii, Red Sea, South Africa

Balistidae

⇦ ***Rhinecanthus aculeatus*** (Linnaeus) ⇩

[*Balistapus aculeatus* (Linnaeus)]

Clear Picasso Triggerfish, Aculeate Triggerfish, Black Barred Triggerfish, Prickly File-fish, White-barred Triggerfish

Diagnosis — Inhabits shallow coral reefs. Quickly hides in small burrows when alarmed. Snout is long, a black band across the eyes. Caudal peduncles have 3 rows of black small spines. Abdominal side has 4 white oblique bands. Anus is black. First dorsal fin has 3 spines, second has 21–26 rays. Anal fin has 21–22 rays. Edible but has records of ciguatera poisoning. Aquarium fish. Up to 30cm in total length.

Distribution — Japan, Ryukyu, Formosa, China, Indo-Pacific, Hawaii, North America, Red Sea, South Africa

Balistidae

Canthidermis maculatus (Bloch) ⇨

[*C. rotundatus* (Procé)]

Ocean Triggerfish, Ocean Turbot, Rotund File Fish, Rough Triggerfish, Spotted Triggerfish, White-spotted Triggerfish

Diagnosis — Found in schools in middle layers of offshore waters. Body is oblong, profusely dotted with pale color. Dorsal and anal fin rays are triangular. A groove is present on surface anterior to eyes. First dorsal fin has 3 spines, second has 23–27 rays. Anal fin has 20–27 rays. Up to 42cm in total length.

Distribution — Japan, Ryukyu, Formosa, Indo-Pacific-Atlantic, Hawaii, South Africa

Balistidae

Xanthichthys mento (Jordan et Gilbert) ⇨

[*X. gotonis* Tanaka]

[*X. purus* Tanaka]

Blueface Triggerfish, Blue-throated Triggerfish, Long-chinned Trigger Fish

Diagnosis — Inhabits coral reefs. Body is oblong, snout is rounded. Cheeks have 5 dark grooves. Dorsal, anal, caudal fin margins are orange. First dorsal fin has 3 spines, second has 29–32 rays, anal fin has 25–29 rays, bases are black. Edible. Up to 30cm in total length.

Distribution — Japan, Izu Islands, Ryukyu, Indo-Pacific, Hawaii, Red Sea, South Africa

▲ Male

Balistidae

⇦ ***Odonus niger*** (Rüppell)

[*O. erythrodon* (Günther)]

Redtooth Triggerfish, Red-toothed Filefish, Redfang, Triggerfish

Diagnosis — Found in schools about outer slope of coral reefs. Body is bluish-purple, head is greenish. Caudal lobes are extended long, caudal peduncles have small processes. Teeth are canine-like and reddish. First dorsal fin has 3 spines, second has 32–36 rays. Anal fin has 28–31 rays. Edible but has records of ciguatera poisoning. Up to 50cm in total length.

Distribution — Japan, Ryukyu, Formosa, Indo-Pacific, Red Sea, South Africa

Balistidae

⇦ ***Xanthichthys lineopunctatus*** (Hollard)

[*X. ringens* (Linnaeus)]

[*Balistes ringens* Linnaeus]

Sargassum Triggerfish, Red-tail Triggerfish, Brown-lined Triggerfish

Diagnosis — Inhabits coral reefs. Resembles *X. mento*, but distinguished by 3 grooves on cheeks, dark stripes and spots on body. Caudal margin is truncated, has a broad orange band. First dorsal fin has 3 spines, second has 27–29 rays. Anal fin has 25–27 rays. Edible. Up to 30cm in total length.

Distribution — Izu Islands, Ryukyu, Indo-Pacific-Atlantic, Hawaii, Mexico, South Africa

▲ Male

Balistidae

⇦ ***Xanthichthys auromarginatus*** (Bennett) ⇨

Gilded Triggerfish

Diagnosis — Inhabits coral reefs. Resembles *X. mento*, but distinguished by processes on scales, forming rows of pale spots. Cheeks have 5 grooves. In male, cheeks are bluish, a broad yellow band on caudal margin. First dorsal fin has 3 spines, second has 27–30 rays. Anal fin has 25–27 rays. Edible. Up to 20cm in total length.

Distribution — Ryukyu, Indo-West Pacific, Hawaii

▼Female

Monacanthidae

⇦ ***Thamnoconus modestus*** (Günther) ⇩

[*Navadon modestus* (Günther)]

Black Scraper, Scraper, File Fish

Diagnosis — Found in schools in muddy-sand bottoms of shoreline waters. Body is compressed and oblong with thick skin. Snout is pointed, mouth is small. Rapidly changes color and blotch when alarmed. First dorsal fin has 1 long spine, second has 36–40 rays. Anal fin has 33–37 rays. Fins are bluish gray. Spawns between May and August. Edible. Up to 30cm in total length.

Distribution — Japan, Formosa, Korea, China, South China Sea, South Africa

Monacanthidae

Stephanolepis cirrhifer (Temminck et Schlegel) ⇨ ⇩

Threadsail Filefish, Filefish, Porky, Skin-peeler

Diagnosis — Inhabits shoreline reefs and sandy bottoms. Body is compressed and deep, color varies according to habitat. Mouth is small, snout is pointed. First dorsal fin has 1 short spine, second has 31–35 rays. Feeds on benthic invertebrate. Spawns between May and August. Edible. Up to 40cm in total length.

Distribution — Japan, Ryukyu, Formosa, China, Indo-West Pacific, East Africa

Monacanthidae

Rudarius ercodes Jordan et Fowler ⇨

Japanese File-fish, Net Work Filefish

Diagnosis — Inhabits seaweed bed of shoreline waters. Body is rhombic form and yellowish-brown. In male, caudal peduncles covered with a patch of bristles. First dorsal fin has 1 curved spine with developed fin membrane, second has 25–28 rays. Anal fin has 23–28 rays. Spawns on seaweed in summer. Up to 10cm in total length.

Distribution — Japan, Formosa, Korea

▲ Body has sharp-pointed spines.

Monacanthidae
Amanses scopas (Cuvier)

Brush-sided Leatherjacket, Broom Filefish

Diagnosis — Inhabits singly shoreline reefs and coral reefs. Body is oval. Resembles *Cantherhines dumerili*, but distinguished by posterior part of male body having a patch of 5−6 sharp-pointed spines. In female, these spines change to bristles. First dorsal fin has 1 large stiff spine, second has 26−29 rays. Anal fin has 22−25 rays. Edible. Up to 30cm in total length.

Distribution — Southern Japan, Ryukyu, Formosa, China, Indo-Pacific, Red Sea, South Africa

Monacanthidae
Cantherhines dumerili (Hollard)

Dumeril's File Fish, Brown Leather Jacket, Leopard Leatherjacket, Whitestar Filefish, Sandwich Islands Filefish, Wire-netting Leatherjacket

Diagnosis — Inhabits coral reefs. Body is oval and dark brown. Caudal fin is rounded with 2 rows of spines on caudal peduncles. First dorsal fin has 1 long spine, second has 33−38 rays. Anal fin has 29−33 rays. Edible but may cause ciguatera poisoning. Up to 40cm in total length.

Distribution — Japan, Ryukyu, Formosa, Indo-Pacific, Hawaii, Mexico

Monacanthidae
Paramonacanthus choirocephalus (Bleeker)

Pig-faced Leather-jacket

Diagnosis — Resembles *P. japonicus* of male, but body has 2 dark oblique bands, caudal fin has 2 dark bands. Pelvic fins have 1 spine, fin membrane is somewhat developed. Upper caudal rays are extended like thread. First dorsal fin has 1 long spine, second has 30 rays. Anal fin has 31 rays. Up to 10cm in total length.

Distribution — Formosa, Indo-West Pacific

▼Male

Monacanthidae
Cantherhines pardalis (Rüppell)

[*Amanses pardalis* (Rüppell)]

[*A. sandwichiensis* (Quoy et Gaimard)]

Leopard Leather Jacket, Spotted Leatherjacket, Wire-net Filefish

Diagnosis — Inhabits shoreline reefs. Body is oval, pale, has broad abdominal membrane and reticulated patterns. Caudal margin is yellow. First dorsal fin has 1 large stiff spine, second has 33−35 rays. Anal fin has 29−31 rays. Edible. Up to 30cm in total length.

Distribution — Japan, Ryukyu, Indo-Pacific, Red Sea, South Africa

▼Female

Monacanthidae
Paramonacanthus japonicus (Tilesius)

[*P. oblongus* (Temminck et Schlegel)]

[*Stephanolepis japonicus* (Tilesius)]

Filefish, Striped Filefish, Hair-finned Leatherjacket, Oblong Filefish

Diagnosis — Inhabits shoreline reefs and sandy bottoms. In male, body is elongated, upper caudal lobe is extended like thread. In female, body is oval, caudal margin is roundish. First dorsal fin has 1 spine, second has 24−30 rays. Anal fin has 24−30 rays. Up to 16cm in total length.

Distribution — Japan, Formosa, Indo-West Pacific

Monacanthidae

⇦ ***Aluterus monoceros*** (Linnaeus) ⇩

[*Aluteres monoceros* (Linnaeus)]

[*Alutera monoceros* (Linnaeus)]

File-fish, Unicorn Filefish, Leatherjacket, Unicorn Fish, Unicorn Leatherjacket, Yellow-finned Leatherjacket

Diagnosis — Found in schools in shoreline reefs and sandy bottoms. Body is remarkably compressed, oblong, gray with dark blotches on back. Head is roundish. Caudal peduncles are slightly long. Pelvic fins are absent. First dorsal fin has 1 slender spine, second has 46−52 rays. Anal fin has 48−52 rays. Edible but may cause ciguatera poisoning. Up to 75cm in total length.

Distribution — Japan, Ryukyu, Formosa, Indo-Pacific-Atlantic, South Africa

Monacanthidae

Aluterus scripus (Osbeck) ⇧ ⇨

[*Aluteres scriptus* (Forster)]

[*Alutera scripta* (Osbeck)]

[*Osbeckia scripta* (Osbeck)]

Scrawled Filefish, Triggerfish, Longtail Filefish, Unicorn Fish, Figured Leatherjacket, Scribbled Leatherjacket, Unicorn Filefish

Diagnosis — Inhabits shoreline reefs and coral reefs. Swims with head down. Body is remarkably compressed, has black spots and blue stripes. Snout and tail are extended. Caudal fin is longer than head length. First dorsal fin has 1 long spine, second has 44−50 rays. Anal fin has 46−52 rays. Edible but viscera may contain virulent palytoxin. Up to 1m in total length.

Distribution — Japan, Ryukyu, Formosa, Indo-Pacific-Atlantic, Hawaii, South Africa

Monacanthidae

Chaetodermis penicilligerus (Cuvier) ⇨

[*C. spinosissimus* (Quoy et Gaimard)]

[*Chaetoderma penicilligera* (Cuvier)]

Prickly Leatherjacket

Diagnosis — Inhabits shoreline reefs. Body is rhombic. Snout and caudal peduncles are extended. Numerous flaps are present over body. Scales are large. First dorsal fin has 1 short spine with flaps, second has 25−26 rays. Anal fin has 23−24 rays. Up to 30cm in total length.

Distribution — Japan, Formosa, Indo-West Pacific

Monacanthidae

⇧ ***Pervagor spilosoma*** (Lay et Bennett)

[*Stephanolepis spilosomus* (Lay et Bennett)]

Fan-tail File Fish

Diagnosis — Inhabits reef slopes of coral reefs. Body is oblong, profusely dotted with black. Snout is slightly pointed. Caudal margin is black. First dorsal fin has 1 long spine, second has 38 rays. Anal fin has 34 rays. Pelvic fins have 1 pointed spine. Up to 13cm in total length.

Distribution — Hawaii

Monacanthidae

Oxymonacanthus longirostris (Schneider) ⇨

Beaked Leatherjacket

Diagnosis — Inhabits in pairs coral reefs. Snout is remarkably projected, mouth is small. Body is greenish-blue with orange blotches. Caudal fin has a black blotch. First dorsal fin has 1 long spine, second has 31−33 rays. Anal fin has 30−32 rays. Feeds on coral polyp and small invertebrate. Aquarium fish. Up to 10cm in total length.

Distribution — Japan, Ryukyu, Formosa, Indo-Pacific, East Africa

Triacanthodidae
Paratriacanthodes retrospinis Fowler ⇨

Threeline Spikefish

Diagnosis — Inhabits waters more than 400m deep. Body is somewhat deep, light-rosy with indistinct pale stripes. Eyes are large. Snout is pointed. Dorsal and anal fin rays, caudal fin are yellow. Anus is black. First dorsal fin has 4 long spines, second has 14–15 rays. Anal fin has 13 rays. Up to 15cm in total length.

Distribution — Southern Japan, Formosa, East China Sea, Hong Kong, East Africa

Ostraciidae
Tetrosomus gibbosus (Linnaeus) ⇩

[*Rhinesomus gibbosus* (Linnaeus)]

Black-blotched Turretfish, Boxfish, Hunchback Boxfish, Thornback Boxfish, Thornback Trunkfish

Diagnosis — Inhabits sandy bottoms. Body is triangular in section. Resembles *T. concatenatus*, but distinguished by a cone-shaped spine on back. Small spines are present on margin of abdomen. Interorbital space has a small spine. Tail has indistinct black bands. Dorsal and anal fins have 9 rays. Skin secretes ichthyocrinotoxin. Up to 30cm in total length.

Distribution — Southern Japan, Indo-West Pacific, Red Sea, East Africa

Ostraciidae
Tetrosomus concatenatus (Bloch et Schneider) ⇨ ⇩

[*Rhinesomus concatenatus* (Bloch et Schneider)]

Boxfish, Lined Boxfish, Linked Boxfish, Triangular Boxfish

Diagnosis — Inhabits rather deep rocky reefs and coral reefs. Body is triangular in section, back is high, ridged with 2 small spines. Spines anterior to eyes are absent. Dorsal and anal fins have 9 rays. Skin secretes ichthyocrinotoxin. Up to 30cm in total length.

Distribution — Japan, Ryukyu, Formosa, Indo-West Pacific, South Africa

Ostraciidae
Lactoria diaphana (Bloch et Schneider) ⇩

[*L. diaphanus* (Bloch et Schneider)]

Spiny Boxfish, Spiny Cowfish, Cowfish, Diaphanous Boxfish, Diaphanous Cowfish, Boxfish

Diagnosis — Inhabits shallow shoreline reefs. Resembles *Ostracion immaculatus*, but distinguished by pentagonal section of body, a spine on back. Both sides of abdomen are angular, posterior end forming a spine. A pair of spines are present anterior to eyes. Dorsal and anal fins have 9 rays. Feeds on benthic invertebrate. Viscera poisonous but not virulent. Skin secretes ichthyocrinotoxin. Up to 30cm in total length.

Distribution — Japan, Ryukyu, Formosa, Indo-Pacific, Hawaii, Red Sea, South Africa

Ostraciidae
⇧ *Lactoria fornasini* (Bianconi) ⇧ ⇩

Shortspined Cowfish, Five-horned Cow Fish, Backspine Cowfish, Boxfish, Cowfish

Diagnosis — Inhabits shoreline reefs and coral reefs. Resembles *L. diaphana*, but distinguished by blunt body with blue stripes and blotches. Back has a large spine directed backward. Dorsal and anal fins have 9 rays. Skin secretes ichthyocrinotoxin. Up to 15cm in total length.

Distribution — Japan, Ryukyu, Indo-Pacific, Hawaii, South Africa

Ostraciidae

Lactoria cornuta (Linnaeus) ⇨⇩

[*L. cornutus* (Linnaeus)]

Longhorned Cowfish, Horned Trunkfish, Horned Boxfish, Boxfish, Cowfish

Diagnosis — Inhabits shallow shoreline reefs. Resembles *L. diaphana*, but distinguished by extended body, longer caudal fin with scattered dark spots. Spines on head and posterior end of abdominal margin are large and stiff. Spines are absent on back. Dorsal fin has 9 rays, anal fin has 8−9 rays. Skin secretes ichthyocrinotoxin. Up to 50cm in total length.

Distribution — Japan, Ryukyu, Formosa, Korea, Indo-Pacific, Hawaii, South Africa

Ostraciidae

Ostracion immaculatus Temminck et Schlegel ⇩

⇦⇩ [*O. tuberculatus* Linnaeus]

Bluespotted Boxfish, Spotted Trunkfish, Box Puffer, Boxy, Trunkfish, Ocellated Boxfish, Black-spotted Boxfish

Diagnosis — Inhabits shoreline reefs. Tetragonal body in section is covered with hard carapace, has hexagonal reticulate patterns. Eyes are large, slightly protruded. Mouth is directed downward. Caudal peduncles move freely. Dorsal and anal fins have 9 rays. Flesh is edible but may cause ciguatera poisoning. Viscera poisonous but not virulent. Skin secretes ichthyocrinotoxin. Up to 45cm in total length.

Distribution — Japan, Ryukyu, Formosa, Korea, Indo-Pacific, Hawaii, Red Sea, South Africa

Ostraciidae

⇩ *Ostracion meleagris meleagris* Shaw

[*O. meleagris* Shaw]

[*O. sebae* Bleeker]

Box-fish, Spotted Trunkfish, Blue Boxfish

Diagnosis — Resembles *O. immaculatus*, but distinguished by somewhat roundish body. In male, body is profusely blotched with orange on blue skin. In female, body is profusely dotted with white on black skin. Dorsal fin has 7−8 rays, anal fin has 8 rays. Caudal margin is rounded. Edible but may cause ciguatera poisoning. Up to 20cm in total length.

Distribution — Japan, Ryukyu, Formosa, Indo-Pacific, Hawaii, East Africa

▲ Body is covered with hard hexagonal carapace.

Ostraciidae

Ostracion whiteleyi Fowler ⇨

[*O. ornatus* Hollard]

Whiteley's Box Fish

Diagnosis — Inhabits shallow coral reefs. Resembles *O. immaculatus*, but male has blue stripes on brown skin with 2 circular blotches on abdominal side. In female, body has many white bands and spots. Dorsal and anal fins have 9 rays. Skin secretes ichthyocrinotoxin. Up to 13cm in total length.

Distribution — Hawaii, Polynesia

Triacanthidae

Triacanthus biaculeatus (Bloch) ⇨

[*T. brevirostris* Temminck et Schlegel]

Black-finned Triple-spine, Triplespine, Hollow-snouted Tripod-fish, Queensland Triple-spine, Short-nosed Tripod-fish, Silver Leather Jacket, Tripod Fish

Diagnosis — Adult inhabit rather deep offshore waters, but young are found in bays. Body is silvery-white, tail is extended. First dorsal and pelvic fins form sharp-pointed spine. Caudal margin is falcated. First dorsal fin has 6 black spines, second has 21−26 rays. Anal fin has 17−22 rays. Edible. Up to 30cm in total length.

Distribution — Japan, Ryukyu, Formosa, China, Indo-West Pacific, Persian Gulf

Diodontidae

Diodon liturosus Shaw ⇧

[*D. bleekeri* Günther]

Bleeker's Porcupinefish

Diagnosis — Inhabits shallow rocky reefs and coral reefs. Resembles *D. holocanthus*, but distinguished by fewer movable spines on body, white-margined black blotches on back. Dorsal and anal fins have 14−16 rays. Edible. Viscera poisonous but not virulent. Up to 60cm in total length.

Distribution — Japan, Ryukyu, Indo-Pacific

Diodontidae

⇦ *Diodon holocanthus* Linnaeus ⇨

Balloon Fish, Spiny Puffer, Long-spined Porcupinefish, Balloon Porcupinefish, Freckled Porcupinefish, Lesser Porcupinefish

Diagnosis — Found in schools in shallow waters. Body is covered with hundreds of movable spines. Eyes are large. Snout is short. Caudal margin is roundish, spines are absent on caudal peduncles. When warned, sucks water and air into stomach, swells, erects spines. Spawns in Kuroshio areas. Dorsal and anal fins have 13−15 rays. Edible. Viscera poisonous but not virulent. Up to 50cm in total length.

Distribution — Japan, Ryukyu, Indo-Pacific-Atlantic, South Africa

Diodontidae

Diodon hystrix Linnaeus ⇧ ⇨

[*Tetraodon hystrix* (Linnaeus)]

Spotted Porcupinefish, Porcupinefish, Sea-hedgehog, Common Porcupinefish, Longspine Porcupinefish

Diagnosis — Inhabits singly shallow coral reefs. Resembles *D. liturosus*, but distinguished by absence of black blotches on body, also by body profusely dotted with dark color. Snout and tail are somewhat long. Swells body and erects spines. Dorsal and anal fins have 14−16 rays. Edible but has records of ciguatera poisoning. Viscera poisonous but not virulent. Up to 90cm in total length.

Distribution — Japan, Ryukyu, Formosa, Indo-Pacific-Atlantic, Hawaii, Red Sea, South Africa

Diodontidae

⇦ ***Chilomycterus affinis*** Günther ⇩

[*C. californiensis*]

Pacific Burrfish, Giant Burrfish, Porcupine Fish, Spiny Puffer, Balloon Fish, Tiger Puffer, Swelltoad, Hairy Boxfish, Globefish, Cucumberfish, Spiny Boxfish

Diagnosis — Inhabits singly shoreline reefs and coral reefs. Spines are sparse and short, bases branched out into 3−4 pieces so that spines do not erect when body is swollen. Jaws have a pair of large tooth like plate. Dorsal and anal fins have 12 rays. Juvenile feed on zooplankton. Edible but may cause ciguatera poisoning. Viscera poisonous but not virulent. Up to 65cm in total length.

Distribution — Japan, Ryukyu, Pacific, Hawaii, Galapagos Islands, California

Molidae

⇦ ***Mola mola*** (Linnaeus) ⇨

[*Tetraodon mola* (Linnaeus)]

[*T. lunae*]

Ocean Sunfish, Headfish, King of the Mackerels, Round-tailed Sunfish, Common Head-fish, Common Mola, Common Sunfish, Giant Sunfish, Globe-fish, Trunkfish, Mole-but

Diagnosis — Migrates singly or in small schools in upper and middle layers of oceanic waters. Body is oval, disk-shaped, scaleless. Pelvic and caudal fins are absent, dorsal and anal fins are developed. Eyes are closable. Jaws have a pair of beak-shaped teeth. Feeds on jellyfish and plankton. Produces about 7 hundred million spawns in one spawning. Edible. Up to 4m in total length, 2.3 tons in weight.

Distribution — Japan, Ryukyu, Formosa, Indo-Pacific-Atlantic, Mediterranean, South Africa

Tetraodontidae

⇦ ***Takifugu chrysops*** (Hilgendorf)

[*Fugu chrysops* (Hilgendorf)]

Redeye Puffer

Diagnosis — Body is somewhat thick with smooth skin. Back is yellowish-brown, profusely dotted with dark color. Eyes are red. Dorsal fin has 11−13 rays, anal fin has 10−11 rays. Ovary, viscera and skin are poisonous. Testis and flesh are not poisonous. Up to 28cm in total length.

Distribution — Japan, Ryukyu

Tetraodontidae

⇦ ***Takifugu xanthopterus*** (Temminck et Schlegel) ⇧

[*Fugu xanthopterum* (Temminck et Schlegel)]

Striped Puffer, Yellowfin Puffer

Diagnosis — Inhabits middle layers of shoreline waters. Body is covered with small spines with white oblique bands on back surface. Fins are yellow. Dorsal fin has 16−18 rays, anal fin has 14−16 rays. Spawns between April and May. Edible. Ovary and viscera are poisonous. Flesh, skin, testis are not poisonous. Up to 50cm in total length.

Distribution — Japan, Ryukyu, Formosa, China, Korea, Yellow Sea, East China Sea

Tetraodontidae

⇦ ***Takifugu niphobles*** (Jordan et Snyder) ⇩

[*Fugu niphobles* (Jordan et Snyder)]

Grass Puffer

Diagnosis — Found in schools in shoreline waters. Body is slightly elongated, covered with small spines. Back is dark-brown, profusely dotted with white. A black blotch is present on posterior surface of pectoral fins. Dorsal fin has 12−14 rays, anal fin has 10−12 rays. Spawns between May and August. Ovary, viscera and skin are poisonous. Testis and flesh are also slightly poisonous. Up to 20cm in total length.

Distribution — Japan, Ryukyu, Formosa, Korea

Tetraodontidae

Takifugu chinensis (Abe) ⇩

[*T. pseudommus* (Chu)]

[*Fugu chinensis* (Abe)]

Eyespot Puffer

Diagnosis — Resembles *T. rubripes*, but distinguished by purplish-black back, black anal fin. Body is covered with small spines. A balck blotch is present on posterior surface of pectoral fins. Dorsal fin has 17−18 rays, anal fin has 14−15 rays. Edible. Ovary and viscera are poisonous. Flesh, skin, testis are not poisonous. Up to 60cm in total length.

Distribution — Japan, China, Yellow Sea, East China Sea

Tetraodontidae ⇧

Takifugu oblongus (Bloch) ⇧

[*Torquigener oblongus* (Bloch)]

Oblong Blowfish, Oblong Toadfish, Blaasop, Toby

Diagnosis — Inhabits rather deep waters. Body is covered with small spines and brown bands on back. Dorsal fin has 13 rays, anal fin has 11 rays. Viscera are poisonous. Testis and flesh are also slightly poisonous. Up to 36cm in total length.

Distribution — Formosa, South China Sea, Indo-Pacific, Hawaii, South Africa

Tetraodontidae

Takifugu rubripes (Temminck et Schlegel) ⇧⇨

[*Fugu rubripes rubripes* (Temminck et Schlegel)]

Tiger Puffer, Ocellate Puffer

Diagnosis — Body is covered with small spines. A large black blotch is present on posterior surface of pectoral fins. Anal fin is white, caudal fin is black. Dorsal fin has 16−19 rays, anal fin has 13−16 rays. Spawns between March and June. Good for breeding. Edible. Ovary and viscera are poisonous. Flesh, skin, testis are not poisonous. Up to 80cm in total length.

Distribution — Japan, Ryukyu, Yellow Sea, East China Sea

Tetraodontidae
Takifugu radiatus (Abe) ⇧
[*Fugu vermicularis radiatus* (Abe)]
Pear Puffer
Diagnosis — Resembles *T. vermicularis*, but distinguished by brown body, a dark blotch on posterior surface of pectoral fins. Small spines are absent on body. Lower part of caudal fin and anal fin are white. Dorsal fin has 13–15 rays, anal fin has 10–13 rays. Spawns between May and June. Except testis are poisonous. Up to 25cm in total length.
Distribution — Japan, Korea, Yellow Sea, East China Sea

Tetraodontidae
Takifugu poecilonotus ⇧
(Temminck et Schlegel) ⇧
[*Fugu poecilonotum* (Temminck et Schlegel)]
Finepatterned Puffer
Diagnosis — Body is covered with small spines. Back is dark-brown with numerous pale spots. An indistinct dark blotch is present on posterior parts of pectoral fins. Dorsal fin has 12–15 rays, anal fin has 10–13 rays, both fins are yellowish. Spawns in April. Ovary, testis, viscera, skin are poisonous. Flesh is also slightly poisonous. Up to 30cm in total length.
Distribution — Japan, Ryukyu, Korea, China, East China Sea, Malay

▼Young Adult

Tetraodontidae
⇧ ***Takifugu vermicularis*** (Temminck et Schlegel) ⇩
[*Fugu vermiculare vermiculare* (Temminck et Schlegel)]
Vermiculated Puffer
Diagnosis — Inhabits shoreline waters. Body is smooth. Resembles *T. porphyreus*, but anal fin is white. No white blotches are present on posterior surface of pectoral fins. Back is covered closely with blackish-brown spots. Lower part of caudal fin is white. Dorsal fin has 12–15 rays, anal fin has 10–13 rays. Spawns between June and July. Ovary, viscera, skin are poisonous. Flesh is also slightly poisonous. Has largest records of Fugu-poisoning. Up to 35cm in total length.
Distribution — Japan, East China Sea

Tetraodontidae
Takifugu porphyreus (Temminck et Schlegel) ⇧
[*Fugu vermiculare porphyreum* (Temminck et Schlegel)]
[*F. vermicularis porphyreus* (Temminck et Schlegel)]
Purple Puffer, Genuine Puffer
Diagnosis — Body is smooth, has no small spines. Back is dark-brown with a black blotch on upper surface of pectoral fins. Dorsal fin base is also black. Young are found in schools in bays, have pale blotches on body. Dorsal fin has 12–17 rays, anal fin has 11–15 yellowish rays. Edible. Ovary, viscera and skin are poisonous. Up to 47cm in total length.
Distribution — Japan, Korea, East China Sea, Sakhalin

Tetraodontidae
⇦ ***Takifugu pardalis*** (Temminck et Schlegel)
[*Fugu pardale* (Temminck et Schlegel)]
Panther Puffer, Equinox Puffer
Diagnosis — Body has no spines, covered closely with small wart-like processes. Back is light-brown with many dark blotches. Dorsal, pectoral, anal fins are orange. Caudal fin is blackish. Dorsal fin has 11–14 rays, anal fin has 9–12 rays. Spawns in March. Ovary, testis, viscera, skin are poisonous. Flesh is considered as nonpoisonous but has been found poisonous in some indivisuals. Up to 35cm in total length.
Distribution — Japan, Ryukyu, Formosa, Yellow Sea, East China Sea

Tetraodontidae
Arothron meleagris
(Schneider) ⇩

[*Tetraodon meleagris* Lacépède]

[*T. lacrymatus* Quoy et Gaimard]

Guineafowl Puffer, Speckled Balloon Fish

Diagnosis — Inhabits coral reefs. Body is covered with small spines, profusely dotted with white. Nostrils have 2 flaps. Dorsal and anal fins have 12 rays. Liver is poisonous. Up to 45cm in total length.

Distribution — Ryukyu, Bonin Islands, Formosa, Indo-Pacific, Hawaii, Central America, East Africa

Tetraodontidae
⇖ *Arothron mappa* (Lesson) ⇧

[*Tetraodon mappa* Lesson]

Scribbled Toadfish

Diagnosis — Body is deep, covered with small spines. Radial stripes are present around eyes. Base and lower part of pectoral fins and anus are black. Dorsal fin has 11−12 rays, anal fin has 10−11 rays. Viscera and flesh are nonpoisonous, but no data on other organs. Up to 70cm in total length.

Distribution — Japan, Ryukyu, Formosa, Indo-Pacific, East Africa

Tetraodontidae
Arothron hispidus (Linnaeus) ⇧ ⇨

[*Tetraodon hispidus* Linnaeus]

Whitespotted Blowfish, Swellfish, Poison Puffer, Broad-barred Toadfish, Star and Striped Toado, Whitespotted Blaasop, Bristly Puffer

Diagnosis — Inhabits coral reefs. Body is somewhat deep, covered with small spines. Back has white spots, ventral side has striped patterns. Gill slits are black, encircled with a ring-shaped blotch. Nostrils have flaps. Dorsal and anal fins have 10−11 rays. Ovary, viscera, skin are poisonous. Flesh is nonpoisonous. Skin secretes tetrodotoxin. Up to 50cm in total length.

Distribution — Japan, Ryukyu, Formosa, Korea, Indo-Pacific, Panama, Red Sea, South Africa

Tetraodontidae
⇦⇙ *Arothron nigropunctatus*
(Schneider) ⇨⇙

[*A. diadematus*]

[*Tetraodon nigropunctatus* (Bloch et Schneider)]

[*Ovoides nigropunctatus* (Bloch et Schneider)]

Blackspotted Blowfish, Blow Puffer, Blackspotted Toado

Diagnosis — Inhabits coral reefs. Light-brown body is covered with small spines, has scattered large black spots. In some individuals, entire body is yellow. Black flaps are present on nostrils. Dorsal and anal fins have 10−11 rays. Liver is poisonous. Has records of flesh poisoning. Up to 25cm in total length.

Distribution — Japan, Ryukyu, Indo-West Pacific, Red Sea, South Africa

Tetraodontidae

Arothron stellatus (Bloch et Schneider)

[*Tetraodon stellatus* Bloch et Schneider]

Starry Blowfish, Spottedfin Blaasop, Starry Toadfish, Toad, Toby

Diagnosis — Inhabits coral reefs. Body is somewhat deep, covered with small spines, closely with small black spots. Pectoral fin bases are black. Flaps are present on nostrils. Young resemble *A. hispidus*, but distinguished by a black blotch on back. Dorsal fin has 11–12 rays, anal fin has 11 rays. Ovary, testis, viscera, skin are poisonous. Flesh is edible but in some individuals poisonous with records of causing death. Up to 1m in total length.

Distribution — Japan, Ryukyu, Indo-Pacific, Persian Gulf, Red Sea, South Africa

Tetraodontidae

Arothron diadematus (Rüppell)

Masked Puffer

Diagnosis — Inhabits shallow coral reefs. Body is covered with small spines. Resembles *A. nigropunctatus*, but distinguished by dark margins of mouth and eyes, also by white between them. Pectoral, dorsal, anal fins are black. Black flaps are present on nostrils. Dorsal and anal fins have 10 rays. Poisonousness is unknown. Up to 30cm in total length.

Distribution — Ryukyu, Australia, Red Sea

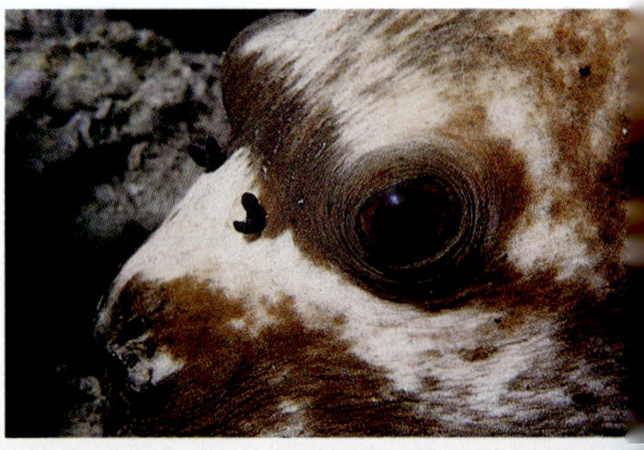

Tetraodontidae

Arothron manillensis (Procé)

[*A. immaculatus* (Bloch et Schneider)]

[*Tetraodon immaculatus* Bloch et Schneider]

Immaculate Blowfish, Plain Toado, Narrow-lined Toadfish

Diagnosis — Inhabits coral reefs and brackish areas. Body is covered with small spines, has dark stripes. Flaps are present on nostrils. Dorsal fin has 10–11 rays, anal fin has 9–10 rays. Poisonousness is unknown. Up to 50cm in total length.

Distribution — Ryukyu, Indo-Pacific, Red Sea, South Africa

Tetraodontidae

Arothron firmamentum (Temminck et Schlegel)

[*Boesemanichthys firmamentum* (Temminck et Schlegel)]

Starry Toado

Diagnosis — Inhabits rather deep waters. Body is slightly elongated, covered with small spines. Body is profusely dotted with white. Dorsal fin has 14 rays, anal fin has 13–14 rays. Flesh and liver are nonpoisonous, but poisonousness of other organs is unknown. Up to 40cm in total length.

Distribution — Southern Japan, South China Sea, New Zealand, Australia

Tetraodontidae

⇦ ***Lagocephalus inermis*** (Temminck et Schlegel)

[*L. laevigatus* (Temminck et Schlegel)]

[*L. laevigatus inermis* (Temminck et Schlegel)]

Unarmed Blaasop, Smoothback Blowfish

Diagnosis — Inhabits rather deep waters. Resembles *L. wheeleri*, but body is somewhat thick. Small spines are absent on back, abdomen has small granular processes. Gill slits and dorsal base are black. Caudal margin is white. Dorsal fin has 13–14 rays, anal fin has 12 rays. In those native in waters around Japan, all organs and parts except liver are nonpoisonous and edible, but the flesh from those of the South China Sea is poisonous. Up to 1m in total length.

Distribution — Southern Japan, South China Sea, Indo-West Pacific

Tetraodontidae

Lagocephalus globeri Abe et Tabeta ⇨

[*L. lunaris* (Bloch et Schneider)]

[*L. lunaris spadiceus* (Richardson)]

[*Gastrophysus lunaris* (Bloch)]

Blowfish, Brown-backed Toadfish, Chestnut Puffer, Pufferfish, Spade Blaasop, Green Rough-backed Blowfish, Toby

Diagnosis — Inhabits middle layers of offshore waters. Resembles *L. wheeleri*, but caudal fin is double emarginated, lobes white, back is dark brown, abdomen is silvery. Dorsal fin has 13 rays, anal fin has 12 rays. Those native in waters around Japan are nonpoisonous and edible. But ovary and viscera are poisonous, also the flesh is slightly poisonous of those from the South China Sea. Up to 40cm in total length.

Distribution — Japan, Ryukyu, Formosa, South China Sea, Indo-Pacific, Red Sea, South Africa

Tetraodontidae

Lagocephalus wheeleri Abe, Tabeta et Kitahama ⇨

[*L. lunaris* subsp.]

Silver Puffer (New Common Name)

Diagnosis — Inhabits middle layers of shoreline waters. Body is somewhat elongated, covered with small spines. Caudal margin is emarginated, lower part is white. Back is yellowish-brown, ventral is silvery. Anal fin is white. Dorsal fin has 13–14 rays, anal fin has 12–14 rays. Spawns between June and July. Edible, liver is slightly poisonous. Up to 35cm in total length.

Distribution — Japan, Formosa, East China Sea

Tetraodontidae

Chelonodon fluviatilis (Buchanan-Hamilton) ⇩

[*Tetrodon fluviatilis* Buchanan-Hamilton]

Estuarine Blowfish

Diagnosis — Inhabits shoreline waters and river-mouths. Body is yellow, profusely blotched with black. Flaps are present on nostrils. Fins are yellowish. Dorsal fin has 14–16 rays, anal fin has 12–13 rays. Poisonousness is unknown. Up to 15cm in total length.

Distribution — Indian Ocean

Tetraodontidae

Amblyrhynchotes hypselogeneion (Bleeker) ⇧

[*A. hypselogenion* (Bleeker)]

Bar-cheeked Toadfish, Harlequin Blaasop, Blaasop, Toby, Orange Banded Blowfish

Diagnosis — Body is brown, spotted closely with pale spots, covered with small spines. Small dorsal fin has 8–10 rays, small anal fin has 7–8 rays. Ovary is poisonous, flesh, skin, liver also slightly poisonous. Up to 18cm in total length.

Distribution — Japan, Ryukyu, Formosa, Indo-Pacific, Hawaii, Red Sea, South Africa

Tetraodontidae
Lagocephalus lagocephalus oceanicus Jordan et Fowler ⇩

Spotted Smooth-backed Blowfish, Blue Blaasop, Toby, Blueback Puffer ⇩

Diagnosis — Inhabits middle layers of offshore waters. Body is elongated. Back is bluish-black with scattered black spots on lower part. Lower part of caudal fin is somewhat long. Dorsal fin has 13–15 rays, anal fin has 11–14 yellowish rays. Poisonousness is unknown. Up to 50cm in total length.

Distribution — Japan, Ryukyu, Indo-West Pacific, Hawaii

Tetraodontidae
Lagocephalus scelaratus (Forster) ⇧

[*Pleuranacanthus scelaratus* (Forster)]

[*Gastrophysus sceleratus* (Gmelin)]

Black Spotted Puffer, Silvercheek Blaasop, Spotted Roughback Blowfish, Giant Toado, Silver-cheeked Toadfish

Diagnosis — Inhabits rather deep waters. Body is elongated, tail is extended. Back is profusely dotted with black, abdomen is covered with small spines. A silvery band is present between snout and caudal peduncles. Gill slits are black. A silvery blotch on surface anterior to eyes. Dorsal fin has 11–13 rays, anal fin has 11–12 rays. Poisonous, has records of causing death. Up to 1.5m in total length.

Distribution — Japan, Ryukyu, Formosa, Indo-Pacific, Persian Gulf, Red Sea, South Africa

Tetraodontidae
Chelonodon patoca (Buchanan-Hamilton) ⇧ ⇨

Gangetic Blowfish, Marbled Toad, Pufferfish, Milkspotted Blaasop, Toby, Milk-spotted Toadfish

Diagnosis — Body is covered with small spines. Back has 4 broad dark bands, also pale spots. Anal fin is white. Caudal margin is black. Dorsal fin has 10 rays, anal fin has 8 rays. Skin and ovary are poisonous, flesh is also slightly poisonous. Up to 40cm in total length.

Distribution — Ryukyu, Formosa, China, Indo-Pacific, Persian Gulf, South Africa

Tetraodontidae
Canthigaster rivulata (Temminck et Schlegel) ⇨

Sharp-nosed Puffer, Scribbled Toby, Purse Puffer, Schlegel's Puffer Fish

Diagnosis — Inhabits shoreline reefs. Body is compressed, somewhat deep, has 2 dark bands. Snout is projected. Abdomen is covered with small spines. Dorsal and anal fins have 9–10 rays. Skin and viscera are poisonous. Ovary and flesh are nonpoisonous. Up to 20cm in total length.

Distribution — Japan, Ryukyu, Bonin Islands, Formosa, Indo-Pacific, Hawaii

Tetraodontidae

Canthigaster jactator (Jenkins) ⇧

Sharp-nosed Puffers, Whitespotted Toby

Diagnosis — Inhabits shallow coral reefs and shoreline reefs. Body is somewhat deep, profusely dotted with bluish-white. Snout is pointed. Eyes are protruded. Caudal fin is transparent. Dorsal and anal fins have 9 rays. Poisonousness is unknown. Up to 8cm in total length.

Distribution — Japan, Izu Islands, Hawaii

Tetraodontidae

⇦ *Canthigaster valentini* (Bleeker) ⇧

[*C. cinctus* (Richardson)]

Saddled Toby, Striped Purse Puffer, Blacksaddled Toby, Blaasop, Black-saddled Puffer, Striped Toby

Diagnosis — Inhabits shallow coral reefs. Body is deep, has 4 dark bands. Snout is pointed. Caudal fin is yellow. Dorsal and anal fins have 9 rays. Poisonousness is unknown. Up to 15cm in total length.

Distribution — Japan, Ryukyu, Bonin Islands, Formosa, Indo-Pacific, Hawaii, South Africa

Tetraodontidae

⇦ *Canthigaster margaritata* (Rüppell)

[*C. solandri* (Richardson)]

Pearl Toby, Ocellated Puffer

Diagnosis — Inhabits shallow coral reefs and shoreline reefs. Resembles *C. jactator*, but radial stripes are present around eyes. Caudal fin is profusely covered with spots. Dorsal fin base has a black blotch. Dorsal and anal fins have 9–10 rays. Poisonousness is unknown. Up to 15cm in total length.

Distribution — Indo-Pacific, Red Sea, South Africa

Triodontidae

Triodon macropterus Lesson ⇨

Threetooth Puffer, Three-toothed Puffer

Diagnosis — Inhabits rather deep waters. Body is elongated. Head is roundish, upper jaw is long, has a pair of beak-shaped tooth. Caudal fin is lunated, tail is slender. Membranous part of abdomen is remarkably developed, covered with laminar scales, has a black blotch. Dorsal fin has 10–12 rays, anal fin has 9–10 rays. In some individuals, first dorsal fin (2 spines) is present. Flesh is edible. Up to 50cm in total length.

Distribution — Japan, Ryukyu, Formosa, South China Sea, Indo-West Pacific

Coelacanth

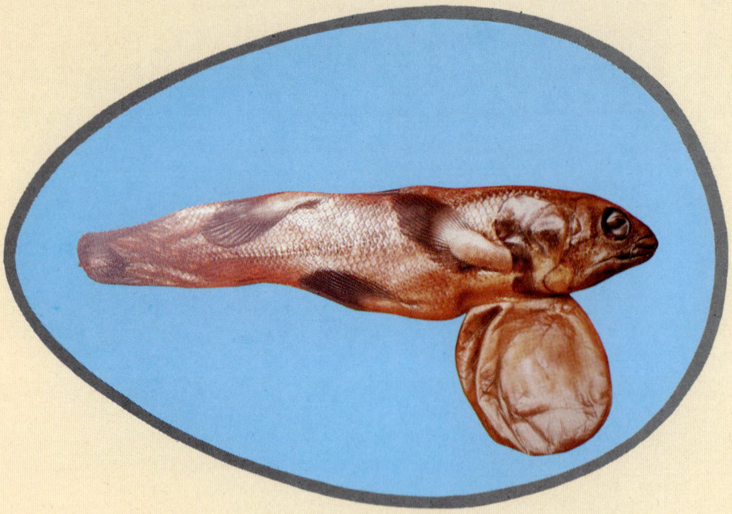

▲ Unborn Coelacanth with yolk sac.
(Courtesy of British Museum, Natural History)

▲ Recent specimen of Coelacanth.
(Courtesy of Zoologisk Museum, Kobenhavn)

▲ Front view of Coelacanth.
(Courtesy of British Museum, Natural History)

Latimeriidae

Latimeria chalumnae Smith

[*Malania anjouanae* Smith]

Coelacanth, Coelacanth Fish, Comoro Coelacanth

Diagnosis — Inhabits bottoms about between 15−400m deep. Body is somewhat compressed, bluish-green, has indistinct pale blotches, covered with large scales. Tail is thick. Pectoral, second dorsal, anal fins are extended with bone. Backbone consists of tubular chorda. Mouth is large with fang-shaped teeth. First dorsal fin has 8 spines, second has 30 rays. Anal fin has 29 rays. Lateral lines are not complete. Anterior part of body has 76 scales, upper part of caudal fin has 23 scales. Well-known fish as an existing species of Crossopterygii, extinct 70 millions years ago. Up to 1.5m in total length, 80kg in weight.

Distribution — Comores Islands, South Africa

Marine Animals of the Indo-Pacific

The Indo-Pacific

The Earth is a sphere with 6,367㎞ in semidiameter and 510,069,100㎢ in surface area. The sea occupies 70.8% of it, 361,128,923㎢.

The Pacific Ocean occupies a surface of 179,800,000㎢ or 49.8% of the whole sea area and 35.2% of the surface of the Earth. The Indian Ocean occupies 74,800,000㎢ or 20.7% of all the waters and 14.7% of the Earth. This means that the combined Indo-Pacific Oceans occupy 70.5% of the sea and about the half of the whole surface of the globe. The percentage tells how vast the Ocean is.

The Indo-Pacific, used in the field of the marine animals, is the tropical and subtropical sea area within the boundaries between thirty degrees to the north and the south. This area coincides with the distribution of coral reefs. The characteristic features of the fauna there are the reef animals including reef-building corals. The Indo-Pacific is also called the "Tropical Indo-Pacific."

The marine fauna of the East Pacific along the western coast of American continent is similar to that of the tropical Atlantic including the Caribbean Sea and different from that of the Hawaiian Islands and Polynesia. The "Indo-Pacific" is said to be the "Indo-West Pacific" bio-geographically. The writer, however, regards the Hawaiian Islands and Polynesia as belonging to the Central Pacific. The West Pacific includes the Caroline Islands such as Yap and Palao, the Philippines, Indonesia, New Guinea and Australia. This is the geographical division adopted in this encyclopedia.

The northern boundary of the Indo-Pacific is the Gulf of Aqaba, northern Red Sea, Persian Gulf, southern Japan, Ryukyu Islands and Hawaiian Islands, while the southern boundary is Mozambique Channel, Mauritius Islands, Lord Howe Island in Eastern Australia, Tubuai Islands of Polynesia, and the Gambier Islands.

In general the sea around Malaysia, the Philippines, and New Guinea show the richest fauna in marine animals, which are appreciated by many scientists. It is regarded as the center of evolution.

The Hawaiian Islands, distant from continents and other islands, are characterized by their "endemic species." The Indian Ocean, recognized as a cradle of many marine animals, has plenty of endemic species, because some reasons impeded them to spread to the Pacific. The salty waters of the Red Sea are also abundant in endemic species.

Marine Vertebrata

The Vertebrata is a group of chordates which have developed vertebras. They have heads, fins or arms and legs, and

The Indo-Pacific Region

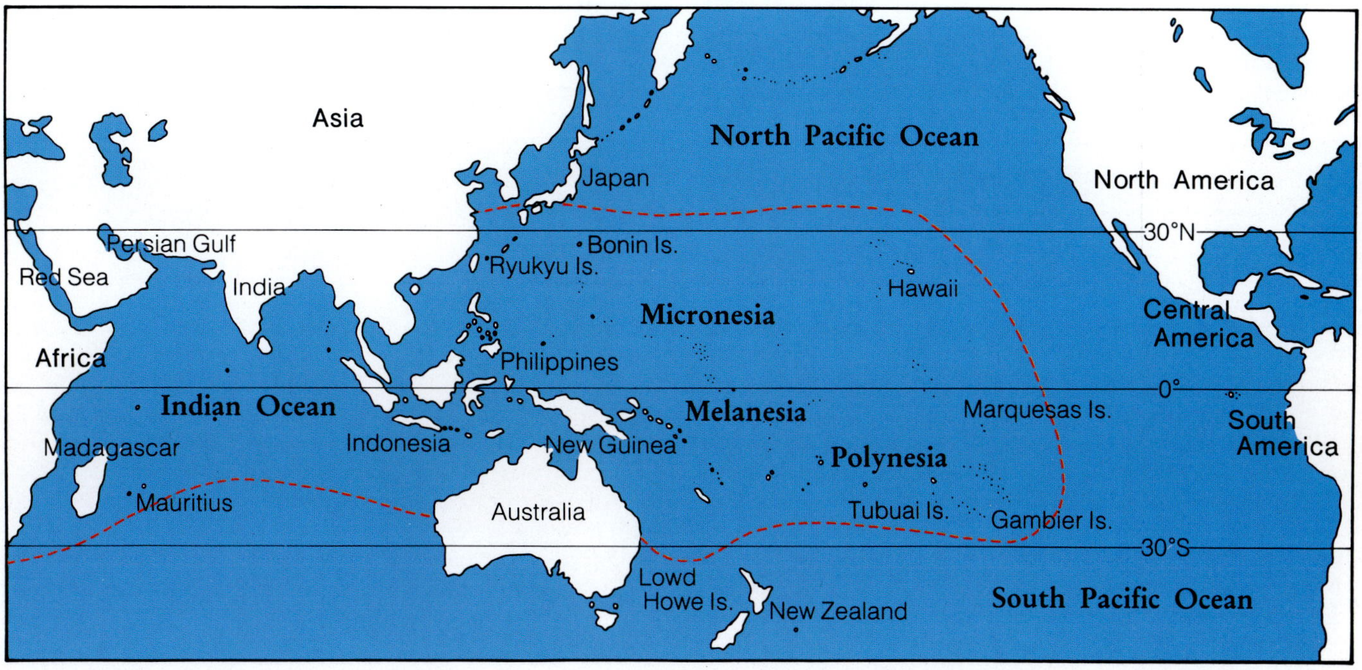

well-developed hearts which are the greatest specific feature of them. Their bodies are divided into three parts, head, trunk, and tail.

The phylum Vertebrata is roughly grouped into eight classes.

(1) **Agnatha** ——— The most primitive group including lampreys and hagfish, etc.
(2) **Acanthodii** ——— Fossil species.
(3) **Placodermi** ——— Fossil species.
(4) **Chondrichthyes** — Fishes with cartilage bones such as sharks, rays and ghost sharks.
(5) **Osteichthyes** ——— Developed fishes including ordinary ones.
(6) **Amphibia** ——— Amphibious animals including frogs and salamanders.
(7) **Reptilia** ——— Land or acquatic animals such as snakes, lizards and turtles.
(8) **Mammalia** ——— Animal group of the highest class, from men to whales.

All the fishes, sea snakes, sea turtles, whales and dolphins are included in the marine animals and belong to either of the five classes.

Marine Mammals

The Mammalia is a group of Vertebrata which nourishes its baby on its own milk. Man belongs to this class. It is the most developed group of all the animals consisting of more than one million species. Fossils uncovered that it was differentiated from the order Theraprida of Reptilia in the Triassic period of the Mesozoic about 200 million years ago. It took the place of the Reptilia which fast declined and developed after the Cenozoic era about 66 million years ago. The present period is called the "Mammalia era."

All the mammals except whales are haired over their whole body. They are distinguished with Aves from the poikilothermic animals such as fishes, frogs and lizards as the isothermal animals by their constant body temperature (35–38℃).

More than half of the 5,000 mammalian species living on the globe belong to the order Rodentia including rats and rabbits. They are fertile and full of vitality.

The Mammalia is grouped into 18 orders, of which Cetacea and Sirenia are marine.

The order Cetacea consists of about 100 species. The species, larger than 4 to 5 meters, are called "whale," while the smaller are named "dolphin." They are commonly, but not scientifically, used names. The dolphin belongs to the suborder Odontoceti which includes the sperm whale.

It is supposed that the whale developed from a land animal, as large as a dog or a cat, which moved to the river for food, and into the sea at last to live in it and that after a long time the animal has become huge.

The sea cow is an acquatic mammals to which the dugong and the manatee belong. Only one species of the dugong lives in the Indo-Pacific. It has a horizontally flat tail but no hind legs, looking like a whale and being close to the elephant from the taxonomical point of view.

Marine Reptiles

The Reptilia is the poikilothermic Vertebrata that live with pulmonary respiration, covered with keratinous scales and laying membranous or calcific eggs.

The Reptilia originated supposingly from the Seymouria which appeared at the end of the Carboniferous period of the Paleozoic era about 280 million years ago. It was in the Masozoic era when the Reptilia flourished. The era (between 230 million years ago and 66 million years ago) is called the "Reptilia or the Dinosaur era." However, the Reptilia which flourished for 200 million years on Earth decayed quickly in the Cenozoic era, and most of them except about 6,000 species, belonging to the orders Chelonia, Crocodilia, Ryncocephalia and Squamata, died out. The surviving orders are grouped into the turtle type with a carapace, the lizard type with arms and legs, and the snake type without them.

The marine reptiles are the sea turtle and the sea snake.

The sea turtle is a group of the Reptilia belonging to the order Chelonia. Originated from the Eunotosaurus which appeared in the Permian period of the Paleozoic era about 250 million years ago. All the species of the order Chelonia had the carapace then, too. At present there live 265 species belonging to 11 families of 2 suborders on Earth. 6 species of the sea turtle belonging to the families Cheloniidae and Dermochelyidae of the suborder Cryptodira are known. Originally the sea turtle is the marine reptiles which lives in the tropical and sub-tropical sea where the coral reef spreads.

The snake is the Reptilia belonging to the suborder Serpentes of the order Squamata, and differentiated from the Eosuchia in the Permian period of the Paleozoic era about 250 million years ago. The oldest fossil ever found was that of the genus *Pachyophis*, 40cm long, that lived in the Cretaceous period of the Mesozoic era about 130 million years ago. Present species are 2,700 belonging 330 genera of 13 families.

The sea snake is a venomous marine snake of which distribution is wide along the coast in the Pacific and the Indian Ocean.

Because of its specific ecological life, it used to be classified in an independent family Hydrophiidae, but in 1967 Dr. Macdowell of the U.S.A. disproved the classification according to the form of the palatines.

He reported that the sea snake of the genus *Laticauda* was not a kind of the family Hydrophiidae, but a snake in the near lineage of genus *Micrurus* distributing over South East Asia and

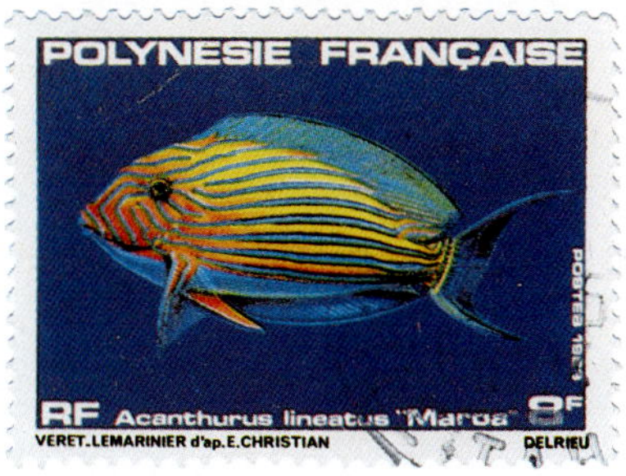

Australia which appeared later than the family Hydrophiidae, and became semi marine snake. Other sea snakes were classified in the group of cobras which had adopted themselves to the life in the sea and which had specialized from the same ancestor as the genus *Demansia* group of the family Elapidae. The former Hydrophiidae was included in the subfamily of the Elapidae. According to his theory, the Hydrophiidae cannot be regarded as an independent group, but this encyclopedia followed the former theory.

The sea snake is venomous with stationary fangs in the front part of the mouth. 52 species are found in the Indo-Pacific. Its distribution is 31 species in the Pacific, 6 in the Indian, and 15 in the Indo-Pacific, that is, 46 (88.5%) in the Pacific and 21 (40.4%) in the Indian Ocean.

It was said that the sea snake was gentle and did not bite at all, but, in fact, there are many which are fierce and aggressive. A large number of deaths caused by their bite has been reported. Those of the fishermen, while they are fishing or diving, are known.

The grooved teeth of the sea snake are furrowed and not as developed as the solenoglyphic teeth of the cobra and the habu. They do not inject much venoms, but it is seventy to eighty times, or one-hundred and fifty times in some kinds, as strong as that of the habu. If a small amount of it goes into a victim's body, it will paralyze his respiratory system and cause him difficult breathing. If it is too heavy, he will die. The sea snake is dangerous.

The bite of the sea snake does not cause pains nor swelling. In 15 minutes to 8 hours after some period of incubation muscles become rigid. They begin to ache. The languor, the paralysis of lips and tongue, and eyelid ptosis occur. The trouble of arms and legs gets worse as time passes and muscular reflex is soon lost. At last the muscles are paralyzed. In a heavy case urine turns brownish red and the victim dies of respiratory arrest in 14 to 18 hours.

Care must be taken because the venom of the snake has allergic antigen so that the victim gets Anaphylaxis after the first bite and has Anaphylactic shock at the second bite. If he does not die, he will suffer from conjunctivitis, nasitis, asthma, and hives.

Fishes

The fishes are the first Vertebrata in the Earth. Its history can be traced back to 500 million years ago (the latter period of the Cambrian). The oldest fossil of it is that of the Ostracodermi which lived in the Ordovician about 440 million years ago. Many of the ancient fishes appeared in the Devonian (400 million to 250 million years ago), so the period is called the "Pisces era."

The fishes means all the animals that belong to the subphylum Pisces. The classification differs according to the scientists. It is classified into four to eleven classes. The writer divides them into five classes.

(A) **Agnatha**

(1) **Agnatha** (jawless fishes) — The most primitive vertebrata groups. Lampreys and hagfishes belong to it. Present species are about 45 species of 11 genera in 2 families.

(B) **Gnathostomata**

(2) **Acanthodii** — Only fossil species.

(3) **Placodermi** (armored fishes) — Only fossil species.

(4) **Chondrichthyes** (cartilaginous fishes) — It appeared in the Devonian. Sharks, rays and ghostsharks are included in this group. It has 575 species of 132 genera in 31 families. It has cartilage but no air bladder. Male has a pair of clasper.

(5) **Osteichthyes** (bony fishes) — It is a group of ordinary fishes, having appeared in the Silurian period. It has about 17,600 species of 3,570 genera in 357 families and 32 orders including 2,000 species of catfish, 3,000 species of carp, 7,500 species of sea bass groups, and other 5,100 species of the 29 orders. It has hard bones. The groups is divided into two: the subclass Choanichthyes and the subclass Teleostomi.

These are 18,220 species of 3,713 genera in 390 families. The number of the species differs from 18,000 to 40,000 according to the scientists. These animals are characterized by the following points.

(1) They live in the water from birth to death.
(2) They breathe on oxygen from water, taking it through the gills.
(3) They have one atrium and one cardiac ventricle.
(4) They are poikilothermic, adopting themselves to the temperature of surroundings.
(5) They have scales covering over the body.
(6) Most of them are oviparous, being externally fertilized.
(7) Most of them have fins.
(8) The body consists of the head, the trunks, and the tail without the neck.

Most fishes are predatory and carnivorous. Some are herbivorous and omnivorous and others are parasitic or symbiotic. Primitive kinds live on land and breathe with the lung.

They live in sea and fresh waters. Some can live in cold water below 0℃, and others live in warm water in hot springs over 38℃. Their distribution is from 5,500 meters above the sea level to 13,000 meters deep under water.

The largest fish is the whale shark in the warm sea. A record reads 20 meters in total length and 40 tons in weight, but the largest fish which has been exactly measured is 12 meters long and 19.3 tons in weight.

The smallest fish is the genus *Mistychtys* a Philippine goby, about 1 centimeter long, the genus *Schindleria* from Samoa

weighs only 6 milligrams. The smallest shark is *Squariolus laticaudatus*, smaller than 15 centimeters. The heaviest fish is Sunfish, about 1 ton in weight. The fastest fish is the swordfishes that can swim a 100 kilometers per hour.

They are ecologically divided into two groups by their distribution; the horizontal distribution in the tropical, the subtropical, the temperate and the frigid zone, and the vertical distribution in the surface, the middle, the deep, and the super-deep stratum.

The largest number of fishes live in the Indo-West Pacific. The sea around South Africa follows it. Between 30 degrees of the northern and the southern latitude in the tropical and the subtropical zone plenty of coral reef fishes live. Most of them are the families Serranidae, Apogonidae, Chaetodontidae, Pomacentridae, Labridae, Scaridae, Acanthuridae in the order Perciformes.

It is a characteristic of the coral reef area that it is rich in species but not in population. The sea areas which are influenced by both the cold and the warm current are also abundant in species. In the places with ascending currents, in particular, the families Gadidae, Clupeidae, Pleuronectidae flourish. They are richer in population than in species, so that they are valued as fisheries resources.

As far as the fishes living deep under water are concerned, if 100 fishes live in waters from 50 to 200 meters, the number will decrease into 80 in 200 to 500 meters, 40 in 500 to 1,000 meters, and 30 below 1,000 meters.

The shape of the fishes are various, but there are generally six.

(1) **Fusiform** ——— The fishes swimming in the open seas are mostly fusiform. They are sharks, yellowtails, bonitoes and mackerels.

(2) **Compressiform** ——— Ordinary fishes have this form, being vertically flat. They live near the shore, inside the bay, deep in the bottom. Jacks, snappers, damselfishes, surgeonfishes and flounders belong to this group.

(3) **Depressiform** ——— The fishes are wide and horizontally flat, living in deep underwater. Rays, flatheads, anglerfishes and guitarfishes are in this group.

(4) **Cylindriform** ——— The fishes are narrow and long, living under the sand. Eels, conger eels, pike congers, and morays belong to this group.

(5) **Spheriform** ——— The fishes of this form have oval cross section. Their movement is slow. They are puffers and porcupinefishes, etc.

(6) **Squariform** ——— The fishes have four angular corners. Boxfishes and spiny boxfishes are of this form.

Basic forms of fishes

① **Fusiform**

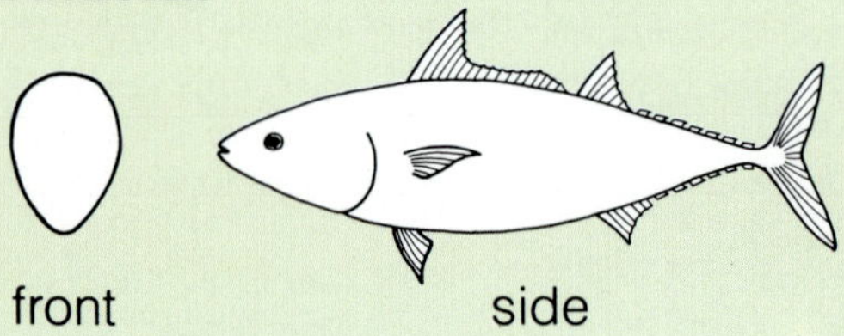

② **Compressiform**

③ **Depressiform**

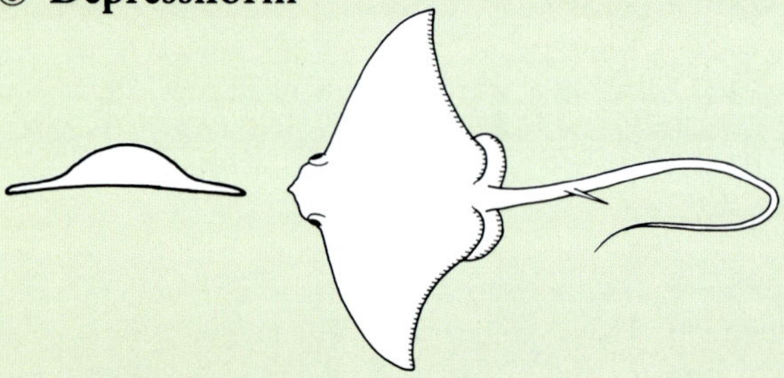

④ **Cylindriform**

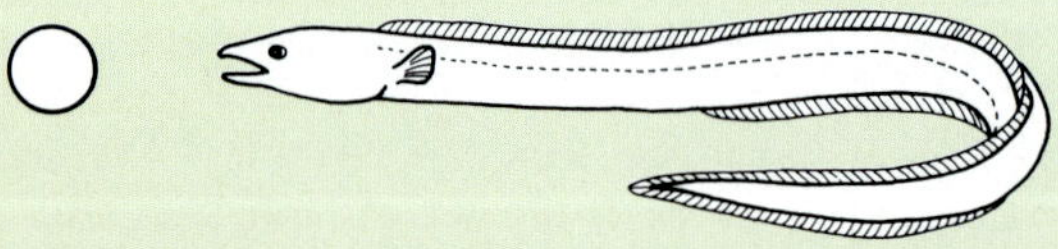

⑤ **Spheriform**

⑥ **Squariform**

BIBLIOGRAPHY

Abe, T., 1939: A List of the Fishes of the Palao Islands, Palao Tropical Stat. Studies, No.4

Abe. T., 1953–58: Figures and Descriptions of the Fishes of Japan. Kazama Shobo, Tokyo

Abe, T., 1957: Illustrated One Thousand Useful Fishes (2). Morikita Shuppan, Tokyo (in Japanese)

Abe, T., 1963: Key to the Marine Fishes of Japan Fully Illustrated in Colors. Hokuryukan, Tokyo (in Japanese)

Abe, T., 1970–82: Fishes of the New Face to Japan. Ito Foundation for Ichthyological Research, Tokyo (in Japanese)

Abe, T., 1976: Angling Marine Fishes. Heibonsha, Tokyo (in Japanese)

Agco, A. G., 1939: Two Rare Pediculate Fishes from the Philippines. Phil. Journ. Sci., Vol. 69, No.2

Akazaki, M., 1962: Studies on the Spariform Fishes, Anatomy, Phylogeny, Ecology and Taxonomy. Misaki Mar. Biol. Inst., Kyoto Univ., Sp. Rep., No.1 (in Japanese)

Akazaki, M., 1972: A Critical Study of Serranid Fishes of the Genus Chelidoperca Found in Japan. Jap. Journ. Ichthyol., Vol. 19, No.4 (in Japanese)

Akazaki, M., 1983: A New Lutjanid Fish, *Lutjanus stellatus*, from Southern Japan and a Related Species, *L. rivulatus* (Cuvier). Jap. Journ. Ichthyol., Vol. 29, No.4

Allen, G. R., J. E. Randall, 1974: Five New Species and a New Genus of Damselfishes (Family Pomacentridae) from the South Pacific Ocean. Trop. Fish Hobbyist, Vol.21, No.9

Allen, G. R., 1974: The Anemonefishes, 2nd Edition. T. F. H. Publ. Inc., Hongkong

Allen, G. R., 1975: Damselfishes of the South Seas, T. F. H. Publ., New Jersey

Allen, G. R., J. E. Randall, 1977: Review of the Sharpnose Pufferfishes (Subfamily Canthigasterinae) of the Indo-Pacific. Rec. Aust. Mus., Vol.30, No.17

Allen, G. R., 1979: Butterfly and Angelfishes of the World. Mergus Publ., West Germany

Allen, G. R., 1980: The Anemonefishes of the World. Aquarium Systems, Ohio

Allyn, R., 1967: Dictionary of Fishes. Great Outdoors Publ. Co., Florida

American Fisheries Society, 1980: A List of Common and Scientific Names of Fishes from the United States and Canada. American Fish. Soc., Maryland

Aoyagi, H., 1943: Coral Fishes. Pt.1, Maruzen Co., Tokyo

Aoyagi, H., 1952: Studies on the Coral Fishes of the Riu-Kiu Islands Ⅷ. Gobiesocidae, Limnichthidae, Parapercidae, Brotulidae, Bothidae and Soleidae. Zool. Mag., Vol.63, No.7 (in Japanese with English abstract)

Aoyagi, H., 1954: Description of One New Genus and Three New Species of Blennidae from the Riu-Kiu Islands. Zool. Mag., Vol.63, No.5 (in Japanese with English abstract)

Aoyagi, H., 1954: Additional Notes on the New and the Rare Fishes of the Family Blennidae from the Riu-Kiu Islands. Zool. Mag., Vol.63, No.6 (in Japanese with English abstract)

Aoyagi, H., 1954: Studies on the Coral Fishes of the Riu-Kiu Islands Ⅸ. Blennidae (1). Zool. Mag., Vol.63, No.7 (in Japanese with English abstract)

Aoyagi, H., 1955: Studies on the Coral Fishes of the Riu-Kiu Islands Ⅹ. Blennidae (2). Zool. Mag., Vol.64, No.3 (in Japanese with English abstract)

Aragaki, S., T. Yoshino, 1984: Color Illustrations of the Angling Fishes of Okinawa. Shinsei-tosho Shuppan, Okinawa (in Japanese)

Asahi Press, 1962: Who's Who of the Sea. Asahi Press, Tokyo (in Japanese)

Asahi Press, 1974: Weekly Encyclopedia of Animals of the World. Asahi Press, Tokyo (in Japanese)

Bagnis, R. A., 1968: Clinical Aspects of Ciguatera (Fish Poisoning) in French Polynesia. Hawaii Med. Jour., Vol.28, No.1

Bagnis, R. A., 1968: A Propos de 10 Cas de Blessures par Requins Chez des Pecheurs Sous-Marins en Polynésie Francaise. Méd. Tropicale, Vol.28, No.3

Bagnis, R. A., 1973: La Ciguatera aux Îles Marquises: Aspects Cliniques et Épidémiologiques. Bull. Org. Mond. Sante, Vol.49

Bagnis, R. A., 1973: Fish Poisoning in the South Pacific. South Pacific Commission, Noumea

Bagnis, R. A., 1979: L'ichtyosarcotoxisme de Type Ciguatera en Nouvelle-Caledonie. Aspects Cliniques et Épidémiologiques. Rev. Épidém. et Sante Publ., Vol.27

Bagnis, R. A., M. Rouanet, 1970: A Propos d'un Cas Mortel de Ciguatera dans les Îles Tuamotu. Rev. Méd. Vét.

Bagnis, R. A., P. Mazellier, J. Bennett, E. Christian, 1972: Fishes of Polynesia. Les Editions du Pacifique, Tahiti

Bagnis, R. A., S. Chanteau, T. Yasumoto, 1977: Un Agent Etiologique Vraisemblable de la Ciguatera. Comptes Rendus des Seances de l' Acad. Sci.

Bagnis, R. A., T. Kuberski, S. Laugier, 1979: Clinical Observations on 3009 Cases of Ciguatera (Fish Poisoning) in the South Pacific. Am. Journ. Trop. Med. Hyg., Vol.28, No.6

Banner, A. H., J. C. Nevenzel, W. R. Hudgins, 1969: Marine Toxins from the Pacific Ⅱ. The Contamination of Wake Island Lagoon. Atoll Res. Bull., No.122

Berry, P. Y., A. Aziz bio Hassan, 1973: Comparative Lethaity of Tissue Extracts from the Malaysian Puffer Fishes, *Lagocephalus lunaris lunais*, *L. l. spadiceus* and *Arothron stellatus*, Toxicon, Vol.11

Bleeker, P., 1862–77: Atlas Ichthyologique des Indes Orientales Neerlandaises. Amsterdam, 9 Vols.

Bloch, M. E., J. G. Schneider, 1801: Systema Ichthyologiae Iconibus ex Illustratum. Post Obitum Auctoris Opus Inchoatum Absolvit, Correxit, Interpolavit Jo. Gottlob Schneider, Sax. Perolini Ⅸ

Boschung, H. T., J. D. Williams, D. W. Gotshall, D. K. Caldwell, M. C. Caldwell, 1983: The Audubon Society Field Guide to North American Fishes, Whales, and Dolphins. Alfred A. Knopf, Inc., New York

Burgess, W. E., 1978: Butterflyfishes of the World. T. F. H. Publ. Inc., New Jersey

Burgess, W. E., H. R. Axelrod, 1972: Pacific Marine Fishes 1, Fishes of Southern Japan and the Ryukyus. T. F. H. Publ. Inc., New Jersey

Burgess, W. E., H. R. Axelrod, 1973: Pacific Marine Fishes 2, Fishes of Southern Japan and the Western Pacific. T. F. H. Publ. Inc., New Jersey

Burgess, W. E., H. R. Axelrod, 1973: Pacific Marine Fishes 3, Fishes of Sri Lanka, Maldive Islands and Mombasa. T. F. H. Publ. Inc., New Jersey

Burgess, W. E., H. R. Axelrod, 1974: Pacific Marine Fishes 4, Fishes of Taiwan and adjacent Waters. T. F. H. Publ. Inc., New Jersey

Burgess, W. E., H. R. Axelrod, 1974: Pacific Marine Fishes 5, Fishes of Taiwan and adjacent Waters. T. F. H. Publ. Inc., New Jersey

Burgess, W. E., H. R. Axelrod, 1975: Pacific Marine Fishes 6, Fishes of Melanesia. T. F. H. Publ. Inc., New Jersey

Burgess, W. E., H. R. Axelrod, 1976: Pacific Marine Fishes 7, Fishes of the Great Barrier Reef. T. F. H. Publ. Inc., New Jersey

Cameron, A. M., R. Endean, 1973: Epidermal Secretions and the Evolution of Venom Glands in Fishes, Toxicon, Vol.11

Carcasson, R. H., 1977: A Field Guide to the Reef Fishes of Tropical Australia and the Indo-Pacific Region. William Collins Sons & Co., Glasgow

Carr, A., 1969: Reptiles. Time Life International Inc., Tokyo (in Japanese)

Chang, K. H., S. C. Lee, 1968: Additions to the Notes on the Fishes Found in the Waters Around the Coastal Line of the Southern-most Part of Taiwan. Ann. Rep. Nat. Taiwan Mus., Taipei (in Chinese with English abstract)

Chang, K. H., S. C. Lee, 1971: Five Newly Recorded Gobies of Taiwan.

Bull. Inst. Zool., Academia Sinica, Vol.10, No.1
Chang, K. H., S. C. Lee, J. C. Lee, C. P. Chen, 1973: Ecological Study on Some Intertidal Fishes of Taiwan. Bull. Inst. Zool., Academia Sinica, Vol.12, No.2
Chang, K. H., C. P. Chen, 1974: The Newly Recorded Sandborers, *Sillago parvisquamis* Gill of Taiwan. Bull. Inst. Zool., Academia Sinica, Vol.13, No.1
Chang, K. H., S. C. Lee, J. C. Lee, W. L. Wu, 1977: Fishes of Reef Limestone Platform at Maopitou, Taiwan: Diversity and Abundance. Bull. Inst. Zool., Academia Sinica, Vol.16, No.1
Chang, K. H., S. C. Lee, K. T. Shao, 1978: A List of Forty Newly Recorded Coral Fishes in Taiwan. Bull. Inst. Zool., Academia Sinica, Vol.17, No.1
Chang, K. H., K. T. Shao, C. S. Hua, 1980: Coral-reef Fishes of Taiwan. Du-jia Press, Taipei (in Chinese)
Chang, K. H., R. Q. Jan, C. S. Hua, 1982: Fishes of Itu Aba Island in South China Sea. Encyclopedia Sinica Inc., Taipei (in Chinese)
Chang, K. H., R. Q. Jan, K. T. Shao, 1983: Community Ecology of the Marine Fishes of Lutao Island, Taiwan, Bull. Inst. Zool. Academia Sinica, Vol.22, No.2
Chen, J. T. F., 1963: A Review of the Sharks of Taiwan. Tunghai Univ. Taichung (in Chinese)
Christian, E., R. A. Bagnis, 1977: Underwater Guide of Tahiti. Les Editions du Pacifique, Tahiti
Clark, E., 1974: The Red Sea's Sharkproof Fish. Nat. Geograph. Mag., Vol. 146, No.5
Coleman, N., 1980: Australian Sea Fishes, South of 30°S. Doubleday Australia Pty. Ltd. , Sydney
Coleman, N., 1981: Australian Sea Fishes, North of 30°S. Doubleday Australia Pty. Ltd., Sydney
Cuvier, G., A. Valenciennes, 1828: Histoire Naturelle des Poissons. 22 Vols., Paris
Day, F., 1865: The Fishes of Malabar, London
Day, F., 1875: The Fishes of India: Being a Natural History of the Fishes Known to Inhabit the Seas and Fresh Waters of India, Burma, and Ceylon, London
Deraniyagala, P. E. P., 1949: Some Vertebrate Animals of Ceylon. Nat. Mus. Ceylon, Colombo
Dunson, W. A., 1975: The Biology of Sea Snakes. University Park Press, Maryland
Edwards, H., 1975: Sharks and Shipwrecks. Lansdowne Press, Melbourne
Ellis, R., 1982: Dolphins and Porpoises. Alfred A. Knopf, Inc., New York
Far Seas Fisheries Laboratory, 1972: Colored Illustrations of Bottomfishes Collected by Japanese Trawlers. Jap. Deep Sea Trawlers Ass., Tokyo (in Japanese)
Far Seas Fisheries Laboratory, 1976: Colored Illustrations of Bottomfishes Collected by Japanese Trawlers, Vol. Ⅱ. Jap. Deep Sea Trawlers Ass., Tokyo (in Japanese)
Faughnan, V. R., 1980: Sharks of Hawaii. Undersea Resources, Ltd., Honolulu
Fouques, M., J. Lagraulet, J. Tapu, R. Vidal, 1972: Traumatismes et Blessures par les Poissons en Polynesie Francaise. La Nouvelle Presse Médicale, No.47
Fourmanoir, P., P. Laboute, 1976: Poissons de Nouvelle Calédonie et des Nouvelles Hébrides. Les Editions du Pacifique, Tahiti
Fowler, H. W., 1967: The Fishes of Oceania. Mem. Bishop Mus., Vol. Ⅹ, Honolulu
Gakushu-kenkyusha, 1974: Learning Encyclopedia, Fishes and Shells, Gakushu-kenkyusha, Tokyo (in Japanese)
Gakushu-kenkyusha, 1975: Gakken School's Encyclopeida, Fishes. Gakushu-kenkyusha, Tokyo (in Japanese)
Gakushu-kenkyusha, 1981: Gakken Biological Encyclopedia, Animals. Gakushu-kenkyusha, Tokyo (in Japanese)
Goodson, G., 1973: The Many Splendored Fishes of Hawaii. Marquest Colorguide Books, California
Gosline, W. A., V. E. Brock, 1960: Handbook of Hawaiian Fishes. Univ. Press Hawaii, Honolulu
Grant, E. M., 1978: Guide to Fishes. Dept. Harbours & Marine, Queensland
Greenberg, J., I. Greenberg, 1981: Sharks and Other Dangerous Sea Creatures. Seahawk Press, Florida
Günther, A., 1859–70: Catalogue of the Fishes in the British Museum, 8 Vols. British Mus., London
Gushiken, S., 1972: Fishes of Okinawa in Colour. Ryukyu Fisheries Ass., Okinawa (in Japanese)
Hattori, J., 1976: The Pharyngeal Bones of Parrotfishes (Scaridae) Excavated from the Shell-Mounds in Okinawa and Amami, Japan. Jap. Journ. Ichthyol., Vol.22, No.4 (in Japanese with English abstract)
Hashimoto, Y., *et al*, 1967: Investigations on the Poisonous Fishes and Molluscus of the South West Islands, Japan Ⅲ., Tokyo Univ. (in Japanese)
Hashimoto, Y., *et al*, 1967: Investigations on the Poisonous Fishes and Molluscus of the South West Islands, Japan Ⅴ., Tokyo Univ. (in Japanese)
Hashimoto, Y., *et al*, 1967: Investigations on the Poisonous Fishes and Molluscus of the South West Islands, Japan Ⅵ., Tokyo Univ. (in Japanese)
Hashimoto, Y., N. Fusetani, 1968: A Preliminary Report on the Toxicity of an Amberjack, *Seriola aureovittata*. Bull. Jap. Soc. Sci. Fish., Vol.34, No.7
Hashimoto, Y., *et al*, 1969: Investigations on the Poisonous Fishes and Molluscus of the South West Islands, Japan Ⅺ., Tokyo Univ. (in Japanese)
Hashimoto, Y., H. Kamiya, 1969: Occurrence of a Toxic Substance in the Skin of a Sea Bass, *Pogonoperca punctata*. Toxicon, Vol.7
Hashimoto, Y., S. Konosu, T. Yasumoto, H. Kamiya, 1969: Ciguatera in the Ryukyu and Amami Islands. Bull. Jap. Soc. Sci. Fish., Vol.35. No.3
Hashimoto, Y., T. Yasumoto, H. Kamiya, T. Yoshida, 1969: Occurrence of Ciguatoxin and Ciguaterin in Ciguatoxic Fishes in the Ryukyu and Amami Islands. Bull. Jap. Soc. Sci. Fish., Vol.35, No.3
Hashimoto, Y., *et al*, 1970: Investigations on the Poisonous Fishes and Molluscus of the South West Islands, Japan Ⅻ., Tokyo Univ. (in Japanese)
Hashimoto, Y., T. Noguchi, 1971: Occurrence of a Tetrodotoxin-like Substance in a Goby, *Gobius criniger*. Toxicon, Vol.9
Hashimoto, Y., *et al*, 1973: Investigations on the Poisonous Fishes and Molluscus of the South West Islands, Japan ⅩⅤ., Tokyo Univ. (in Japanese)
Hashimoto, Y., *et al*, 1973: Investigations on the Poisonous Fishes and Molluscus of the South West Islands, Japan ⅩⅣ., Tokyo Univ. (in Japanese)
Hashimoto, Y., 1977: Marine Toxins. Univ. Tokyo Press, Tokyo (in Japanese)
Hayashi, J., 1968: Common Animals of the World in Colour Vol.1. Hoikusha, Osaka (in Japanese)
Heibonsha, 1974: Coloured Encyclopedia of Animals. Heibonsha, Tokyo (in Japanese)
Heibonsha, 1980: Grand Encyclopedia of the World. Heibonsha, Tokyo (in Japanese)
Helfrich, P., A. H. Banner, 1968: Ciguatera Fish Poisoning Ⅱ. General Paterns of Development in the Pacific. Occa. Pap., Bishop Mus., Vol.23, No.14
Helfrich, P., T. Piyakarnchana, P. S. Miles, 1968: Ciguatera Fish Poisoning 1. The Ecology of Ciguateric Reef Fishes in the Line Islands. Occa. Pap. Bishop Mus., Vol. 23, No.14
Herre, A. W. C. T., 1923: A Review of the Eels of the Philippine Archipelago. Phil. Journ. Sci., Vol.23, No.2
Herre, A. W. C. T., 1924: Poisonous and Worthless Fishes. Phil. Journ. Sci., Vol.25, No.4
Herre, A. W. C. T., 1925: Notes on Philippine Sharks Ⅱ. Phil. Journ. Sci., Vol.26, No.1
Herre, A. W. C. T., 1927: Gobies of the Philippines and the China Sea. Bur. Sci., Manila
Herre, A. W. C. T., 1927: A New Genus and Three New Species of Philippine Fishes. Phil. Journ. Sci., Vol.32, No.3
Herre, A. W. C. T., H. R. Montalban, 1927: The Philippine Species of Khuliidae. Phil. Journ. Sci., Vol.33, No.2
Herre, A. W. C. T., H. R. Montalban, 1927: Philippine Sparoid and

Rudder Fishes. Phil. Journ. Sci., Vol.33, No.4
Herre, A. W. C. T., H. R. Montalban, 1927: The Philippine Butterfly Fishes and Their Allies. Phil. Journ. Sci., Vol.34, No.1
Herre, A. W. C. T., 1927: Philippine Surgeon Fishes and Moorish Idols. Phil. Journ. Sci., Vol.34, No.4
Herre, A. W. C. T., H. R. Montalban, 1928: The Philippine Siganids. Phil. Journ. Sci., Vol.35, No.2
Herre, A. W. C. T., H. R. Montalban, 1928: The Goatfishes, or Mullidae, of the Philippines. Phil. Journ. Sci., Vol.36, No.1
Herre, A. W. C. T., 1928: The Philippine Gars or Needlefishes. Phil. Journ. Sci., Vol.36, No.2
Herre, A. W. C. T., 1939: The Philippine Blennies. Phil. Journ. Sci., Vol.70, No.4
Herre, A. W. C. T., E. S. Herald, 1950: Noteworthy Additions to the Philippine Fish Fauna with Descriptions of a New Genus and Species. Phil. Journ. Sci., Vol.79, No.3
Herre, A. W. C. T., 1951: A Review of the Scorpaenoid Fishes of the Philippines and Adjacent Seas. Phil. Journ. Sci., Vol.80, No.4
Herre, A. W. C. T., 1954: The Guitarfishes, Family Rhinobatidae of the Philippine and Adjacent Sea. Phil. Journ., Sci., Vol.83, No.4
Hirosaki, Y., T. Honma, 1979: One Hundred Species of the Angling Fishes. Hokuryukan, Tokyo (in Japanese)
Hiyama, Y., F. Yasuda, 1972: Edible and Poisonous Fishes of the Pacific Ocean. Kodansha, Tokyo (in Japanese)
Hiyama, Y., 1943: Survey Report on the Poisonous Fishes of South Seas. Nissan Fisheries Laboratory, Odawara (in Japanese)
Hobson, E. S., E. H. Chave, 1972: Hawaiian Reef Animals. Univ. Press Hawaii, Honolulu
Hoese, D. F., J. E. Randall, 1982: Indo-Pacific Fishes No.1, Revision of the Gobiid Fish Genus *Stonogobiops*. Bishop Mus., Honolulu
Ichthyological Society of Japan, 1981: Dictionary of Japanese Fish Names and Their Foreign Equivalents. Sanseido, Tokyo (in Japanese)
Imai, S., 1957: The Characteristics and Keys for the Determination of the Species of Flying-fishes Obtained in Japan. Journ., Fac. Fish., Kagoshima Uviv., Vol.5 (in Japanese with English abstract)
Imaizumi, Y., 1960: Coloured Illustrations of the Mammals of Japan. Hoikusha, Osaka (in Japanese)
Imaizumi, Y., *et al*, 1970: Animal Life. Japan Mail Order, Tokyo (in Japanese)
Imamura, T., 1961: New Classification of the Animal Kingdom, Hokuryukan, Tokyo (in Japanese)
Institute Oceanology, Academia Sinica, 1962: Economic Animals of China, Marine Fishes. Science Press, Beijing (in Chinese)
Institute Zoology, Institute Oceanology, Academia Sinica, Shanghai Fisheries College, 1962: The Fishes of South China Sea. Science Press, Beijing (in Chinese)
Ishiyama, R., 1967: Fauna Japonica, Rajidae, Biogeogr. Soc. Japan
Iwai, T., 1985: Fisheries Vertebrata Ⅱ. Pisces. Kouseisha-Kouseikaku, Tokyo (in Japanese)
Japan Fisheries Resource Conservation Association, 1982: Fishes of the Kyushu-Palao Ridge and Tosa Bay. Jap. Fish. Res. Conserv. Ass., Tokyo
Japan Fisheries Resource Conservation Association, 1984: Fishes of the Okinawa Trough and the Adjacent Waters (Ⅰ) Jap. Fish. Res. Conserv. Ass., Tokyo
Japan Fisheries Resource Conservation Association, 1985: Fishes of the Okinawa Trough and the Adjacent Waters (Ⅱ). Jap. Fish. Res. Conserv. Ass., Tokyo
Japan Marine Fishery Resource Research Center, 1977: Fishes of Indian Ocean. Jap. Mar. Fish. Res. Resear. Center, Tokyo
Japan Marine Fishery Resource Research Center, 1982: Fishes of the South China Sea. Jap. Mar. Fish. Res. Resear. Center, Tokyo
Johnson, R. H., 1978: Sharks of Polynesia. Les Editions du Pacifique, Tahiti
Jones, R. S., 1971: Two Nonfatal Shark Attacks in the Truk District, Eastern Caroline Islands. Micronesia, Vol.7
Jordan, D. S., B. W. Evermann, 1973: The Shore Fishes of Hawaii., C. E. Tuttle Co., Tokyo
Jordan, D. S., R. E. Richardson, 1910: Check-list of the Species of Fishes Known from the Philippine Archipelago. Dept. Interior, Bur. Sci., Manila
Kainuma, M.,1975: Book of Puffer Fishes. Shibata Shoten, Tokyo (in Japanese)
Kamohara, T., 1937: A Review of the Triacanthodid Fishes Found in the Water of Japan. Annot. Zool. Jap., XVI
Kamohara, T., 1938: Gemphylidae of Japan. Annot. Zool. Jap., XVII
Kamohara, T., 1940: Fauna Nipponica, Zeoidei, Rhegnopteri, Holconoti. Sanseido, Tokyo (in Japanese)
Kamohara, T., 1940: Fauna Nipponica, Scombroidei, Sec. Percomorphi. Sanseido, Tokyo (in Japanese)
Kamohara, T., 1940: Fauna Nipponica, Sclerodermi. Sanseido, Tokyo (in Japanese)
Kamohara, T., 1950: Description of the Fishes from the Province of Tosa and Kishu, Japan. Kochi Pref. Bunkyo Ass., Kochi (in Japanese)
Kamohara, T., 1950: Deep Sea Fishes. Nihon Shuppansha, Osaka (in Japanese)
Kamohara, T., 1952: Studies of the Family Peristediidae Found in Japan. Jap. Journ. Ichthyol., Vol.2, No.1 (in Japanese with English abstract)
Kamohara, T., 1955: Coloured Illustrations of the Fishes of Japan (1). Hoikusha, Osaka (in Japanese)
Kamohara, T., 1961: Coloured Illustrations of the Fishes of Japan (2). Hoikusha, Osaka (in Japanese)
Kamohara, T., 1962: Notes on Six Additions to the Marine Fish Fauna of Kochi Prefecture, Japan. Rep. Usa Mar. Biol. Stat., Kochi Univ., Vol.9, No.2
Kamohara, T., 1963: On the Fishes of the Family Scaridae of Japan, Including the Riu Kiu Islands. Rep. Usa Mar. Biol. Stat., Kochi Univ., Vol.10, No.1
Kamohara, T., T. Yamakawa, 1965: Fishes from Amami-Oshima and Ajacent Regions. Rep. Usa Mar. Biol. Stat., Kochi Univ., Vol.12, No.2
Kamohara, T., 1966: Standard Colour Illustrations of the Fishes. Hoikusha, Osaka (in Japanese)
Kamohara, T., T. Yamakawa, 1967: On Some Fishes from the Waters of Okinawa and Yaeyama. Rep. Usa Mar. Biol. Stat., Kochi Univ., Vol.14, No.1
Kamohara, T., T. Yamakawa, 1968: Additional Records of Marine Fishes from Amami. Rep. Usa Mar. Biol. Stat., Kochi Univ., Vol.15, No.1
Kamohara, T., T. Yamakawa, 1968: Additional Records of Marine Fishes from Amami (Ⅱ). Rep. Usa Mar. Biol. Stat., Kochi Univ., Vol.15, No.2
Kataoka, T., S. Yagi, S. Manabe, M. Sekido, 1966: Butterflyfishes of Okinawa. Mie Seibutu, No.16 (in Japanese)
Katayama, M., 1959: Studies on the Serranid Fishes of Japan (Ⅰ). Bull. Fac. Edu., Yamaguchi Univ. 8, Pt.2
Katayama, M., 1960: Fauna Japonica, Serranidae. Biogeogr. Soc. Japan
Katayama, M., 1960: Studies on the Serranid Fishes of Japan (Ⅱ). Bull. Fac. Edu., Yamaguchi Univ., 9, Pt.2
Katayama, M., 1963: A New Genus and Species of Anthinid Fish from Sagami Bay, Japan. Bull. Fac. Edu., Yamaguchi Univ., 13, Pt.2
Katayama, M., 1973: Serranid Fishes of the Okinawa Islands (Ⅰ). Bull. Fac. Edu., Yamaguchi Univ., 23, Pt.2
Katayama, M., 1974: On the Anthinid Fish, *Mirolabrichthys tuka pascarus* Jordan et Tanaka, Obtained from Okinawa. Bull. Fac. Edu., Yamaguchi Univ., 24, Pt.2 (in Japanese)
Katayama, M., 1974: Serranid Fishes of the Okinawa Islands (Ⅱ), Bull. Fac. Edu., Yamaguchi Univ., 24, Pt.2
Katayama, M., 1975: Serranid Fishes of the Okinawa Islands (Ⅲ). Bull. Fac. Edu., Yamaguchi Univ., 25, Pt.2
Katayama, M., 1978: Serranid Fishes of the Okinawa Islands (Ⅳ). UO No.29.
Katayama, M., T. Taki, 1984: *Lates japonicus*, a New Centropomid Fish from Japan. Jap. Journ. Ichthyol., Vol.30, No.4
Kenny, N. T., 1968: Sharks: Wolves of the Sea. Nat. Geograph. Mag., Vol.133, No.2

Kimura, S., 1979: Ecology of Marine Fishes. Kankaido Shuppan, Chiba (in Japanese)
Kobayashi, T., *et al*, 1954: Palaeontology (1). Asakura Shoten, Tokyo (in Japanese)
Kosaki, T. I., B. J. Stephens, H. H. Anderson, 1968: Marine Toxins from the Pacific Ⅲ. Comparative Bio-assay of Ciguatoxin(s) in the Mouse and Chicken. Proc. West. Pharm. Soc., Vol.11
Kosaki, T. I., H. H. Anderson, 1968: Marine Toxins from the Pacific Ⅳ. Pharmacology of Ciguatoxin(s). Toxicon, Vol.6
Kuroda, N., 1951: A Nominal List with Distribution of the Fishes of Suruga Bay, Inclusive of the Freshwater Species Found Near the Coast. Jap. Journ. Ichthyol., Vol.1, No.5 (in Japanese)
Kuronuma, K., Y. Abe, 1972: Fishes of Kuwait. Kuwait Inst. Sci. Resear., Kuwait
Laboute, P., Y. Magnier, 1979: Underwater guide to New Caledonia. Les Editions du Pacifique, Tahiti
Lacépède, B. G., 1798–1803: Histoire Naturelle des Poissons. 5 Vols., Paris
Lagraulet, J., J. Tapu, G. Guzon, M. Fouques, 1973: Piqûres par les Poissons Venimeux en Polynésie Francaise. La Nou. Pres. Med. 14
Lee, S. C., H. C. Yang, 1983: Fishes of the Suborder Scombroidei of Taiwan. Bull. Inst. Zool. Academia Sinica, Vol.22, No.2
Leis, J. M., D. S. Rennis, 1949: The Larvae of Indo-Pacific Coral Reef Fishes. Univ. Press Hawaii, Honolulu
Lindberg, G. U., A. S. Heard, T. S. Rass, 1980: Multilingual Dictionary of Names of Marine Food-fishes of World Fauna. Ministry of Fish., USSR
Lindsey, C. C., 1963: Guide to Families of Malaysian Fishes. Dept. Zool., Univ. Singapore, Singapore
Mainichi Press, 1975: The Fishes of Shikoku Island. Shikoku Mainichi Koukokusha, Takamatsu (in Japanese)
Martin, C., H. R. Montalban, 1934: Philippine Sillaginidae. Phil. Journ. Sci., Vol.55, No.3
Martin, C., H. R. Montalban, 1935: Philippine Parapercidae. Phil. Journ. Sci., Vol.56, No.2
Masuda, H., T. Abe, 1972: Sea Fantasy, Wonders of Fish Life in Japan's Coastal Waters, Mainichi Press, Tokyo (in Japanese)
Masuda, H., C. Araga, T. Yoshino, 1975: Coastal Fishes of Southern Japan. Tokai Univ. Press, Tokyo (in Japanese)
Masuda, H., 1984: Field Encyclopedia on Marine Fishes. Tokai Univ. Press, Tokyo (in Japanese)
Masuda, H., K. Amaoka, C. Araga, T. Uyeno, T. Yoshino, 1984: The Fishes of the Japanese Archipelago. Tokai Univ. Press, Tokyo (in Japanese and English)
Matsubara, K., 1955: Fish Morphology and Hierarchy Ⅰ-Ⅲ. Ishizaki Shoten, Tokyo (in Japanese)
Matsubara, K., A. Ochiai, T. Iwai, 1970: Ichthyology (1). Kouseisha-Kouseikaku, Tokyo (in Japanese)
Matsui, I. 1977: The Book of Eel. Shibata Shoten, Tokyo (in Japanese)
Matsuura, K., M. Toda, 1981: First Records of Two Pufferfishes, *Arothron mappa* and *A. reticularis*, from Japan. Jap. Journ. Ichthyol., Vol.28, No.1
McGraw-Hill·Kodansha, 1977: Encyclopedia of Science and Technology. Kodansha, Tokyo (in Japanese)
Mcwhirter, N., 1985: The Guinness Book of Records, Guinness Superlatives Ltd., England
Mendis, A. S., 1954: Fishes of Ceylon. Bull. No.2, Fish. Resear. Stat., Ceylon
Miyaji, D., H. Kawanabe, N. Mizuno, 1963: Coloured Illustrations of the Freshwater Fishes of Japan. Hoikusha, Osaka (in Japanese)
Montilla, J. R., 1935: A Review of Philippine Menidae and Gerridae. Phil. Journ. Sci., Vol.58, No.2
Montilla, J. R., 1938: A Review of Philippine Holocentridae. Phil. Journ. Sci., Vol.67, No.2
Montalban, H. R., 1972: Pomacentridae of the Philippine Islands. Bur. Sci., Manila
Munro, I. S. R., 1955: The Marine and Fresh Water Fishes of Ceylon. Dept. Ext. Affairs, Canberra
Munro, I. S. R., 1967: The Fishes of New Guinea, Dept. Agr., Stock Fish., Port Moresby
Nakamura, I., R. Nakamura, 1982: New Records of Two Species of *Scomberomorus* from Japan. Jap. Journ. Ichthyol., Vol.28, No.4
Nakamura, K., S. Uyeno, 1963: Japanese Reptiles and Amphibians in Colour. Hoikusha, Osaka (in Japanese)
National Geographic Society, 1965: Wondrous World of Fishes. Nat. Geograph. Soc., Washington, D. C.
Nelson, J. S., 1984: Fishes of the World, Second Edition. John Wiley & Sons, Inc., Toronto
Obunsha, 1967: Colored Encyclopedia, Fishes and Shells. Obunsha, Tokyo (in Japanese)
Obunsha, 1976: Learning Encyclopedia, Fishes. Obunsha, Tokyo (in Japanese)
Ochiai, A., 1968: Ecology of Fishes. Soyosha, Tokyo (in Japanese)
Ochiai, A., 1970: Fishes of Tosa. Marunouchi Shuppan, Tokyo (in Japanese)
Ogura, Y., J. Nara, T. Yoshida, 1968: Comparative Pharmacological Actions of Ciguatoxin and Tetrodotoxin, a Preliminary Account. Toxicon, Vol.6
Okada, Y., 1938: A Catalogue of Vertebrates of Japan, Maruzen Co., Tokyo
Okada, Y., *et al*, 1950: Animals of Japan, Vertebrata (Ⅰ). Nihon Shuppansha, Tokyo (in Japanese)
Okada, Y., K. Matsubara, 1953: Checklist of the Bibriographies on Japanese Fishes. Mie Pref. Univ. (in Japanese)
Okada, Y., 1959: Checklist of the Animals in Ryukyus. Okinawa Biol. Ass. for Education (in Japanese)
Okada, Y., *et al*, 1965: New Outlines of Biology (3), Vertebrata (A). Nakayama Shoten, Tokyo (in Japanese)
Okada, Y., *et al*, 1965: New Illustrated Encyclopedia of the Fauna of Japan. Hokuryukan, Tokyo (in Japanese)
Okada, Y., H. Masuda, 1971: Marine Creatures, Fishes. Yamatokeikokusha, Tokyo (in Japanese)
Okinawa Biological Association for Education, 1976: Creatures of Okinawa. Shinsei-tosho, Okinawa (in Japanese)
Oshima, M., 1925: A Review of the Carangoid Fishes Found in the Waters of Formosa. Phil. Journ. Sci., Vol.26, No.3
Oshima, M., 1940: Systematic Vertebrates, Fishes. Sanseido, Tokyo (in Japanese)
Pepper, S. J., 1975: Sharks. Inform. Bull. No.1, Air Training Command, Combat Crew Training Wing Environmental Information Division, Alabama
Pope, P. E., 1949: Dictionary of Sharks. Great Outdoors Publ. Co., Florida
Randall, J. E., 1955: Fishes of the Gilbert Islands. Atoll Res. Bull., 47
Randall, J. E., 1955: An Analysis of the Genera of Surgeon Fishes (Family Acanthuridae). Pac. Sci., Vol.9, No.3
Randall, J. E., 1955: A Revision of the Surgeon Fish Genera *Zebrasoma* and *Paracanthurus*. Pac. Sci., Vol.9, No.4
Randall, J. E., 1956: A Revision of the Surgeon fish Genus *Acanthurus*. Pac. Sci., Vol.10, No.2
Randall, J. E., 1960: A New Species of *Acanthurus* from the Caroline Islands, with Notes on the Systematics of Other Indo-Pacific Surgeonfishes. Pac. Sci., Vol.14, No.3
Randall, J. E., 1961: Two New Butterflyfishes (Family Chaetodontidae) of the Indo-Pacific Genus *Forcipiger*. Copeia, 1961, 1
Randall, J. E., D. K. Caldwell, 1970: Classification of the Species of the Butterflyfish Genus *Forcipiger*. Copeia, 1970, 4
Randall, J. E., K. Aida, T. Hibiya, N. Mitsuura, H. Kamiya, Y. Hashimoto, 1971: Grammistin, the Skin Toxin of Soapfishes, and its Significance in the Classification of the Grammistidae. Publ. Seto Mar. Bio. Lab., Vol.14, Nos.2–3
Randall, J. E., 1972: The Hawaiian Trunkfishes of the Genus *Ostracion*. Copeia, 1972, 4
Randall, J. E., 1972: A Revision of the Labrid Fish Genus *Anampses*. Micronesia, Vol.8, Nos.1–2
Randall, J. E., 1975: A Revision of the Indo-Pacific Angelfish Genus *Genicanthus*, with Descriptions of Three New Species. Bull. Mar. Sci., Vol.25, No.3
Randall, J. E., K. Matsuura, A. Zama, 1978: A Revision of the Triggerfish Genus *Xanthichthys*, with Description of a

New Species. Bull. Mar. Sci., Vol.28, No.4
Randall, J. E., S. C. Shen, 1978: A Review of the Labrid Fishes of the Genus *Cirrhilabrus* from Taiwan, with Description of a New Species. Bull. Inst. Zool., Academia Sinica, Vol.17, No.1
Randall, J. E., 1978: A Revision of the Indo-Pacific Labrid Fish Genus *Macropharyngodon*, with Descriptions of Five New Species. Bull. Mar. Sci., Vol.28, No.4
Randall, J. E., 1979: A Review of the Serranid Fish Genus *Anthias* of the Hawaiian Islands, with Descriptions of Two New Species. Los Angeles Count. Mus., Contr. Sci., No.302
Randall, J. E., 1980: A Survey of Ciguatera at Eniwetok and Bikini, Marshall Islands, with Notes on the Systematics and Food Habits of Ciguatoxic Fishes. Fish. Bull., Vol.78, No.2
Randall, J. E., 1980: Revision of the Fish Genus *Plectranthias* (Serranidae: Anthiinae) with Descriptions of 13 New Species. Micronesia, Vol.16, No.1
Randall, J. E., H. Ida, J. T. Moyer, 1981: A Review of the Damselfishes of the Genus *Chromis* from Japan and Taiwan, with Description of a New Species. Jap. Journ. Ichthyol., Vol.28, No.3
Randall, J. E., P. Struhsaker, 1981:*Naso maculatus*, a New Species of Acanthrid Fish from the Hawaiian Islands and Japan. Copeia, 1981, 3
Randall, J. E., 1981: Waterproof Underwater Guide to Hawaiian Reef Fishes. Harrowood Books, Pennsylvania
Randall, J. E., T. Shimizu, T. Yamakawa, 1982: A Revision of the Holocentrid Fish Genus *Ostichthys*, with Descriptions of Four New Species and a Related New Genus. Jap. Journ. Ichthyol., Vol.29, No.1
Randall, J. E., A. Ben-Tuvia, 1983: A Review of the Groupers (Pisces: Serranidae: Epinephelinae) of the Red Sea, with Description of a New Species of *Cephalopholis*. Bull. Mar. Sci., Vol.33, No.2
Randall, J. E., 1983: Red Sea Reef Fishes. Immel Publ., London
Read, K. R. H., 1971: Nonfatal Shark Attack, Palau Islands. Micronesia, Vol.7
Ricciuti, E. R., 1973: Killers of The Sea. Macmillan Publ. Inc., New York
Robins, C. R., 1980: A List of Common and Scientific Names of Fishes from the United States and Canada. Am. Fish. Soc., Sp. Publ. 12, Maryland
Romashko, S., 1974: Sharks, Lord of the Sea. Windward Publ. Inc., Florida
Roxas, H. A., 1934: A Review of Philippine Mugilidae. Phil. Journ. Sci., Vol.54, No.3
Roxas, H. A., A. G. Agco, 1941: A Review of Philippine Carangidae. Phil. Journ. Sci., Vol.74, No.1
Scheuer, P. J., W. Takahashi, J. Tsutsumi, T. Yoshida, 1967: Ciguatoxin: Isolation and Chemical Nature. Science, Vol. 155
Schroeder, R. E., 1980: Philippine Shore Fishes of the Western Sulu Sea. Nat. Media Production Center, Manila
Seale, A., 1909: New Species of Philippine Fishes. Phil. Journ. Sci., Vol.4, No.6
Seale, A., 1910: Fishes of Borneo, with Descriptions of Four New Species. Phil. Journ. Sci., Vol.5, No.4
Seale, A., 1912: Some Poisonous Philippine Fishes. Phil. Journ., Sci., Vol.7, No.4
Seale, A., 1914: Fishes of Hongkong. Phil. Journ. Sci., Vol.9, No.1
Sengoku, S., 1979: Color Illustrated Amphibia and Reptilia of Japan. Ie-no-hikari Ass., Tokyo (in Japanese)
Senou, H., T. Suzuki, 1980: The Inland Water Fishes of the Yaeyama Islands, Okinawa Prefecture, Japan 1. Nanki Seibutu, Vol.22 (in Japanese)
Senou, H., T. Suzuki, 1980: The Inland Water Fishes of the Yaeyama Islands, Okinawa Prefecture, Japan 2. Nanki Seibutu, Vol.22 (in Japanese)
Senou, H., T. Suzuki, 1981: The Inland Water Fishes of the Yaeyama Islands, Okinawa Prefecture, Japan 3. Nanki Seibutu, Vol.23 (in Japanese)
Shen, S. C., 1984: Coastal Fishes of Taiwan. Dept. Zool., Nat. Taiwan Univ., Taipei
Shibuzawa, K., 1958: Compilation of the Common Names on the Fishes of Japan. Kadokawa Shoten, Tokyo (in Japanese)
Shiino, S. M., 1972: List of English Names of Japanese Fishes with Proposition of New Names. Sci. Rep. Shima Marineland, No.1
Shiino, S. M., 1976: List of Common Names of Fishes of the World, Those Prevailing among English-speaking Nations. Sci. Rep. Shima Marineland, No.4
Shimizu, T., T. Yamakawa, 1979: Review of the Squirrelfishes (Subfamily Holocentrinae: Order Beryciformes) of Japan, with a Description of a New Species. Jap. Journ. Ichthyol., Vol.26, No.2
Shinohara, S., 1960: Review of the Luthanidae (Snappers) Found in the Waters of Ryukyu (1). Journ. Univ. Ryukyus, Ser. Sci., No.4 (in Japanese with English abstract)
Shinohara, S., 1962: Review of the Luthanidae (Snappers) Found in the Waters of Ryukyu (2). Journ. Univ. Ryukyus, Ser. Sci., No.5 (in Japanese with English abstract)
Shinohara, S., 1963: Review of the Lethrinidae Found in the Waters of Ryukyus. Journ. Univ. Ryukyus, Ser. Sci., No.6 (in Japanese with English abstract)
Shinohara, S., 1964: On the Distribution of the Luthanid Fishes. Biol. Mag. Okinawa, Vol.1, No.1 (in Japanese)
Shinohara, S., 1966: Studies on Morphology, Taxonomy and Distributions of the Luthanid Fishes Found in the Waters of Ryukyus. Journ. Univ. Ryukyus, Ser. Sci., No.9 (in Japanese with English abstract)
Shirano, K., 1883: Color Illustrated Fishes of the Kagoshima Waters. Shimazu Shuppan-kai, Kagoshima (in Japanese)
Shirai, S., 1974: Marine Animals of Okinawa, Fudokisha, Okinawa (in Japanese)
Shirai, S., 1977: Ecological Encyclopedia of the Marine Animals of the Ryukyu Islands. Shinsei-tosho, Okinawa (in Japanese)
Shirai, S., Y. Sano, 1977: Nature of Okinawa, Fishes. Shinsei-tosho, Okinawa (in Japanese)
Shirai, S., 1982: Color Encyclopedia of the Poisonous and Injurious Animals and Plants of the Ryukyu Islands. Shinsei-tosho, Okinawa (in Japanese)
Shirai, S., 1982: Morphology of Fishes. Iwasaki Shoten, Tokyo (in Japanese)
Shirai, S., 1985: Fishes of Kagoshima Prefecture. Shin Nippon Kyoiku Tosho, Tokyo (in Japanese)
Smith, J. L. B., 1949: The Sea Fishes of Southern Africa. Central News Agency, Cape Town
Smith, J. L. B., 1956: The Parrotfishes of the Family Callyodontidae of the Western Indian Ocean. Ichthyol. Bull., Dept. Ichthyol., Rhodes Univ., 1
Smith, J. L. B., 1959: The Identity of *Scarus gibbus* Rüppell, 1828 and of Other Parrotfishes of the Family Callyodontidae from the Red Sea and the Western Indian Ocean. Ichthyol. Bull., Dept. Ichthyol., Rhodes Univ., 16
Smith, J. L. B., 1960: Fishes of the Family Gobiidae in South Africa. Ichthyol. Bull., Dept. Ichthyol., Rhodes Univ., 18
Smith, J. L. B., 1960: Coral Fishes of the Family Pomacentridae from the Western Indian Ocean and the Red Sea. Ichthyol. Bull., Dept. Ichthyol., Rhodes Univ., 19
Smith, J. L. B., 1962: The Moray Eels of the Western Indian Ocean and the Red Sea. Ichthyol. Bull., Dept. Ichthyol., Rhodes Univ., 23
Smith, J. L. B., M. M. Smith, 1969: The Fish of Seychelles. Smith Inst. Ichthyol., S-Africa
Smith, J. L. B., 1967–70: Studies in Carangid Fishes. 1–5, Occas. Pap., Dept. Ichthyol., Rhodes Univ., 12–15
Smith, J. L. B., 1977: Smith's Sea Fishes. Trust. Sea Fish Southern Africa Book Fund., South Africa
Snyder, J. O., 1904: A Catalogue of the Shore Fishes Collected by the Steamer "Albatross" About the Hawaiian Islands in 1892. Bull. U. S. Fish. Comm., Vol.22
Snyder, J. O., 1907: A Review of the Mullidae, Surmullets or Goatfishes of the Shore of Japan. Proc. U. S. Nat. Mus., Vol.32
Snyder, J. O., 1908: Descriptions of Eighteen New Species and Two New Genera of Fishes from Japan and the Riu Kiu Islands. Proc. U. S. Nat. Mus., Vol.35

Snyder, J. O., 1909: Descriptions of New Genera and Species of Fishes from Japan and the Riu Kiu Islands. Proc. U. S. Nat. Mus., Vol.36

Snyder, J. O., 1911: Descriptions of New Genera and Species of Fishes from Japan and the Riu Kiu Islands. Proc. U. S. Nat. Mus., Vol.40

Snyder, J. O., 1912: Japanese Shore Fishes Collected by the United States Bureau of Fisheries Steamer "Albatross" Expedition of 1906. Proc. U. S. Nat. Mus., Vol.42

Snyder, J. O., 1912: The Fishes of Okinawa, One of the Riu Kiu Islands. Proc. U. S. Nat. Mus., Vol.42

South China Sea Fisheries Institute, 1979: The Fishes of the Islands in the South China Sea. Science Press, Beijing (in Chinese)

Steene, R. C., 1978: Butterfly and Angelfishes of the World, Vol.1 Australia. Reed Pty Ltd., Sydney

Staenes, W. C., K. Mochizuki, 1982: Occurrence of the Percichthyid Fish *Neoscombrops pacificus* Near Samoa. Jap. Journ. Ichthyol, Vol.29, No.3

Strasburg, D. W., 1966: New Fish Records from Hawaii: *Hime, Pikea,* and *Omobranchus*. Pac. Sci., Vol.20, No.1

Suvatti, C., 1981: Fishes of Thailand. Royal Inst. Thailand, Bangkok

Suyehiro, Y., 1951: Text Book of Ichthyology. Iwanami Shoten, Tokyo (in Japanese)

Suyehiro, Y., 1968: Colour Books, Fishes of Japan. Hoikusha, Osaka (in Japanese)

Suyehiro, Y., T. Abe, T. Kanno, 1973: Encyclopedia of Fishes and Shells. Shogakukan, Tokyo (in Japanese)

Suzuki, K., K. Kobayashi, 1972: Fauna of the Shallow Water Fishes Around South and West Izu Peninsula. Rep. Resear. Shizuoka Marine Parks (in Japanese)

Sylva, D. P. de, 1963: Systematics and Life History of the Great Barracuda, Studies in Trop. Oceanography No.1. Univ. Miami Press, Florida

Takada, W., T. Uyeno, 1978: A Record of the Filefish, *Amanses scopas*, Collected from the Amami Islands, Japan. Jap. Journ. Ichthyol., Vol. 25, No.2 (in Japanese with English abstract)

Takemura, Y., F. Yasuda, 1965: The Sea Fishes of Japanese Coastal Waters. Hokuryukan, Tokyo (in Japanese)

Tanaka, S., 1905: On Two New Species of *Chimaera*. Journ. Coll. Sci. Imp. Univ. Tokyo, Vol.20, Art.11

Tanaka, S., 1908: Descriptions of Eight New Species of Fishes from Japan. Annot. Zool. Jap. Vol.7

Tanaka, S., 1908: Notes on Some Japanese Fishes, with Descriptions of Fourteen New Species. Journ. Coll. Sci. Imp. Univ. Tokyo, Vol.23

Tanaka, S., 1908: Notes on Some Rare Fishes of Japan, with Descriptions of Two New Genera and Six New Species. Journ. Coll. Sci. Imp. Univ. Tokyo, Vol.23

Tanaka, S., 1909: Descriptions of One New Genus and Ten New Species of Japanese Fishes. Journ. Coll. Sci. Imp. Univ. Tokyo, Vol.27

Tanaka, S., 1911–30: Figures and Descriptions of the Fishes of Japan, Including Riukiu Islands, Bonin Islands, Formosa, Kurile Islands, Korea and Southern Sakhalin. No.1–48, Tokyo (in Japanese and English)

Tanaka, S., 1926: The Edible Fishes. Nankosha, Tokyo (in Japanese)

Tanaka, S., 1931: Dr. Tanaka's Japanese Fishes in Life Colours. Daichi Shoin, Tokyo (in Japanese)

Tanaka, S., 1942: Deep Sea Fishes. Sangabou, Tokyo (in Japanese)

Tanaka, S., T. Abe, 1955: Illustrated One Thousand Useful Fishes of Japan (2). Morikita Shuppan, Tokyo (in Japanese)

Tani, I., 1945: Toxicological Studies of the Japanese Puffer Fishes. Teikoku Shoin, Tokyo (in Japanese)

Tani, I., 1948: The Puffer Fishes. Sougensha, Osaka (in Japanese)

Taniuchi, T., 1974: Three Species of Hammerhead Sharks in the Southwestern Waters of Japan. Jap. Journ. Ichthyol., Vol.21, No.3

Taniuchi, T., 1976: Sharks. Diving World-sha, Tokyo (in Japanese)

Taniuchi, T., 1979: Species and Distribution of Pelagic Sharks Caught by Tuna Longline. Aquabiology, Vol.4, Nos.1–4 (in Japanese)

Taniuchi, T., H. Tachikawa, Y. Kurata, Y. Nose, 1985: Galapagos Sharks, *Carcharhinus galapagensis*, from the Ogasawara Islands, Japan. Jap. Journ. Ichthyol., Vol.31, No.4 (in Japanese)

Tateishi, A., 1973: Corals and Coral Fishes of Okinawa. Fudokisha, Okinawa (in Japanese)

Tayama, J., 1981: The Story of Tuna. Kyoritsu Shuppan, Tokyo (in Japanese)

Teng, H. T., 1962: Taxonomic Studies and Distributions of the Serachii Fishes of Taiwan (in Japanese)

Tinker, S. W., C. J. DeLuca, 1973: Sharks and Rays. C. E. Tuttle Co., Tokyo

Tinker, S. W., 1978: Fishes of Hawaii. Hawaiian Service, Inc., Honolulu

Tomiyama, I., T. Abe, T. Tokioka, 1958: Encyclopedia Zoologica Illustrated in Colours Ⅱ. Hokuryukan, Tokyo (in Japanese)

Uchida, S., *et al*, 1947: Illustrated Encyclopedia of the Fauna of Japan. Hokuryukan, Tokyo (in Japanese)

Uchida, T., 1965: Basis of Systematic Animal Taxonomy. Hokuryukan, Tokyo (in Japanese)

Ui, H., 1939: Fishes of Kii Peninsula. Yodoya Shoten, Osaka (in Japanese)

Van der Elst, R., 1981: A Guide to the Common Sea Fishes of Southern Africa. C. Struik Publ. Ltd., Cape Town

Wakiya, Y., 1924: The Carangoid Fishes of Japan. Ann. Carneg. Mus., Vol. XV, Nos.2–3

Webb, J. E., J. A. Wallwork, J. H. Elgood, 1981: Guide to Living Fishes. Macmillan Press Ltd., London

Weber, M., L. F. de Beaufort, 1923–29: The Fishes of the Indo-Australian Archipelago, 7 Vols. E. J. Brill, Leiden

White, A. W., M. A. Barwani, 1980: Common Sea Fishes of the Arabian Gulf and Gulf of Oman. Trucial States Council, Oman

Whiteley, G. P., 1927: A Check List of Fishes Recorded from Fijian Waters. Journ. Pan-Pacific Res. Inst., Ⅱ, No.1

Wood, L. G., 1982: Guiness World, Animals. Kodansha, Tokyo (in Japanese)

Yamada, U., 1976: On Young Stages of *Pampus echinogaster* from the East China Sea. UO, No.27 (in Japanese)

Yamada, U., T. Nakabou, 1983: Morphology and Ecology of *Satyrichthys rieffeli* (Family: Peristediidae) from Japan. UO, No.33 (in Japanese with English abstract)

Yamaguchi Prefecture, 1982: Puffer Fishes. Yamaguchi Pref. Food Hyg. Ass. (in Japanese)

Yamakawa, T., 1966: On the Damsefish, *Teixeirichthys mossambicus* Smith, New to the Fish Fauna of Japan. Jap. Journ. Ichthyol., Vol.14, Nos.1–3 (in Japanese with English abstract)

Yamakawa, T., 1969: Additional Records of Marine Fishes from Amami (Ⅲ). Rep. Usa Mar. Biol. Stat., Kochi Univ., Vol.16, No.2

Yang, H. J., 1967: Poisonous and Injurious Fishes of Taiwan. Ann. Rep. Nat. Taiwan Mus., No.10, Taipei (in Chinese)

Yang, H. J., 1975: Illustrated Poisonous Fishes of the Pacific. Bur. Fish., Dept. Agr., Taiwan, Taipei (in Chinese)

Yasuda, F., 1966: Coral Fishes in Colour, their Ecology and Breeding Methods. Asahi Press, Tokyo (in Japanese)

Yasuda, F., Y. Tominaga, 1970: Two New Long-tailed Pomacanthine Fishes from Miyake-jima and Okinawa-jima, Japan. Jap. Journ. Ichthyol., Vol.17, No.4

Yasuda, F., 1975: The Fishes of Coral-reefs, Okinawa Islands. Diving World-sha, Tokyo (in Japanese)

Yasuda, F., 1975: The Fishes of Coral-reefs, Ogasawara Islands. Diving World-sha, Tokyo (in Japanese)

Yasumoto, T., P. J. Scheuer, 1969: Marine Toxins of the Pacific Ⅷ. Ciguatoxin from Moray Eel Livers. Toxicon, Vol.7

Yoshino, T., S. Nishijima, S. Shinohara, 1975: Catalogue of Fishes of the Ryukyu Islands. Bull. Sci. Eng., Univ. Ryukyus, Math. Nat. Sci. No.20 (in Japanese with English abstract)

Zama, A., K. Fujita, 1977: An Annotated List of Fishes from the Ogasawara Islands. Journ. Tokyo Univ. Fish., Vol.63, No.2

Zhu, Y. D., 1960: Serachii Fishes of China. Science Press, Beijing (in Chinese)

Zhu, Y. D., C. L. Chang, Q. T. Cheng, 1963: The Fishes of East China Sea. Science Press, Beijing (in Chinese)

INDEX
(Scientific Name)

A

B

C

D

E

F

G

H

I

J

K

L

M

N

O

P

R

S

T

U

V

W

X

Y

Z

INDEX
(Common Name)

A

B

C

D

E

F

G

H

I

J

K

L

M

N

O

P

Q

R

S

T

U

V

W

Y

Z